HOMEWORK HELPERS

Chemistry

GREG CURRAN

CBS PUBLISHERS & DISTRIBUTORS

NEW DELHI • BANGALORE

CBS Publisher ISBN : 81-239-1250-1
Career Press ISBN : 1-56414-721-5

First Indian Reprint : 2005

Publishing Director : Vinod K. Jain

Published by :
Satish Kumar Jain for CBS Publishers & Distributors,
4596/1-A, 11 Darya Ganj, New Delhi - 110 002 (India)
E-mail : cbspubs@del3.vsnl.net.in • Website : www.cbspd.com

Branch Office :
2975, 17th Cross, K.R. Road, Bansankari 2nd Stage, Bangalore - 70
Fax : 080-26771680 • E-mail : cbsbng@vsnl.net

Printed at :
India Binding House, Noida (U.P.)

Dedication

I would like to dedicate this book to my entire family, especially my wife, Rosemarie, who suffered with me during the process of writing it and for reminding me when it was time to work; my children, James, Amanda, and Jessica, for reminding me when it was time to play; my mother-in-law, Annette, who provided me with enough babysitting to finish it; my cousin Chris, who always encouraged me to try my hand at writing; my mother, Kathleen, who passed on her love of reading to all of her children; and my father, Peter, who continues to serve as the only role model that I will ever need for every stage of life.

Acknowledgments

I would like to thank Jessica Faust and BookEnds-Inc., for making this project possible. I would like to recognize the hard work of Eileen Dow Munson, Stacey Farkas, Jodi Brandon, Michael Pye, Michael Lewis, and everyone at Career Press who had a hand in turning my crude looking manuscript into a polished book.

I would also like to acknowledge my colleagues and students, past and present, each of whom have contributed to my growth and development as a teacher. Special thanks to John Haag, Matt Distefano, Mike Curtin, Dennis Ahern, Bob Gomprecht, Fr. Mickey Corcoran, and Jeff Butkowski.

Contents

Preface

Welcome to *Homework Helpers: Chemistry*!

Warning! Studying chemistry will change you! More specifically, it will change the way that you look at the world around you. A person who learns chemistry often begins to see things that others don't see, because he or she comes to understand and appreciate the interactions of the invisible atoms that make up the world around us. As you learn chemistry, try to make connections between the course material and the real world. You may start to find yourself imagining air molecules being displaced as you watch a leaf fall from a tree. You may be thinking about the forces exerted by the molecules of water as you make iced tea. Watching a stick burning in a fire might call to mind the laws of conservation of mass and energy. Don't be afraid of such changes; they are signs that you are growing and that your view of the world is becoming more sophisticated.

Science is concerned with studying the laws of nature. Chemistry is the science that involves the study of atoms, the building blocks of matter. It is concerned with how these atoms interact with each other in chemical reactions, to produce new substances. Chemists study the structure of matter and try to make connections between the structure and the properties of substances. Studying the characteristics of the elements and compounds that are found in nature has allowed chemists to create other substances with desirable properties, some of which make our lives longer and easier.

Chemistry is a rich, vibrant, and diverse subject, and you can't learn it all in one course, one year, or even one lifetime. Just the history of chemistry alone contains countless fascinating stories, about characters such as

John Dalton, J.J. Thomson, and Antoine Lavoisier. The only limit to how much you can learn is how much time and effort you put into the subject. In reality, there is enough material to keep you interested and engaged for a lifetime.

Learning chemistry, as is the case with all learning, is all about attitude. Don't resent the people who are trying to help you learn, and don't resist the learning process. Many people struggle with a subject when they are young but find that as they mature they come to enjoy it. If all students could develop a love of learning at an early age, both teaching and learning would be much easier, and review books such as this one might become obsolete.

Homework Helpers: Chemistry has been written with the student in mind. With this book, we hope to simulate the feel of one-on-one tutoring sessions with a teacher. Throughout this book you will find tables and images that illustrate important points in the text. Numerous review questions have also been provided, and each question has an answer and explanation.

I hope to make learning chemistry easier, by providing clear explanations of the topics that students struggle with, including the many types of calculations they will encounter. Numerous practice questions have been provided, and they all come with answers and explanations to help the student become more familiar with the material and more confident about the subject.

The book can be used to supplement a chemistry course textbook or as a review book for a standardized test, such as the SAT II in chemistry. It can serve as a stand-alone text for anyone who wants to review some of the chemistry that he or she might have forgotten over the years. The book can be read from cover to cover or used as a reference to review only specific types of problems, such as molarity or gas equations. I hope you enjoy the book and, more importantly, the subject of chemistry.

An Introduction to Chemistry: The Science of Chemistry

Lesson 1–1: Classification of Matter

When a person is confronted with a large number of objects or ideas, it is only natural to want to classify and organize them into groups. The advantage of grouping items is that you will end up with a smaller number of groups than objects. In your day-to-day life, you may group baseball cards according to teams or positions. You may organize your books according to titles or authors. At the very least, you probably have a sock drawer. Do you organize your clothes into drawers, according to the type of item? This allows you to remember where you keep your shirts, rather than memorizing where a specific shirt may have been placed.

A good classification scheme will allow you to memorize the characteristics of the groups, which you can then apply to all of the objects in the groups. In other words, rather than memorizing the characteristics of millions of organisms, a biologist will memorize the characteristics of the different kingdoms. If an organism is known to belong to a certain kingdom, then the biologist will know some of the characteristics of the organism, based on the known characteristics of the kingdom.

It is important to realize that these classification schemes are man-made, which means that we make up the categories and classes. The classification schemes can change, if someone comes up with a system that scientists like better than the present system. The present classification scheme for chemistry will probably seem very simple and elegant to you, especially if you have recently studied the classification system of biology.

As you can see from the graphic on page 14, all matter can be divided up into four main categories. Of course, there are other ways to classify matter, but this system is the one that seems to be generally recognized

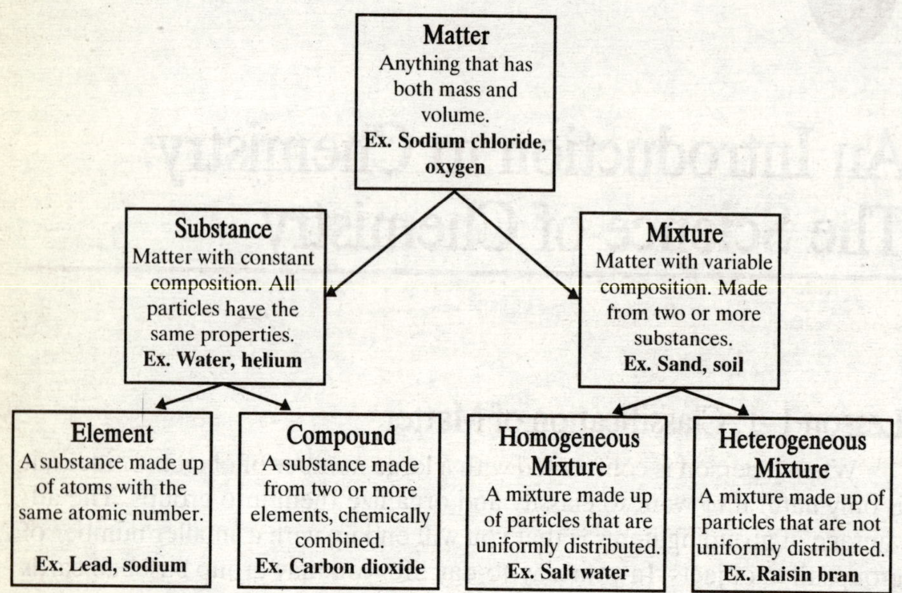

right now. Although the diagram is concise, it may not be completely clear unless I explain the categories in more detail. Once you understand the categories, the chart should be all that you need to review.

Matter is anything that is made up of atoms, and because all atoms have mass and volume, so does all matter. Even colorless gases, which you can't see, contain atoms that have both mass and volume. If you doubt that invisible air has volume, blow up a balloon and see how much space the air takes up. All objects that you encounter in your day are examples of matter.

A ***substance*** is a type of matter that has a consistent composition. What I mean is that no matter where you find a specific substance, its composition will be the same. In other words, a molecule of water from India has the same composition as a molecule of water from Canada. There is no real variation in the composition of a substance (except, perhaps, on the subatomic level, as you will learn when we discuss ***isotopes***). There are two major types of substances: elements and compounds.

Elements are substances that are made up of only one type of atom. By "type of atom," I mean that all of the atoms in this type of substance have the same atomic number, which is the number of protons. For example, all samples of pure oxygen (O_2) are only made up of atoms of oxygen, with 8 protons in their nuclei. So oxygen is a substance, because it has a consistent composition (it is the same, wherever you find it) and it is an element because it is only made up of one type of atom (atoms with 8 protons). Other common examples of elements include hydrogen (H_2), carbon (C), gold (Au), iron (Fe), and copper (Cu).

Have you ever noticed that when oxygen is written in a chemical equation, say the formula for photosynthesis that you studied in biology, it is written with a small subscript "2" (O_2)? The reason for this subscript is that oxygen is a diatomic element. Most of the elements exist as ***monatomic*** forms, which means that the smallest individual complete parts of these samples of pure elements are single atoms. Seven of the known elements are called ***diatomic elements***, because they are found in nature in their elemental form as two-atom molecules. Examination of a sample of pure chlorine, for example, would reveal two-atom molecules of the element. For this reason, when chlorine (Cl_2) is represented in a chemical equation as a pure element, it is also given the subscript "2." The seven diatomic elements are chlorine (Cl_2), fluorine (F_2), bromine (Br_2), iodine (I_2), hydrogen (H_2), oxygen (O_2), and nitrogen(N_2).

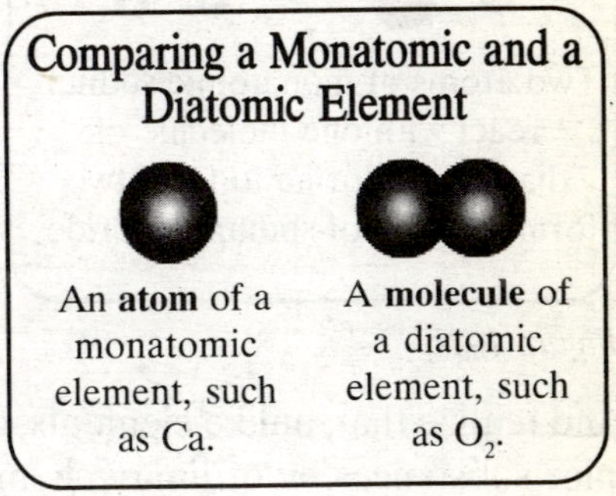

Comparing a Monatomic and a Diatomic Element

Figure 1–1a

One important characteristic of elements is that they cannot be broken down by ordinary chemical means. Chemical reactions can break more complex substances down into elements, but elements can only be broken down further by nuclear reactions. As of this writing, there are presently around 110+ known elements. I am being somewhat vague on purpose here, because the number of known elements changes over time, and some of the "man-made" elements are awaiting confirmation. A branch of nuclear chemistry deals with trying to produce atoms of undiscovered elements, and many of these heavier elements are made to "exist" for fractions of a second, until they break apart into smaller parts. All of the known elements, whether natural or man-made, are represented on the ***Periodic Table of Elements***, where they are organized into similar groups.

Compounds are substances that are made up of two or more elements **chemically** combined. The key to this definition is that the elements must be chemically combined. If you physically mix together two elements, let's say iron and copper, you get a mixture, which can be physically separated. Compounds are formed by chemical reactions, where the individual elements lose their individual properties and take on the new properties of the compound that is formed.

One clear example of this can be found on your kitchen table. Elemental chlorine is a yellowish-green poisonous gas that was used as a weapon in World War I. Elemental sodium is a highly reactive metal, which will react violently with water, often with explosive results. When these two potential deadly elements are combined in a chemical reaction, you get sodium chloride or table salt, which you can put on your French fries and safely eat. You see, elements that combine chemically to form a compound lose their individual properties in the chemical reaction. The new substance will have its own unique properties. See the reaction that follows.

Sodium and chlorine are elements, and they react to form sodium chloride, which is a compound. The reaction in Figure 1–1b shows that two parts of the monatomic element sodium will react with one part of the diatomic element chlorine to form two parts of sodium chloride. Notice that diatomic chlorine is represented as **Cl_2**, whereas monatomic sodium is represented as **Na**, without a subscript. Chlorine does not get a subscript when it is part of the compound, NaCl, because when we say that an element is diatomic, we mean in its elemental form.

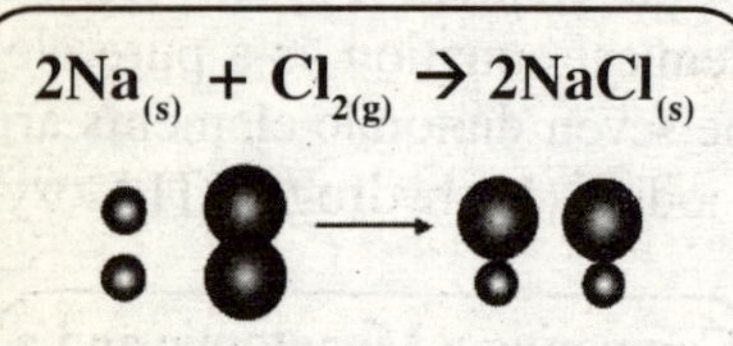

Two atoms of monatomic sodium react with one molecule of diatomic chlorine to form two formula units of sodium chloride.

Figure 1–1b

You can also look at Figure 1–1b and realize that, unlike elements, compounds can be decomposed into simpler substances by ordinary chemical means. Following this paragraph is the reverse reaction for the decomposition of the compound sodium chloride into its constituent elements, sodium and chlorine.

$$2NaCl_{(s)} \rightarrow 2Na_{(s)} + Cl_{2(g)}$$

sodium chloride yields sodium + chlorine

Do you know what the subscript "(s)" and "(g)" represent in the equation? They indicate the phase in which the substance in question is found in. The sodium and the sodium chloride are found in a solid phase, as represented by the "(s)" subscript. The chlorine is followed by a "(g)," because it is found in a gas phase. If any of the substances were found in the liquid phase, they would be followed by a subscript "(l)."

Other examples of compounds, which are chemical combinations of two or more elements, include water (H_2O), carbon dioxide (CO_2), and glucose ($C_6H_{12}O_6$). Compounds that are only made up of two elements, such as water (H_2O) and carbon dioxide (CO_2), are called ***binary compounds***. Compounds that contain three elements, such as glucose ($C_6H_{12}O_6$), are called ***ternary compounds***.

Whereas a compound is made from the chemical combination of two or more substances, a ***mixture*** is made from two or more substances that are **physically** combined. You don't need a chemical reaction to make a mixture, nor do you need one to separate a mixture. If you had a handful of copper coins and a handful of silver coins, and you mixed them together in a purse, you would have a mixture. Notice that, unlike in the example of compounds, the substances that make up a mixture do not lose their individual properties. The copper coins don't stop being copper coins just because they are physically combined with silver ones.

There are many different types of mixtures, but they all have these common characteristics: They represent physical combinations of two or more substances. The individual substances in a mixture do not lose their original properties, and the substances in a mixture can be separated by physical means. Another characteristic that all mixtures share is that their composition is variable. By this I mean that the substances that are found in the mixture can be mixed in with varying proportions or concentrations. If you mixed in five silver coins and three copper coins, or four silver coins and seven copper coins, you still end up with a mixture of copper and silver coins.

Heterogeneous mixtures are mixtures that **do not** have a consistent, or uniform, composition throughout the entire sample. A good example of this would be chicken soup. When you make chicken soup, the denser materials will sink to the bottom. If you scoop out soup to fill two bowls, it will make a difference from which area of the pot you take from. If you take a scoop from near the surface of the pot you will end up with mostly broth and the less dense items that float, such as pasta. If you take from the bottom of the pot, you will end up with more of the dense items, such as chicken and carrots. No two samples taken from the same pot will be all that similar, because this type of mixture is not uniform. Other examples of heterogeneous mixtures include a bowl of mixed nuts, a handful of sand, a chocolate chip cookie, and a scoop of rocky road ice cream.

Unlike heterogeneous mixtures, ***homogeneous mixtures*** **do** have a uniform composition. By this I mean that the composition of an entire sample is consistent. If you were to take one scoop out of one area of a homogenous mixture and a second scoop out of a different area of the same sample, the composition of each scoop would be essentially the same.

Homogeneous mixtures are also called ***solutions***, and the uniformity found in samples of homogeneous mixtures can be illustrated by a glass of salt water. If you dissolve a spoonful of salt in a glass of water, the salt becomes uniformly dispersed in the water. By the time the salt is fully dissolved, the concentration of the salt solution will be the same in each area of the sample. It is not as if the top of the glass will contain only pure water and the bottom of the glass will have really salty water; rather, the distribution of the salt will be uniform. Other common examples of solutions (homogeneous mixtures) would include metal alloys (such as bronze), the mixture of gases that we breathe, carbonated water, and mixtures of water and alcohol.

There is another important symbol that is used in chemistry that will help you to recognize solutions. When a substance such as salt is dissolved in water, its chemical formula will be followed by the subscript "(aq)," which stands for ***aqueous***. An ***aqueous*** substance is one that is dissolved in water (in other words, a solution). You need to get in the habit of paying attention to the symbols, so that you can differentiate between the symbolic representations of substances.

$NaCl_{(s)}$— This formula represents a single substance, the compound NaCl in its solid form.

$NaCl_{(aq)}$— This formula represents a homogeneous mixture of NaCl and water.

Another important way that mixtures can be differentiated from compounds is that the actual composition of a mixture is not a fixed thing, although the ratio by which elements combine in a compound is always the same. For example, water is always made from two atoms of hydrogen and one atom of oxygen; yet there are many ways to make a solution of salt water. If you add a tiny amount of salt to water, you get a salt-water solution; if you add much more salt, you still have a salt-water solution. Mixtures come in varying ***concentrations***.

If you drink coffee, you know the difference between a "strong" cup of coffee and a "weak" cup of coffee. If you don't drink coffee, I am sure you can recall having a glass of "weak" iced tea, or some other powdered drink mix. When we talk about the "strength" of one of these drinks, we are really talking about the ***concentration*** of a solution. We will be learning more about concentration later on, but for now you should remember that you could vary the concentration of a mixture but not of a compound.

Try the following practice questions and check your answers at the end of the chapter, before moving on to the next lesson.

Lesson 1–1 Review

1. A substance that is made up of only one type of atom is called a(n)____________.
2. Two or more substances physically combined form a(n)____________.
3. Two or more substances chemically combined are called a(n)____________.
4. Sand is a good example of a(n) ____________ mixture.
5. ____________ elements, such as oxygen (O_2), are found in nature in their elemental form as two-atom molecules.
6. Which of the following substances cannot be broken down by ordinary chemical reactions?
   a) water c) carbon dioxide
   b) gold d) glucose
7. Which of the following substances does not possess uniform composition?
   a) pure water b) salt water c) helium gas d) salad
8. Which of the following is made up of only one type of atom?
   a) NO b) He c) CO d) MgS
9. Which of the following represents a mixture?
   a) $NaCl_{(s)}$ b) $NaCl_{(l)}$ c) $NaCl_{(g)}$ d) $NaCl_{(aq)}$
10. Which of the following shows a ternary compound?
    a) CO b) NaCl c) NH_3 d) $NaNO_2$

Lesson 1–2: Phases of Matter

All of the matter that we encounter in our normal day is found in one of the three phases: solid, liquid, or gas. Have you ever stopped to think about the difference between these forms of matter? Why are some substances solid at room temperature whereas others are liquids or gases? Why do gases spread out and escape from containers that are able to hold solids and liquids? We will explore the difference between the phases in this lesson.

Most of the elements on the Periodic Table of Elements exist as solids at standard temperature and pressure. Much of the matter that you encounter on a daily basis is also found in the solid phase. Probably the two most interesting things about solids are that they are mostly made up of empty space and that even their molecules are in constant motion.

Solids aren't really solid? They are mostly empty space? When we discuss the structure of an atom in Chapter 3 you will learn that the size of the nucleus is actually very tiny compared to the size of the atom. The space that an atom effectively occupies really has to do with the fact that the electrons are moving very quickly and that the repulsive forces between electrons of different atoms keep them a certain distance apart. If you could strip the electrons off an atom and neutralize the charge of the nucleus, the atom would occupy only a fraction of its original size. To picture the size change, if the original size of the atom was about the size of a baseball stadium, the altered atom would be about the size of a marble. Just think: The "solid" floor that you stand on is almost completely empty space!

Did you know that the molecules of a solid are in constant motion? We tend to think of solids as static, but the truth is that they are made up of particles that are moving. In a solid, the motion of these molecules is restricted to mainly vibrational motion, meaning that each particle stays in the same position relative to each other, but they are still vibrating in place. What we call "temperature" is a measure of the average kinetic energy of the particles of a substance, so if the temperature of an object is above zero on the Kelvin scale, neither the mass, nor the average velocity of its particles, could be zero.

The limitations on the motion of the particles of solids account for the fact that they are said to have definite shape and definite volume. By definite shape, we mean that they do not take the shape of a container, the way that gases and liquids will. By definite volume, we mean that they do not expand or contract to occupy the entire container. This does not mean, however, that the volume of a solid is a constant. Solids do expand and contract in response to temperature changes. That is why, when you look up the density of a solid, it will always indicate the temperature at which the value for density is listed.

Another interesting thing about solids is that all true solids have what is called crystalline structure. By this, we mean that their particles are arranged in a three-dimensional, orderly pattern. Some things that most people think of as true solids, such as glass, are not considered true solids to chemists, because they don't possess this crystalline structure.

Solids will undergo phase changes when they encounter energy changes. ***Fusion***, commonly called melting, is the process by which a solid becomes a liquid. ***Solidification***, or freezing, is the process by which a liquid becomes a solid. ***Sublimation*** is an unusual process by which a solid goes directly to the gas phase, without turning to a liquid first. Sublimation is seen in substances such as carbon dioxide, which have relatively high vapor pressure and relatively low intermolecular forces. ***Deposition***, the opposite of sublimation, is when a gas vapor goes directly into the solid phase without becoming a liquid first.

Very few of the elements listed on the Periodic Table of Elements exist as liquids at standard temperature and pressure. On the other hand, approximately three-fourths of our planet is covered with the liquid known as water, so you should be very familiar with the properties of liquids. Unlike solids, liquids do not have definite shape. If you pour a liquid from a cylindrical bottle into a square container, it changes shape to match the container. This is possible because the motion of the individual particles within the liquid is much less restricted than in a solid. The particles are not locked into fixed positions, and they push past each other, allowing the liquid sample to flow. Some liquids, such as water, flow readily, whereas other liquids, such as molasses, are said to be viscous and flow slowly. The ***viscosity*** of a liquid is its relative resistance to flow. Regardless of how fluid a liquid is, the space that a liquid occupies is more fixed, and it will not expand to occupy an entire vessel the way a gas will.

When you change the temperature and/or the pressure of a liquid, you get phase changes to occur. ***Vaporization***, or boiling, is the process by which a liquid becomes a gas. The temperature and pressure at which a substance undergoes vaporization depend upon the intermolecular forces between its particles. When a substance such as gasoline evaporates at a relatively low temperature, it is an indication that the forces between its molecules are not as strong as those between water molecules. Vaporization takes place when the vapor pressure of the liquid is equal to the pressure of the atmosphere on its surface. ***Condensation*** is the process by which gas becomes a liquid. We see condensation form on the outside of a cold glass, as water vapor in the air turns into a liquid.

An interesting question to explore at this point involves another process that you are familiar with, called evaporation. ***Evaporation*** is when the liquid on the surface of a sample changes to the gas phase. If the normal boiling point of water is 100 °C or 373 K, how is it possible for water to evaporate at room temperature? The answer to this question can be found in the definition of temperature. Remember that we define temperature as the **average** kinetic energy of the particles of a substance. Even in a sample of water with a temperature of 22 °C, there are some molecules of water with exceptionally high kinetic energy, just as you can score a 100% on your chemistry examination even if the class **average** is only 70%. When a molecule of water that is close to the surface gains enough individual kinetic energy to escape the molecular attraction of its neighbors, it can enter the gas phase, even when the temperature of the sample is below the boiling point.

Can you see why having sweat or other water evaporate off the surface of your skin can cool you off? What would happen to the average grade in your chemistry class if your teacher removed all of the "A" grades from the average? The class average, deprived of the highest grades, would go down. What happens to the sample of water, when the molecules with the highest temperature escape as gas? The average temperature of the sample, deprived of the "hottest" molecules, goes down, leaving you feeling cooler.

Gases have no definite shape and no definite volume. They will expand to fill the container that they are made to occupy. Ten molecules of a gas can occupy as much space as ten million molecules of a gas, as they move freely around a container at relatively high speeds. The intermolecular forces of attraction between molecules in the gas phase are so weak that the particles are allowed to travel quite far away from each other. Gases differ significantly from the other phases of matter in that, the particles of a gas are often separated by very great distances. You already know that most of an atom is made up of empty space. So, too, most of a gas sample is made up of empty space, to the point that the size of a gas sample typically has little to do with the size of the actual particles.

For example, if you look at a party balloon filled with helium, the actual helium atoms would occupy a very tiny fraction of that space if you could get them to stop moving around. The balloon is really held open by the pressure exerted by the helium atoms, as they crash into the walls of the balloon at high speed. You can get a better understanding of this by leaving a balloon in a freezer for a few hours. You will notice that the size of the balloon decreases significantly, not because the atoms in the balloon shrink or escape, but because the atoms of helium have lost some of their kinetic energy and are not exerting as much pressure on the inside of the balloon. As the balloon returns to room temperature, its volume will increase, as the atoms of helium gain kinetic energy from the room.

If you take the same balloon and hold it under hot water for a period of time, the balloon's volume will increase. This happens because, as the helium atoms gain kinetic energy, they crash into the walls of the balloon faster, exerting more pressure on the inside of the balloon and causing it to expand. (You will learn much more about gases in Chapter 8.)

As you continue to learn more about chemistry, pay attention to the phase of each of the substances around you. Now, try the following review questions and check your answers at the end of the chapter before moving on to the next lesson.

Lesson 1–2 Review

1. ______________ is the process by which a gas becomes a liquid.
2. ______________ is the process by which a liquid becomes a gas.
3. The normal boiling point of water is _________.
4. The ___________ of a liquid is its resistance to flow.
5. ______________ is the process by which a solid turns directly to a gas.
6. ______________ have definite shape and definite volume.
7. ______________ is the process by which a solid turns to a liquid.
8. ______________ is the process by which a gas turns directly to a solid.
9. ______________ is the process by which a liquid becomes a solid.
10. True solids are said to possess _____________ structure.

Lesson 1–3: Chemical and Physical Properties

With more than one hundred known elements and millions of compounds in the universe, how do we distinguish between all of the kinds of matter out there? We can identify substances according to their chemical and physical properties. Given just two samples, you could probably tell the difference between a piece of gold and a piece of copper from their colors, which is an example of a physical property. Pyrite, however, is a mineral that looks so much like gold that it has been called "fool's gold." To distinguish between pyrite and the real thing requires knowledge of more properties.

Physical properties are the characteristics of a material that can be observed without carrying out a chemical reaction on it. Some physical properties are easy to observe, such as color, size, texture, and shine. Other physical properties may have to be measured or even calculated, such as mass and density. Still other physical properties can't be observed unless you physically change the substance. ***Malleability*** is a physical property of metals that allows them to be hammered into thin sheets, such as aluminum foil. In order to see if a particular substance were malleable, you would need to try to flatten it out.

Chemical properties are properties that can't readily be observed. In order to see if an unknown substance has a particular chemical property it is necessary to try to carry out a chemical reaction on it, which will, of course, produce a new substance. How something reacts to acid, for example, would be a chemical property. To see if a particular metal reacts with a particular acid, you would need to try the reaction. You would pour some acid on the metal and look for evidence of a chemical reaction. By the time that you are done testing the metal, it has combined with part of the acid to make a salt. That is the defining characteristic of a chemical property: In order to observe one you must carry out a chemical reaction and produce a different substance.

Examples of Physical and Chemical Properties

Physical Properties	Chemical Properties
Color, texture, density, freezing point, electrical conductivity, luster, hardness, mass, weight.	Flammability, reactivity with acid, reactivity with oxygen, ability to decompose into specific elements.

Another useful way to characterize properties is according to whether or not the size of the sample will affect the property.

Intensive properties are not affected by the size of the sample. Picture a piece of pure copper. It doesn't matter how big the piece of copper is; it will still have that distinctive orange-brown copper color. It will still be a good conductor of electricity. It will still have the ability to be flattened into a thin sheet. It will have a certain density for a given temperature. Color, electrical conductivity, malleability, and density are all intensive properties.

Extensive properties are characteristics that depend on the size of the sample in question. Picture two lumps of gold, one much larger than the other. In some ways, the samples of gold are both the same. For example, they have the same color and shine. These, then, are intensive properties. The ways that the two samples differ would be extensive properties. For example, they would have different sizes and different masses. Mass and volume, which do depend on the size of the sample, are extensive properties.

Examples of Intensive and Extensive Properties

Intensive Properties	Extensive Properties
Density, freezing point, malleability, ductility, electrical conductivity.	Mass, volume, weight, length.

Try the following practice questions and check your answers at the end of the chapter before moving on to the next lesson.

Lesson 1–3 Review

1. _____________ properties depend upon the size of the sample.
2. _____________ is a property of metals which allows them to be hammered into thin sheets.
3. _____________ properties can only be observed by carrying out a chemical reaction.
4. _____________ properties don't depend on the size of the sample.
5. _____________ properties can be observed without carrying out a chemical reaction.
6. Which of the following is **not** an example of an extensive property?
   a) density b) mass c) volume d) weight
7. Which of the following is **not** an example of a physical property?
   a) color c) freezing point
   b) flammability d) luster
8. Which of the following is an example of an extensive property?
   a) length b) melting point c) malleability d) ductility
9. Which of the following is **not** an example of an intensive property?
   a) length b) color c) density d) luster
10. How do intensive properties differ from extensive properties?

Lesson 1–4: Chemical and Physical Changes

Matter can undergo many types of changes. In nuclear changes, atoms can be split apart or fused together with other atoms, and thus transformed into other elements. (We will learn more about these types of nuclear changes in Lesson 6–5.) As you study chemistry, you are more likely to observe matter undergoing various chemical and physical changes. The ***Law of Conservation of Mass*** states that matter cannot be created or destroyed; it can only change form. This law, as it is stated here, does not apply to nuclear changes, yet it holds true with regard to physical and chemical changes. It means that if you start off with 10.0 g of matter before a physical or chemical reaction, you will end up with 10.0 g of matter. These changes don't create or destroy atoms; they only change how the atoms are connected to each other. Even if we burn a piece of wood in a fire, we are not destroying atoms; we are just breaking bonds between atoms and forming new ones.

Physical changes are changes that do not result in the production of a new substance. If I break a glass beaker, I won't be able to use it anymore, but the material does not change into a new element or compound. If I crumple up a ball of aluminum foil, I have changed its shape and size, but I still have aluminum foil. Even a significant physical change, such as grinding a piece of wood into sawdust, is not a chemical change, because it does not result in the production of a new substance.

The "change-of-phase operations," discussed in Lesson 1–2, are very common physical changes. Imagine a beautiful ice sculpture in the shape of a swan, on a table at an outdoor party. The sculpture is made up of frozen water, which can be indicated with the notation $H_2O_{(s)}$, where the subscript "s" stands for "solid." As the day wears on and the ice sculpture continues to absorb heat from its surroundings, much of the ice will melt, producing a puddle of liquid water, $H_2O_{(l)}$. Now, the liquid water will continue to absorb energy from its surroundings, and, over time, much of it will evaporate, becoming gaseous water, $H_2O_{(g)}$. Despite the fact that the water has existed in three different phases, it has remained water, H_2O. Because none of these changes resulted in the production of a new substance, they are each examples of a physical change.

Sometimes, you will see a change-of-phase operation illustrated in a form that you may associate with chemical reactions, such as the equation for melting ice that you see following this paragraph. Don't be confused by the format, it is still showing a physical change.

$$H_2O_{(s)} + \text{energy} \rightarrow H_2O_{(l)}$$

You are probably familiar with most change-of-phase operations. In Lesson 1–2 we discussed the process of sublimation, which you may not have been familiar with. Recall that ***sublimation*** is the process of a substance changing from a solid phase directly into a vapor phase. Even if you don't know the term, you have probably seen sublimation taking place. When "dry ice," which is frozen carbon dioxide, appears to give off a white "smoke," you see evidence of the solid carbon dioxide turning directly into vapor. Carbon dioxide is actually an invisible gas. The white "smoke" is actually water vapor condensing because of the cold gas coming off the solid block of dry ice.

Sublimation happens more readily with "dry ice" than frozen water because the forces between carbon dioxide molecules are not as strong as the forces between water molecules. An example of sublimation can be represented by the following equation:

$$\mathbf{CO_{2(s)} + energy \rightarrow CO_{2(g)}}$$

Experimentally, we can determine the freezing point or boiling point of a pure substance by determining at what temperature the substance stops getting hotter or colder for a period of time. If we graph our data for heating or cooling, the "plateaus" or flat areas represent where a change-of-phase operation is occurring. It is also important to note that, for a pure substance, the freezing point and melting point are exactly the same. For example, the normal freezing point of pure water is 0 °C, and the normal melting point for pure ice (frozen water) is also 0 °C. The same relationship holds true for the boiling point and condensation point of a pure substance.

Examples of Physical Changes

Melting wax, cutting wood, breaking glass, boiling water, bending metal, freezing carbon dioxide, dissolving salt, cutting hair.

Chemical changes, also called chemical reactions, are more complex, and they tend to be harder to reverse than physical changes. There are many different types of chemical changes. Some happen very quickly, such as an explosion, and some happen slowly, such as rusting, but they all result in the production of one or more new substances. Look at the equation for the decomposition of water, shown on page 28, and compare it to the equation for melting water, shown previously.

This reaction, called the electrolysis of water, produces two new substances. Hydrogen and oxygen are both colorless gases, and they both have properties that are different from the properties of water. When liquid

$$2H_2O_{(l)} + \text{energy} \rightarrow 2H_{2(g)} + O_{2(g)}$$

Two water molecules are broken down into two hydrogen molecules and one molecule of oxygen.

Figure 1–4a

water evaporates to form water vapor, it only requires a temperature change to bring it back to liquid water. In the reaction shown in Figure 1–4a, the hydrogen and oxygen won't join to make water again unless another chemical reaction is carried out.

Examples of Chemical Changes

Photosynthesis:	$6O_2 + C_6H_{12}O_6 \rightarrow 6H_2O + 6CO_2$
Zinc reacting with an acid:	$Zn + 2HCl \rightarrow ZnCl_2 + H_2$
Methane gas burning:	$CH_4 + 2O_2 \rightarrow CO_2 + 2H_2O$
Synthesis of magnesium oxide:	$Mg + O_2 \rightarrow 2MgO$

You may have noticed that chemical equations are sometimes written with more information than at other times. For example, I don't always include the subscripts (for example, (s) or (g)), which indicate the state of the matter at the time of the reaction. Also, I don't always indicate whether energy is present as a reactant or as a product. When you use these notations depends upon the type of information that you want to convey at a given time. In general, it is always okay to show more information than you need to. In the next lesson you will learn how to show energy changes in chemical equations.

Try the following practice questions and check your answers at the end of the chapter before moving on to the next lesson.

Lesson 1–4 Review

1. _____________ changes produce new substances, with new properties.
2. The Law of _____________ of Mass states that matter cannot be created or destroyed.

3. ______________ is the process of a solid changing directly to a gas, without passing through the liquid phase.
4. ______________ changes do not result in the production of new substances.
5. Melting is an example of a change-of- ______________ operation.
6. Which of the following is not an example of a physical change?
   a) breaking glass c) cutting wood
   b) melting gold d) burning wood
7. Which of the following represents matter in a solid form?
   a) $CO_{2(g)}$ b) $H_2O_{(s)}$ c) $CCl_{4(l)}$ d) $O_{2(g)}$
8. Which of the following correctly represents the process of sublimation?
   a) $CO_{2(g)} \rightarrow CO_{2(s)}$ c) $CO_{2(g)} \rightarrow CO_{2(l)}$
   b) $CO_{2(s)} \rightarrow CO_{2(g)}$ d) $CO_{2(l)} \rightarrow CO_{2(s)}$
9. Explain how physical changes differ from chemical changes.
10. Does dissolving sugar in water represent a physical change or a chemical change? Explain your answer.

Lesson 1–5: Energy and Chemical Reactions

In the last lesson you read about the Law of Conservation of Mass. Energy has its own conservation law, appropriately called the ***Law of Conservation of Energy***. This law states that the total amount of energy in the universe is conserved. In other words, energy is not created or destroyed; it only changes form. Einstein combined these two conservation laws into the ***Law of Conservation of Mass-Energy***, with his formula $E=mc^2$, to illustrate what happens in a nuclear reaction. When dealing with only physical and chemical changes, as we are now, we apply the original conservation laws.

Let's suppose a bowling ball fell off a shelf and landed on your foot, breaking one of your toes. When the bowling ball hit your foot, it transferred energy into it. Where did the bowling ball get the energy in the first place? It certainly didn't create it. The person who put the bowling ball on the shelf did work lifting it up, transferring energy into it, and the bowling ball stored that energy as potential energy.

Energy is often defined as the ability to do work. ***Work***, in this case, is defined as a force exerted over a distance. Energy, then, is the ability to exert a force over a distance. There are many forms of energy, including heat, mechanical, chemical, electrical, solar, and nuclear. According to the Law of Conservation of Energy, we can convert one form into another. Energy is measured in units called joules, or J for short.

Gravitational Potential Energy:

The energy that an object possesses due to its mass and its height above a reference point.

Gravitational P.E. =

mass × acceleration due to gravity × height or P.E. = mgh

When the bowling ball was lifted onto the shelf, it was given gravitational potential energy. ***Gravitational potential energy*** is the stored energy that an object possesses due to its mass and its position above a reference point, in this case your foot. When the bowling ball fell off the shelf, its height above the floor decreased, decreasing its gravitational potential energy. When the bowling ball hit your foot, its height above it was zero, so its gravitational potential energy, with reference to your foot, was zero. Was its potential energy destroyed? No—it just changed into another form of energy called kinetic energy.

Kinetic Energy:

The energy that an object possesses due to its mass and velocity.

Kinetic energy = 1/2 mass × velocity^2 or K.E. = 1/2 mv^2

Gravitational Potential Energy is Converted to Kinetic Energy

The ball has high gravitational potential energy (mgh), because its height is high. It has low kinetic energy (1/2mv^2), because its velocity is low.

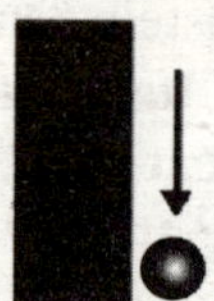

The ball has low gravitational potential energy (mgh), because its height is low. It has high kinetic energy (1/2mv^2), because its velocity is high.

Figure 1–5a

Kinetic energy, which is often called the energy of motion, is the energy that an object has due to its mass and velocity (think speed). When the bowling ball was sitting on the shelf, its velocity was zero, so its kinetic energy was zero. As it was falling, it was accelerating toward your foot, and its kinetic energy continued to increase. Where did the kinetic energy come from? As the bowling ball lost potential energy, because its height above your foot decreased, it gained kinetic energy as its velocity increased. So, the energy didn't just "appear." The gravitational potential energy was simply converted to kinetic energy.

Just like macroscopic (large) objects, microscopic (tiny) objects such as atoms can store energy due to their positions. We call the energy that atoms store ***chemical energy***. This chemical energy is stored within atoms as their potential to form bonds with other atoms, and it is released when new bonds are made. As in the example with the bowling ball, the energy released will take on a different form than the stored energy. Most often, the chemical energy is released in the form of heat.

When wood is burned in a fireplace, energy is released as the atoms from the wood combine with the atoms of oxygen in the air to make new compounds, such as carbon dioxide and carbon monoxide. However, before the wood will burn, you need to put energy into the wood to get it started. This initial energy, perhaps from burning kindling, is used to break bonds between atoms in the wood so that they can make new bonds. If the energy that we put into getting the wood burning was less than the energy that we got from the wood while it was burning, it would be an inefficient way to try to heat a room.

Reactions that release more energy than they take in are called ***exothermic reactions***. In this type of reaction, the potential energy of the products is lower than the potential energy of the reactants, with the extra energy being released, usually in the form of heat. For this reason, exothermic reactions will heat the area around them. An example of an equation for an exothermic reaction is shown here:

$$\mathbf{CH_{4(g)} + 2O_{2(g)} \rightarrow CO_{2(g)} + 2H_2O_{(g)} + 890.4\ kJ\ of\ energy}$$

As you can see from the equation, energy is released and can be thought of as a product of the reaction. What the equation doesn't show is that a certain amount of energy is required to get this reaction started. The energy required to get a chemical reaction to start is called ***activation energy***. When you rub a match across a strip of sandpaper, the friction generates the heat that acts as the activation energy to get the burning reaction started.

Examples of Exothermic Reactions

$$C_{(s)} + O_{2(g)} \rightarrow CO_{2(g)} + 393.5 \text{ kJ}$$

$$N_{2(g)} + 3H_{2(g)} \rightarrow 2NH_{3(g)} + 91.8 \text{ kJ}$$

$$C_6H_{12}O_6 + 6O_2 \rightarrow 6CO_2 + 6H_2O + 2804 \text{ kJ}$$

Other reactions, called ***endothermic reactions***, take in more energy than they release. Energy can be thought of as a necessary reactant in this type of reaction, as shown in this reaction:

$$\mathbf{N_{2(g)} + O_{2(g)} + 180.6 \text{ kJ} \rightarrow 2NO_{(g)}}$$

Endothermic reactions will cool the area around them by absorbing the necessary energy from the surroundings. As the result of an endothermic reaction, the products end up with more potential energy than the reactants. You may be familiar with chemical cold packs. When the chemicals in a cold pack are allowed to mix, they absorb energy from the surroundings to carry out an endothermic reaction. More examples of endothermic reactions are shown here.

Examples of Endothermic Reactions

$$2H_2O_{(g)} + 483.6 \text{ kJ} \rightarrow 2H_{2(g)} + O_{2(g)}$$

$$CO_{2(g)} + 2H_2O_{(l)} + 890.8 \text{ kJ} \rightarrow CH_{4(g)} + 2O_{2(g)}$$

$$N_{2(g)} + 2O_{2(g)} + 66.4 \text{ kJ} \rightarrow 2NO_{2(g)}$$

Both exothermic and endothermic reactions are often represented graphically with potential energy diagrams, as seen in Figure 1–5b. Maybe you can see why endothermic reactions are sometimes called "uphill reactions," as they need energy to be continuously added in order to continue.

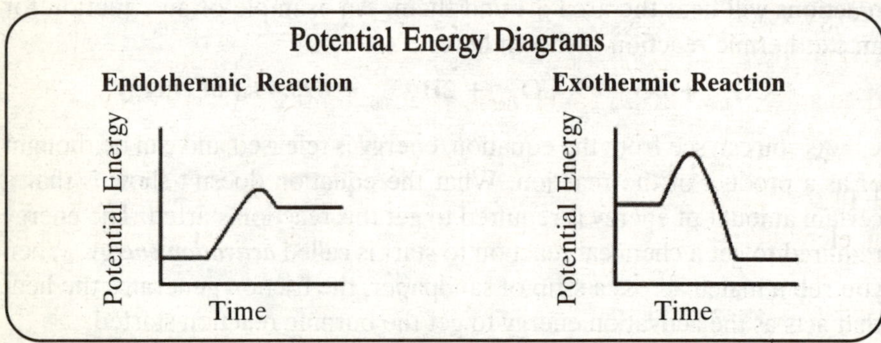

Figure 1–5b

As you continue to study chemistry, you will have the opportunity to observe many chemical reactions. Use these opportunities to apply the information that you have learned in this chapter.

Now, try the following practice questions and check your answers at the end of the chapter, before moving on to the next lesson.

Lesson 1–5 Review

1. ______________ energy is the energy required to get a chemical reaction started.
2. ______________ reactions result in products at a higher energy level than the initial reactants.
3. ______________ is defined as the ability to do work.
4. ______________ energy is the energy of matter in motion.
5. Select the correct formula for gravitational potential energy.
   a) $1/2mv^2$ b) 2mv c) mvh d) mgh
6. Is it possible for a baseball to have more kinetic energy than a truck?
7. Is it possible for a baseball to have more gravitational potential energy than a truck?

Chapter 1 Examination

Part I—Matching

Match the following terms to the definitions that follow. Not all of the terms will be used.

a. substance	h. chemical property	p. exothermic reaction
b. matter	i. physical change	q. endothermic reaction
c. compound	j. chemical change	r. activation energy
d. mixture	k. intensive property	s. ternary compound
e. aqueous	l. extensive property	t. binary compound
f. physical property	m. work	u. monatomic element
g. element	n. energy	v. diatomic element
	o. kinetic energy	

_____ 1. A type of reaction that releases energy.

_____ 2. A change that results in the production of a new substance.

_____ 3. Two or more substances chemically combined.

_____ 4. A substance that is dissolved in water.

_____ 5. A force exerted over a distance.

_____ 6. A substance made up of only one type of atom.

_____ 7. A property that depends on the size of the sample.

_____ 8. A compound that is made up of three elements.

_____ 9. Anything that is made up of atoms.

_____ 10. The energy of matter in motion.

Part II—Multiple Choice

For each of the following questions, select the best answer.

11. Which of the following represents a homogeneous mixture?
    a) $CuCl_{2(s)}$ b) $Hg_{(l)}$ c) $CO_{2(g)}$ d) $C_6H_{12}O_{6(aq)}$

12. Which of the following can be broken down by a chemical change?
    a) hydrogen c) lithium
    b) carbon dioxide d) bromine

13. Which of the following change-of-state operations is exothermic?
    a) gas to liquid c) liquid to gas
    b) solid to liquid d) solid to gas

14. Which of the following is not an element?
    a) gold b) bronze c) copper d) iron

15. Which of the following is a diatomic element?
    a) neon b) iodine c) barium d) calcium

16. Which of the following does not represent a compound?
    a) H_2O b) He c) CO d) KI

17. Which of the following elements is malleable?
    a) copper b) hydrogen c) neon d) iodine

18. Which of the following represents a compound?
    a) Co b) He c) Ar d) NO
19. Which of the following is made up of more than one type of atom?
    a) gold b) table salt c) silver d) oxygen
20. Which of the following is an example of a chemical property?
    a) conductivity c) luster
    b) hardness d) reactivity to acid
21. Which of the following is an extensive property?
    a) volume b) melting point c) density d) ductility
22. Which of the following is an example of a chemical change?
    a) breaking glass c) melting metal
    b) rusting iron d) cutting wood
23. Which of the following is an example of a physical change?
    a) photosynthesis c) sublimation of carbon dioxide
    b) burning gasoline d) zinc reacting with acid

Part III—Short Essay

Answer the following questions in a few sentences.

24. Compare and contrast mixtures and compounds.
25. Explain the Law of Conservation of Energy.

Answer Key

The actual answers **will be shown in brackets**, followed by the explanation. If you don't understand an explanation that is given in this section, you may want to go back and review the lesson that the question came from.

Lesson 1–1 Review

1. [element]
2. [mixture]—You might have said "solution, "homogeneous mixture," or "heterogeneous mixture," which are also true answers. "Mixture" is the most general of the possible correct answers.
3. [compound]—Remember: The key is that the elements in a compound are chemically combined. If the question asked about things that are physically combined, the answer would have been a "mixture."

4. [heterogeneous]—Sand is composed of particles of different substances, mixed together in a non-uniform distribution.
5. [diatomic]—"Diatomic" means "two-atom," so think of O_2 or H_2 when you see it.
6. [B. gold]—Gold is an element, and elements can't be broken down into simpler substances by ordinary chemical means.
7. [D. salad]—We toss a salad to try to make its composition more uniform, but it never becomes truly uniform.
8. [B. He]—NO, CO, and MgS all represent compounds. We can tell because they all contain more than one capital letter and, therefore, more than one elemental symbol.
9. [D. $NaCl_{(aq)}$]—The "(aq)" indicates that the salt has been dissolved in water, making it a mixture of two or more substances physically combined.
10. [D. $NaNO_2$]—A ternary compound contains three different elements. Sodium nitrite ($NaNO_2$) contains the elements sodium, nitrogen, and oxygen, so it fits the definition.

Lesson 1–2 Review

1. [condensation]—If you go outside on a cool morning, you may find condensation on many cool surfaces.
2. [vaporization]—The steam that you see coming from a teapot is the result of vaporization.
3. [100 °C or 373 K]—You will learn more about the Kelvin scale in Chapter 2.
4. [viscosity]—Liquids with a great deal of viscosity are very resistant to flow.
5. [sublimation]
6. [solids]—Solids don't take the shape and size of their containers.
7. [fusion]—The term fusion is also used for a nuclear process that we will discuss in Chapter 6.
8. [deposition]—When water vapor "freezes" onto a cold surface, deposition is taking place.
9. [solidification]—This is the process that we normally call "freezing." Many people associate the term "freezing" with "cold," but the process of solidification can happen at relatively high temperatures for some substances.
10. [crystalline]

Lesson 1–3 Review

1. [extensive]—Think of the word *extent.*
2. [malleability]

3. [chemical]—To know if something is flammable, we must try to burn it.
4. [intensive]—Think of these properties as being "in" the substance (**in**tensive), such as color and density. They don't change due to the size of the sample.
5. [physical]—You can see that a metal is shiny without trying to carry out a chemical reaction on it.
6. [A. density]—The density of a material can be used to identify it. It won't change due to the size of the sample.
7. [B. flammability]—We only observe the flammability of a substance when we are carrying out a chemical change (combustion) on it.
8. [A. length]—The length of an object will certainly depend on the size of the sample.
9. [A. length]—See answer 8.
10. [Intensive properties don't depend on the amount of the substance (sample size) present, whereas extensive properties do.]—The extensive properties of a substance will change with the size of the sample. For example, the color (intensive property) of table salt is the same, regardless of how much of it you have. The weight (extensive property) of your salt sample will certainly depend upon the size of your sample. (The wording of your answer may vary from mine.)

Lesson 1-4 Review

1. [chemical]—Chemical changes, which are also called chemical reactions, result in the production of new substances, with their own properties.
2. [conservation]—The Law of Conservation of Mass is very important to the study of chemistry.
3. [sublimation]—Dry ice is called "dry" because it never melts into the liquid phase.
4. [physical]—If I cut a wooden board in two pieces, both pieces are still made of wood.
5. [state or phase]
6. [D. burning wood]—If we completely burn a piece of wood in a fire, we won't have wood anymore.
7. [B. $H_2O_{(s)}$]
8. [B. $CO_{2(s)} \rightarrow CO_{2(g)}$]
9. [Physical changes don't result in the production of a new substance; chemical changes do.]

10. [physical change; Dissolving sugar in water does not result in the production of a new substance so it represents a physical change. The sugar can be recovered through evaporation, which is a physical process.]

Lesson 1–5 Review

1. [activation]—Just think of it as the energy needed to "activate" the reaction, to get things started.
2. [endothermic]—Endothermic reactions take energy in, which allows the products to end up with more potential energy than the initial reactants.
3. [energy]
4. [kinetic]
5. [D. mgh]—Think "mass, gravity, height."
6. [yes]—It's possible when the baseball is flying through the air and the truck is standing still. Remember: If the formula for kinetic energy is $1/2mv^2$, what is the kinetic energy of the truck when its velocity is zero?
7. [yes]—If our reference point is the street, the baseball will have more gravitational potential energy than the truck when the ball is up in the air and the truck is sitting on the street. If the formula for gravitational potential energy is mgh, what happens when the height of the truck is zero?

Chapter 1 Examination

1. [p. exothermic reaction]—Think of "exo" as "out," as in the word, *exoskeleton*. Energy comes out of an exothermic reaction.
2. [j. chemical change]—New substances are produced in chemical reactions, also called chemical changes.
3. [c. compound]—When two or more substances (say, for example, oxygen and hydrogen) combine chemically, you get a compound (in this case, water).
4. [e. aqueous]—As in the word *aquarium*, the prefix "aqua-" refers to water.
5. [m. work]—To do work, in this sense of the word, you must exert a force over a distance. Do you see why a weight lifter that is holding a barbell over his head is doing no work, while the barbell is motionless?
6. [g. element]
7. [l. extensive property]
8. [s. ternary compound]
9. [b. matter]—Matter is the most general category in our classification scheme, including mixtures, compounds, and elements.
10. [o. kinetic energy]

11. [D. $C_6H_{12}O_{6(aq)}$]—A mixture requires two substances. The symbol (aq) tells us that this sugar is dissolved in water, making a solution, which is a homogeneous mixture.
12. [B. carbon dioxide]—Elements cannot be broken down by a chemical change, but compounds can.
13. [A. gas to liquid]—You need to add energy (heat) to get a solid to turn to a liquid (melting) or a liquid to turn to a gas (boiling), but energy is released when a gas turns to a liquid (condensation).
14. [B. bronze]—You won't find bronze on the periodic table, because it is an alloy, not an element.
15. [B. iodine]—Iodine is, indeed, one of the seven diatomic elements.
16. [B. He]—Compounds will always show two or more capital letters in their formulas.
17. [A. copper]—Copper has the ability to be hammered into sheets, meaning it is malleable.
18. [D. NO]—Notice that NO is the only formula shown that has two or more capital letters.
19. [B. table salt]—Table salt contains the elements sodium and chlorine and is represented by the formula NaCl.
20. [D. reactivity to acid]—Reactivity to acid would only be observable in a chemical reaction.
21. [A. volume]—The volume of an object is a measure of how much space it occupies. This will surely change if the size of the sample changes.
22. [B. rusting iron]—When iron rusts, it loses its original properties and takes on the properties of the new compound, Fe_2O_3.
23. [C. sublimation of carbon dioxide]—Sublimation is a change of state operation, which is a type of physical change.
24. [Mixtures consist of two or more substances physically combined, whereas compounds consist of two or more substances chemically combined.]
25. [Energy isn't created or destroyed; it only changes form.]

Measurements and Calculations

Lesson 2–1: The International System of Measurements (SI)

A common mistake that many students make when they are beginning their study of chemistry has to do with the use of units. In math class, you may do many calculations during the course of the year where you are only working with numbers. When you perform calculations in chemistry class, however, it is likely that you will always be required to work with units. If you measure the length of an object in the laboratory, you can't simply record the data as "2.35," because no one will know whether you mean 2.35 inches, 2.35 centimeters, or 2.35 feet! In science, you must always work with units.

In order for people to communicate most effectively, they need to speak the same language. This is very true in science. When scientists communicate information, they must be extra careful when dealing with numbers and units. A miscommunication between doctors could lead to the administration of improper doses of medication. A miscommunication between engineers could lead to a building, a device, or a vehicle with structural flaws. As do all people, scientists need a common language in order to communicate most effectively. The ***International System of Measurements*** was designed for just this purpose.

The General Conference on Weights and Measures updated the metric system in 1960 and renamed it the International System of Measurements. The system is commonly referred to as SI, which is short for the French name, Le Systeme International d'Unites. Scientists from all around the world have adopted SI, and there has been a push in many countries to convert the general population to the SI units.

The advantages of SI will seem obvious to any student of science. Let's compare the SI units of length to the English units of length:

Comparing English Units of Length to SI Units of Length	
English Conversions	**SI Conversions**
12 inches = 1 foot	10 centimeters = 1 decimeter
3 feet = 1 yard	10 decimeters = 1 meter
1760 yards = 1 mile	10 meters = 1 dekameter

Figure 2–1a

Do you notice the difference? With the English units, there doesn't appear to be any logic behind the conversions, whereas, with the SI conversions, you are always working with multiples of ten. Need another example? Let's compare some U.S. (United States) units for volume to SI units of volume:

Comparing U.S. Units of Volume to SI Units of Volume	
U.S. Conversions	**SI Conversions**
8 ounces = 1 cup	1000 cubic millimeters = 1 cubic centimeter
2 cups = 1 pint	1000 cubic centimeters = 1 cubic decimeter
2 pints = 1 quart	1000 cubic decimeters = 1 cubic meter
4 quarts = 1 gallon	1000 cubic meters = 1 cubic dekameter

Figure 2–1b

Again, the SI conversions are more uniform and easier to remember. The older system is even more confusing than it appears here when you consider the fact that English cups, pints, ounces, and quarts are different than the U.S. units with the same names. For example, there are 1.2 U.S. gallons in 1.0 English gallon! For this reason alone, American students can probably see the need for a common system.

Another thing that American students might find interesting is that the units "liter" and "milliliter" are absent. When the General Conference on Weights and Measures updated the metric system in 1960, they eliminated the liter! This means that the only "metric" unit that we have really embraced in this country is the one that was declared "outdated" more than 40 years ago!

Why was the liter eliminated from the International System of Measurements? Well, one of the goals of the conference was to cut down on the number of existing units and start off with the fewest number of base units that were needed to measure essentially all of the known physical quantities. Because volume can be expressed in cubed units of length, there really is no reason to support special units of volume. Following are the seven ***SI base units***.

These base units can be combined with prefixes in order to derive larger or smaller quantities. For example, we combine the prefix "kilo-" with the base unit "meter" to get one "kilometer," which has a value of 1000 meters. The ***SI prefixes*** are shown on page 44.

SI Base Units

Property	**Unit**	**Symbol**
length	meter	m
time	second	s
mass	kilogram	kg
temperature	Kelvin	K
electric current	ampere	A
amount of substance	mole	mol
luminous intensity	candela	cd

Figure 2–1c

If you are asked to convert between units of the same quantity, simply pay attention to what the prefixes stand for. If asked to convert 3.45 kilograms to grams, keep in mind that "kilo-" means thousand, so one kilogram must equal one thousand grams. This makes the conversion as simple as the work shown here:

$$\mathbf{3.45\ \cancel{kg} \times 1000\ g/1\ \cancel{kg} = 3450\ g}$$

Notice that we multiply the original quantity (3.45 kg) by a ratio of 1000 g/1 kg. By orienting the ratio in such a way that the symbol "kg" is in the denominator, we're able to "cross-cancel" the unwanted unit (kg), leaving us with the desired unit (g).

You can check your answer by reasoning, "If 1 kilogram is 1000 grams, then 3 kilograms would be 3000 grams." Your answer seems reasonable. If you find this difficult, you may want to skip ahead to Lesson 2–5, where we will be covering conversions in greater detail.

SI Prefixes			
Prefix	**Symbol**	**Multiply the base by**	**Scientific Notation**
exa-	**E**	1 000 000 000 000 000 000	10^{18}
peta-	**P**	1 000 000 000 000 000	10^{15}
tera-	**T**	1 000 000 000 000	10^{12}
giga-	**G**	1 000 000 000	10^{9}
mega-	**M**	1 000 000	10^{6}
kilo-	**k**	1000	10^{3}
hecto-	**h**	100	10^{2}
deca-	**da**	10	10^{1}
deci-	**d**	0.1	10^{-1}
centi-	**c**	0.01	10^{-2}
milli-	**m**	0.001	10^{-3}
micro-	**μ**	0.000 001	10^{-6}
nano-	**n**	0.000 000 001	10^{-9}
pico-	**p**	0.000 000 000 001	10^{-12}
femto-	**f**	0.000 000 000 000 001	10^{-15}
atto-	**a**	0.000 000 000 000 000 001	10^{-18}

Figure 2–1d

Some properties, such as time or length, can be expressed in terms of SI base units. Other properties, such as volume or density, are expressed in SI derived units, which are really made by combining the SI base units. Following are some examples of physical properties and the ***SI-derived units***, which can be used to measure them.

Examples of SI-Derived Units		
Property	**SI-Derived Units**	**Symbol**
volume	cubic centimeters	cm^3
velocity	meters per second	m/s
density	grams per cubic centimeter	g/cm^3

Figure 2–1e

Some properties are measured in units that are derived from such a large combination of base units that scientists have given them new unit names and symbols. These new units and symbols are still based on, and are equivalent to, the original combination of base units. Figure 2–1f shows some examples.

Units and Their SI Equivalents			
Property	SI-Derived Unit	Equivalent Unit	Symbol
force	kg × m/s^2	Newton	N
energy	kg × m^2/s^2	joule	J

Figure 2–1f

Now, try the following practice questions and check your answers at the end of the chapter before moving on to the next lesson.

Lesson 2–1 Review

1. The SI unit for length is the __________.
2. The SI unit called the "ampere" measures __________.
3. The SI unit for temperature is the __________.
4. There are __________ centigrams in one gram.
5. There are __________ centigrams in one kilogram.
6. Convert 3.45 meters to centimeters.
7. Convert 7640 decimeters to kilometers.
8. Which of the following is not an SI base unit?
   a) meter b) candela c) mole d) joule
9. Which of the following units is used to measure energy?
   a) meter b) candela c) mole d) joule
10. Which of the following quantities cannot be measured with a single SI base unit?
    a) speed b) time c) length d) mass

Lesson 2–2: Measuring Matter

Chemistry is a quantitative science, which means that it is concerned with measurements that involve quantities or numbers, such as the amount of space a substance takes up or how much it weighs. Those measurements, which analyze the quantity of matter, are of especial importance to chemists.

As you begin your laboratory work in chemistry, it is likely that you will be asked to find the mass, volume, and density of substances early on.

The ***volume*** of a substance is simply the amount of space that it takes up (in other words, how big it is). The standard method for measuring the volume of a substance depends upon whether or not the substance is a solid, liquid, or gas.

Your laboratory will include several types of vessels that can be used to measure the volume of a liquid. The most accurate measurements can be taken in a vessel called a graduated cylinder, or "graduate," for short. Graduated cylinders come in many sizes, and the accuracy of these instruments depends upon the number of lines or gradations on them. Usually, the smaller graduates are more accurate than the larger ones, and all graduates tend to give more accurate readings than beakers or flasks.

When you need to find the volume or size of a solid, the method that you choose will depend upon whether or not the solid has a regular shape. For example, if you need to find the volume of a uniform block, you can measure the length, width, and height of the object, and then multiply them together using the formula: volume = length × width × height.

If you need to find the volume of a solid that is not a regular shape, you can often use what is called the "water displacement method." Imagine that we wanted to find the volume of a nail. A nail isn't close enough to a perfect cylinder to consider using the formula for the volume of a cylinder, so how would we measure its volume? We would make use of a real cylinder, a graduated cylinder. How can we find the volume of a solid in a vessel that is made to measure the volume of a liquid? We could, theoretically, melt the nail, but we would have to get it too hot and the process would change its volume slightly. We have to come up with another plan.

Have you ever poured a glass of water or soda, and then added ice cubes? Did you ever add enough ice to make the level of the liquid overflow from the glass? This happens because the ice takes up space, and in order for it to fit in the glass it pushes some of the liquid out. Solids take up space, and they will displace liquids that they are submerged in.

Take a graduated cylinder, fill it with some measurable amount of water, and record this volume. Next, carefully place the nail into the graduate, and the water level will rise. In order to be accurate, the nail must be completely submerged. Now, record the new volume. The volume of the nail will be the difference between the two water levels. Look at the formula here:

$$\mathbf{v_f - v_i = v_{nail}}$$

volume final – volume initial = volume of the nail

The Water Displacement Method

Empty graduated cylinder.	Graduate is filled with known amount of water.	Water rises by a volume that is equal to the volume of the object.

Figure 2–2a

Measuring the volume of a gas is a bit different, because a gas will take on the size of the container you put it in. This means that if you measure a gas in a 2.0 L bottle, the volume of the gas will be 2.0 L, but if you put the same sample in a 5.0 L bottle, the volume of the gas will be 5.0 L! As you will learn in Chapter 8, the volume of a gas doesn't tell us much, unless we also know the temperature of the gas and the pressure it exerts.

In Lesson 2–1 I told you that the liter is an outdated unit, but the truth is that it continues to be useful in chemistry class. In your laboratory, you can find graduated cylinders, beakers, and flasks that measure volume in liters and milliliters. To convert between the liter and the more accepted units of cubed lengths, try to commit the following conversion factors to memory.

Important Conversion Factors for Volume

1 ml = 1 cm^3 1 L = 1 dm^3 1000 cm^3 = 1 dm^3 1000 ml = 1 L

Another very important property to be able to measure in the chemistry laboratory is mass. ***Mass*** is a measure of the amount of matter that an object contains. In a sense, measuring the mass of an object is akin to detecting how many atoms are in it, because we can calculate the number of atoms in a pure sample when we know its mass. Mass and volume are often confused, because mass takes on another meaning in English. When you see something that is large, you might say, "That is massive!" In chemistry, however, mass has nothing to do with the size of an object. If you see a large object in chemistry, you should say, "That has a large volume!"

In the chemistry laboratory, you will be using an instrument called a balance to measure the mass of various materials. These balances are sometimes incorrectly referred to as "scales." Sometimes, the process of "massing" an object is improperly called "weighing" an object. Mass and weight are actually two different things. The mass of an object has nothing to do with where it is located. If you measured the mass of a brick in your chemistry lab and found it to be 1.89 kg, then it would have the same mass if you sent it into space or transported it to the moon. The number of atoms that the brick contains wouldn't change just because you moved it around, but its weight would change.

What we call ***weight*** is actually a measurement of the attraction between two objects, due to the force of gravity. If you weigh yourself in your bathroom, you are measuring the force of gravitational attraction between you and the Earth. If you weighed yourself on the moon, you would be measuring the force of gravitational attraction between you and the moon. It makes sense that your weight should change, due to where you actually do the weighing, but your mass doesn't change due to a change in location.

Does this mean that your mass never changes? Your mass is a measure of the amount of matter, or atoms, that you contain. As you eat, you take in atoms, changing your mass. In fact, as you breathe, you take in atoms, slightly changing your mass. Your mass does change, but it doesn't change based on location. A brick, which doesn't eat or breathe, will maintain a constant mass, unless it gets damaged.

Now, try the following practice questions and check your answers at the end of the chapter before moving on to the next lesson.

Lesson 2–2 Review

1. The _____________ of an object is the amount of space that it takes up.
2. The _____________ of an object is the amount of matter it contains.
3. We find the mass of an object with an instrument called a(n) _____________.
4. There are ____ cm^3 in one liter.
5. Convert 3.43 dm^3 to cm^3.
6. Convert 563 ml to L.
7. How many cubic centimeters does 4.3 liters represent?
8. The volume of a liquid can be measured in a _____________ cylinder.

9. Which of the following represents the greatest volume of water?
 a) 1 ml b) 1 dm^3 c) 100 cm^3 d) 0.5 L
10. Which of the following represents the greatest mass of water?
 a) 2.6 cg b) 124 mg c) 0.07 kg d) 450 g

Lesson 2–3: Uncertainty in Measurements

There is no such thing as a "perfect" measurement. Each measurement contains a degree of uncertainty, as a result of the limitations of the instrument used to make the measurement and the care of the person taking the measurement. An instrument, such as a graduated cylinder, only measures to a certain degree of accuracy, and no amount of care on your part will allow you to improve measurements taken from that graduate beyond that limitation. On the other hand, you could use a very sensitive instrument, and the accuracy of your measurements might be hindered by your carelessness. A chemistry student must work with the instruments that he or she has in order to take the most precise measurements possible.

Before we continue talking about measurements, we should discuss two important terms, as they apply to measurements in chemistry. The ***accuracy*** of a measurement refers to how close it is to the "true" or accepted value. The term ***precision***, which has several meanings, is used in chemistry to indicate how close together a group of measurements are to each other. The analogy of a dartboard is often used to illustrate accuracy and precision. Precise measurements are likened to a group of darts that are close together on a dartboard. Regardless of how close the darts fall to the bull's-eye, a group of darts that are close together would represent a high degree of precision. A dart striking the bull's-eye would be considered accurate.

Not all instruments are created equal. Some are designed for relatively crude measurements, whereas others are designed for careful, accurate measurements. An easy example is to compare the scale that you find in your bathroom to the scale that is used to measure fruits and vegetables at your supermarket. Many bathroom scales only have lines or marks for the whole pound. This is because even the most weight-conscious person is not likely to want to check if he or she has lost 1/10th or 1/100th of a pound. Such small changes in weight would probably be regarded as insignificant, and so we tend to round our weights off to the whole pound, or even tens of pounds, if we are not overly concerned with our weight.

In the supermarket, however, where you pay for fruit and vegetables by the pound, we are interested in the 10th or even 100th of a pound, because if you were to round off so many measurements in one day, the little increments would start to add up. For this reason, the supermarket is equipped with a much more sensitive scale than you are likely to have at home. Both scales do the jobs that they were designed to do, but they are not created equal.

As are the instruments that they use, not all students take measurements the same way. Some students take very careful measurements, recording every digit that their instrument allows; others take quicker measurements, mentally rounding off the numbers that they see. This should not be the case. If you weigh yourself on a scale at home and it reads 158 lbs., there is no harm if you round up to the nearest tens place. You can tell people that you weigh 160 lbs., and it is not likely to make any kind of difference, unless you need an accurate weight for a sports team. Chemistry, however, is a science that deals with microscopic atoms, and small variations in mass or volume are often significant. For this reason, every student should take each measurement the same way. You will want to use each instrument correctly and record every digit that the instrument allows. Try to avoid the mental "rounding off" that you might practice when taking measurements in your daily life.

When you take measurements in your chemistry laboratory, record every digit that each instrument allows, and include one additional ***estimation digit***. This is very important, as you will see in the next lesson, because when you perform calculations with these measured values, our answers can only be as accurate as our least accurate measurement. Before you use an instrument for a measurement, you should look at it and determine to what place value it will allow you to measure. Look at the ruler section shown here in Figure 2–3a.

1.0 cm 2.0 cm

Figure 2–3a

Upon examination, you should note that each line or gradation on the ruler represents a tenth of a centimeter. This means that you can accurately measure the length of an object to the tenth of a centimeter. In science, remember, you are allowed to add one additional estimation digit. This means that you can add a digit, which estimates how far between lines an object appears to reach. The place value that you should record to is dictated by the instrument, not the object that you are measuring, and every student who uses this ruler should be recording his or her measurements to the hundredth of a centimeter.

Taking Measurements in the Chemistry Laboratory

The student using the ruler on the right should record the length of the ribbon to be 1.11 cm. The first digits are "exact," because the length is clearly between 1.1 cm and 1.2 cm. The last digit is an "estimation digit," because it involves a judgment.

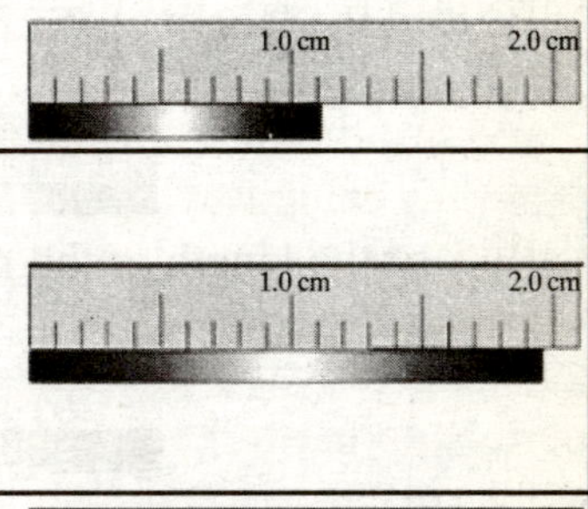

The second measurement should be recorded as 1.95 cm or 1.96 cm. Notice, it is okay to disagree on the value of the estimation digit, because it involves estimation. What you can't disagree on is the number of digits that must be recorded. Always include one and only one estimation digit.

The third ribbon appears to have a length of about 1.31 cm. If the student estimates that the ribbon falls exactly on the 0.3 cm line, he could use his estimation digit to say that, by recording the measurement as 1.30 cm. In this way, the estimation digit is still significant.

Figure 2–3b

Figure 2–3b gives some examples of how to record measurements properly. Imagine that a number of students are using this ruler to measure different strips of magnesium ribbon, such as you may use in the laboratory this year.

Some students show a lack of concern for accuracy when they take their measurements in the chemistry lab. For some, it seems more important to be quick than to be accurate. The problem with this is that the laboratory activities that your instructor is likely to offer depend upon reasonably accurate measurements. If your measurements are poor, your results will be inaccurate, and you miss those "magic" moments when the results of an experiment verify some concept or constant that you have been studying in your text. To get the most out of your learning experiences, put a great deal of care into the measurements that you take in the lab.

Try the following practice questions and check your answers at the end of the chapter before moving on to the next lesson.

Lesson 2–3 Review

1. The term ____________ refers to how close together a set of measurements are to each other.

2. A(n) ________________ measurement is close to the "true" or accepted value.
3. Explain how "estimation digits" are used in chemistry.
4. Record the length of the ribbon shown in the figure below to the proper digit.
5. Record the length of the ribbon shown in the figure below to the proper digit.

Lesson 2–4: Calculating with Significant Digits

Imagine that you wanted to find the volume of a regularly shaped wooden block in your chemistry lab. You measured the length, width, and height of the block and found them to be 4.55 cm, 9.10 cm, and 2.54 cm respectively. To find the volume of the block, you would multiply the three numbers together, as shown in Figure 2–4a.

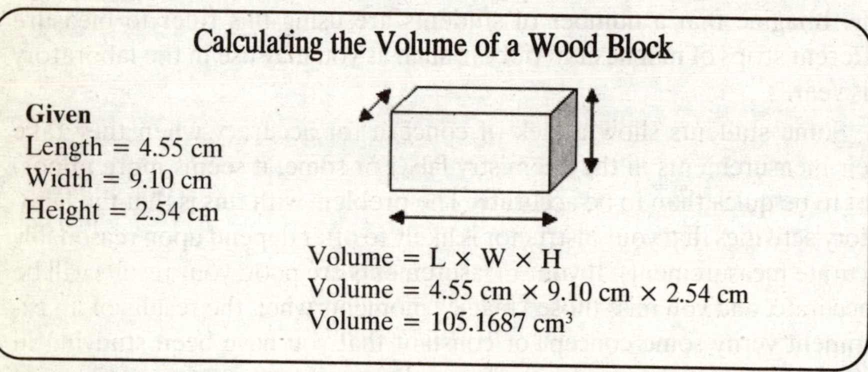

Figure 2–4a

Do you see anything wrong with the reported answer to the calculation? Notice that each of the original measurements was only considered accurate (including the estimation digit for each) to the hundredths place, or 2 digits past the decimal. However, the reported answer suggests that it is actually accurate to 4 digits past the decimal place. This doesn't make any sense! We can't have the result of a calculation that is more accurate than the measurements that the calculation was based on. When doing calculations in science, we must take care not to report answers that claim to be more accurate than the original measurements that the calculations were based on. How do we do this? By following specific rules for rounding.

Each measurement is considered to have a certain number of what are called ***significant digits*** or ***significant figures*** (sometimes called "sig. digs." or "sig. figs." for short). We determine the number of significant digits that a number shows according to the following rules:

Rules for Identifying the Number of Significant Digits in a Number

Rules	Examples
1. Any *nonzero* (digits from 1 to 9) digits are significant.	1. The number 942, with 3 *nonzero* digits, shows 3 significant digits.
2. Any zeros (regardless of how many) found *between* two significant digits are significant.	2. The number 50003, with 2 *nonzero* digits, and 3 zeros *between* significant digits, shows 5 significant digits.
3. Any zeros that are found to the right of ***both*** a significant digit and a decimal place are significant.	3. The number 75.00, with 2 *nonzero* digits, and 2 zeros that are to the right of ***both*** a significant digit and a decimal, shows 4 significant digits.

Now, when you look at a number in chemistry, you should always be aware of how many significant digits you are looking at. Some texts consider numbers such as "500" to be ambiguous examples, because the zeros don't fall under any of the rules for significant digits. Other texts, including this one, will not consider "placeholder" zeros, such as those in the number 500, to be significant. As written, the number 500 only shows 1 significant digit, the nonzero digit. The zeros are not significant, because they are neither between two significant digits nor to the right of both a decimal and a significant digit. There will be times when you get a number like 500, and you will want to show three significant digits. What options do you have for reporting 500 as having three significant digits? Look at the following.

Methods for Showing "Ambiguous" Zeros in the Number 500 as Significant Digits

Method	Example
1. Use **scientific notation** to show the zeros to the right of both a decimal and a significant digit.	1. If we show the number 500 as 5.00×10^2, the value of the number is the same, but the zeros are significant because of Rule 3 on page 53.
2. Place a decimal after the last zero to show that you actually consider the last zero significant.	2. If we show the number as 500. we will consider all three digits significant.
3. Place a bar over the last zero to indicate that the digit is significant.	3. If we show the number as $50\overline{0}$ people will consider all three digits significant.

When we do calculations in science, we must round properly and according to the rules of significant digits. We will follow one rule when adding and subtracting and another when multiplying and dividing. These two important rules are shown here.

- ***Rule for addition and subtraction:*** When adding or subtracting, our answer must show the same number of decimal places as the measurement in the problem with the least number of decimal places.
- ***Rule for multiplication and division:*** When multiplying or dividing, our answer must show the same number of significant digits as the measurement in the problem with the least number of significant digits.

Let us return to our problem with the volume of the wood block and see how we would apply the rules of rounding. (See Figure 2–4b on page 55.)

Calculating the Volume of a Wood Block

Given

Length = 4.55 cm ← shows 3 significant digits
Width = 9.10 cm ← shows 3 significant digits
Height = 2.54 cm ← shows 3 significant digits

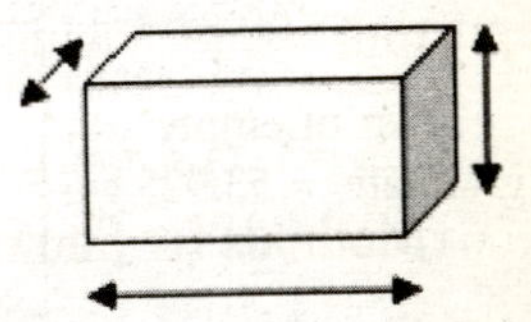

Volume = L × W × H

Notice that each of our measurements shows 3 significant digits. Because we are multiplying, we follow the rule for multiplication and division. Therefore, our answer must show the same number of significant digits as the measurement with the least number of significant digits.

$$\text{Volume} = 4.55 \text{ cm} \times 9.10 \text{ cm} \times 2.54 \text{ cm} = 105.1687 \text{ cm}^3$$

We must round to the ones place. We look to the right of the ones place and we see a "1," which tells us to round down. Our final answer becomes:

$$\text{Volume} = \mathbf{105 \text{ cm}^3}$$

Figure 2–4b

Now, our answer no longer claims to be more accurate than the measurements that it was based on. Figure 2–4b illustrates the rule for rounding after multiplication. Each of the measurements showed only three significant digits, so the calculated volume could only show three significant digits.

Before we move on to the next lesson, let's go over an example of the use of the rule for addition and subtraction. Imagine that you wanted to find the mass of a sample of water in your chemistry laboratory. You can't just pour the water onto the balance, so how would you do it? First, you would find the mass of an empty vessel, perhaps a graduated cylinder. Then, you would add the water to the graduate, and mass it again. Finally, you would subtract the mass of the empty graduate from the mass of the full graduate, and the difference would represent the mass of just the water. When you got your final answer, you would round according to the number of decimal places in your measurements, as demonstrated in Figure 2–4c.

Finding the Mass of a Sample of Water

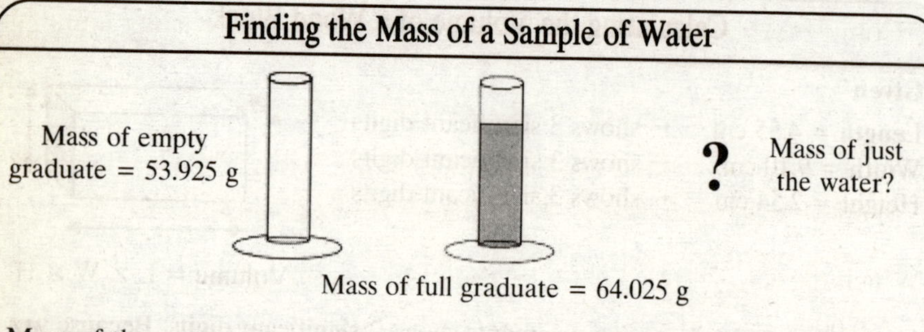

Mass of the water =
mass of the full graduate – mass of the empty graduate

Mass of the water = 64.025 g – 53.925 g = 10.1 g

How interesting! Now, we have the exact opposite of the problem that we encountered when we were calculating the volume of the wood block! Now, we have measurements that are accurate to the thousandth place, but we have an answer that is based on those measurements that only claims accuracy to the tenth place. What do we do? We follow the rule for addition and subtraction, and show an answer with as many decimal places as the measurement with the least number of decimal places. Each of the measurements show three digits past the decimal, so our answer becomes **10.100 g.** Notice, we didn't change the value of the number, we simply reported the correct degree of certainty.

Figure 2–4c

Now, try the following practice questions and check your answers at the end of the chapter before moving on to the next lesson.

Lesson 2–4 Review

1. The number 40 shows _____ significant digit(s).
2. The number 602 shows _____ significant digit(s).
3. The number 0.000043 shows _____ significant digit(s).
4. Complete the following calculation and round to the correct number of significant digits.

3.21 g + 44.100 g + 21.0 g

5. Complete the following calculation and round to the correct number of significant digits.

   2.34 cm × 0.21 cm × 32.4 cm

6. Which of the following shows the number 500 with 3 significant digits?

   a) 500 b) 500. c) 5.0×10^2 d) 5×10^3

7. Which measurement shows a total of four significant digits?

   a) 0.005 g b) 4.0×10^4 g c) 3204 g d) 5400 g

8. How many significant digits should the answer to the calculation below show?

   4.99 m × 4.0 m

   a) 1 b) 2 c) 3 d) 4

9. What is the correct answer, including significant digits, to the following calculation?

   3.42 cm + 2.3 cm + 33.4 cm

   a) 39 cm b) 39.1 cm c) 39.12 cm d) 40 cm

10. What is the correct answer, including significant digits, to the following calculation?

    2.755 cm × 5.0 cm

    a) 14 cm^2 b) 13.8 cm^2 c) 13.78 cm^2 d) 13.775 cm^2

Lesson 2–5: The Factor-Label Method

Unit conversion is such an important part of working with numbers in science that it becomes necessary for us to learn how to make these conversions in a neat and organized fashion. The ***factor-label method***, which is also called ***dimensional analysis***, is designed to do exactly that. Some students find the method intimidating and hard to learn at first, but developing the ability to use it is certainly well worth the effort. Not only will the factor-label method save you a great deal of time in the long run, but it will also improve your grade by preventing you from making careless errors in your calculations.

Let's suppose we asked two different students to convert 4.75 years into seconds. The workspace of a student who doesn't use the factor-label method might look like the space shown below.

Converting 4.75 Years to Seconds, Without the Factor-Label Method

$$4.75 \times 365 = 1733.75 \text{ days}$$

$$1733.75 \times 24 = 41610 \text{ hours}$$

$$41610 \times 60 = 2496600 \text{ minutes}$$

$$2496600 \times 60 = 14979600 \text{ seconds}$$

Now, let's look at the imaginary workspace of an imaginary student who chooses to solve the same problem using the factor-label method.

Converting 4.75 Years to Seconds, With the Factor-Label Method

$$4.75 \cancel{\text{yr}} \times \frac{365 \cancel{\text{dy}}}{1 \cancel{\text{yr}}} \times \frac{24 \cancel{\text{h}}}{1 \cancel{\text{dy}}} \times \frac{60 \cancel{\text{min}}}{1 \cancel{\text{h}}} \times \frac{60 \text{ sec}}{1 \cancel{\text{min}}} = 14979600 \text{ seconds}$$

Comparing the workspaces, you would probably agree that the factor-label method is a more organized way to make conversions. If you use the factor-label method, it is easier to check your work, and you are less likely to make a mistake in the first place, because it is such a structured process. If you are not yet convinced of the advantages of the factor-label method, take my word for it, and you will see it in the months to come.

The key to the factor-label method is the conversion factors. A ***conversion factor*** is a relationship between units that express the same property (for example, 1 hour = 60 minutes). Hours and minutes are both units that are used to express time, and the conversion factor shows how they are related.

You use conversion factors such as this all of the time in your day-to-day life. For example, if a movie has a listed running time of 90 minutes, you may mentally convert minutes to hours. Without really thinking about what you are doing, you divide 90 minutes by 60 minutes/hour to get 1 1/2 hours. Let's use this very simple example as our first demonstration of the factor-label method.

Example 1

Convert 90 minutes to hours.

Step 1. Show what you are given on the left side of your workspace and the units that you want on the right side of your workspace.

$$90 \text{ min} \qquad\qquad = \text{h}$$

Step 2. Insert the required conversion factors to change between units. The conversion factors will be in fraction form, and they will be orientated in such a way that we can cross-cancel the units that we don't want, and keep the units that we do want. In this particular example, we will only need one conversion factor, 1 hour = 60 minutes. We will put the unit for hours on the top, so that we will end with hours on the top. We will put the unit for minutes on the bottom, because we want to cross-cancel the minutes that we already have on the top.

$$90 \text{ min} \times \frac{1 \text{ h}}{60 \text{ min}} = \text{h}$$

Step 3. Cancel the units that appear on both the top and bottom, and solve the math. Notice how we have minutes on both the top and the bottom? We can cross-cancel them, because minutes divided by minutes would give us the number "1".

$$90 \,\cancel{\text{min}} \times \frac{1 \text{ h}}{60 \,\cancel{\text{min}}} = \frac{90 \text{ h}}{60} = 1.5 \text{ h}$$

Now, I am sure that this example didn't convince you of the necessity of the factor-label method, because you can do the conversion in your head. Remember that we started with an easy example, just to show you how it works. As the examples get more difficult, the usefulness of the method will become clearer. Let's try a slightly harder example now.

Example 2

If a snail moves with a constant velocity of 2.38 inches/minute, how many yards could it travel in 1 hour?

Step 1. Show what you are given on the left side of your workspace, and the units that you want on the right side of your workspace.

$$\frac{2.38 \text{ in}}{\text{min}} \qquad = \frac{\text{yd}}{\text{h}}$$

Step 2. Insert the required conversion factors to change between units. The conversion factors will be in fraction form, and they will be orientated in such a way that we can cross-cancel the units that we don't want and keep the units that we do want. In this example, we will need either 2 or 3 conversion factors, depending on how you choose to solve it. If you convert inches to yards directly, you will only need 2 conversion factors. For the sake of clarity, I will use two steps to convert inches to yards and will use a total of 3 conversion factors.

60 minutes = 1hour; 12 inches = 1 foot; 3 feet = 1 yard.

$$\frac{2.38 \text{ in}}{\text{min}} \times \frac{60 \text{ min}}{1 \text{ hr}} \times \frac{1 \text{ ft}}{12 \text{ in}} \times \frac{1 \text{ yd}}{3 \text{ ft}} = \frac{\text{yd}}{\text{h}}$$

Before we move on to the next step, look carefully at how we set up the conversion factors. The actual order that you write the conversion factors doesn't matter at all, so I could have started with 1 foot = 12 inches, or 1 yard = 3 feet. What does matter is that you set it up so that you can cross-cancel the units that you don't want. You know you set it up correctly if you can cross out all of the units, except for the ones that you wanted to keep. Finally, notice that each conversion factor is reducible to "1," because both the denominator and the numerator represent the same value. You would not set up a fraction that reads 12 inches / 60 minutes, because that is not a true conversion.

Step 3. Cancel the units that appear on both the top and bottom, and solve the math.

$$\frac{2.38 \cancel{\text{in}}}{\cancel{\text{min}}} \times \frac{60 \cancel{\text{min}}}{1 \text{ hr}} \times \frac{1 \cancel{\text{ft}}}{12 \cancel{\text{in}}} \times \frac{1 \text{ yd}}{3 \cancel{\text{ft}}} = \frac{142.8 \text{ yd}}{36 \text{ h}}$$

$$= \mathbf{3.97 \text{ yd/h}}$$

You may wonder why I rounded the answer the way that I did. The original problem gave me 3 significant digits, so I rounded the final answer to 3 significant digits. When rounding, we don't take the conversion factors into account, or we would always be rounding to 1 significant digit.

When you are more comfortable with the factor-label method and the International System of Measurements, you will find these conversions very easy. You will not have to write out the names of all of the units; rather, you will just use symbols. Before you move on to the next lesson, look at one final example of how the factor-label method is used. I will give you a problem to try on paper. See if you can do it on your own, and then check your work and your answer against what I wrote here.

Example 3

If an animal were to consume exactly 25 g of food each day, how many kg of food would it consume in one year?

Remember to try it on your own before checking the answer that follows!

$$\frac{25\ \not{g}}{\not{dy}} \times \frac{1\ \text{kg}}{1000\ \not{g}} \times \frac{365\ \not{dy}}{1\ \text{yr}} = \frac{9.125\ \text{kg}}{\text{yr}}$$

which rounds to **9.1 kg/yr**

Once again, the order that you place down the conversion factors doesn't matter. What is important is that you were able to cross-cancel the units that you didn't want and that you got the same final answer. Did you remember to round your answer to 2 significant digits? We rounded that way because we started off with 2 significant digits in the problem.

Now, try the following practice questions and check your answers at the end of the chapter before moving on to the next lesson.

Lesson 2–5 Review

Use the factor-label method to make the conversions in numbers 1–5.

1. 4.53 km to cm
2. 2.34 days to seconds
3. 0.23 yards to inches
4. 45.4 centimeters/second to meters/hour
5. 0.03 milliliters/second to liters/week
6. Which of the following represents a proper conversion factor?

   a) $\frac{4\ \text{feet}}{\text{second}}$ b) $\frac{1\ \text{m}}{100\ \text{cm}}$ c) $\frac{3\ \text{yd}}{\text{ft}}$ d) $\frac{\text{day}}{\text{week}}$
7. Which of the following represents a proper conversion factor?

   a) $\frac{1\ \text{mg}}{1000\ \text{g}}$ b) $\frac{1000\ \text{km}}{1\ \text{m}}$ c) $\frac{100\ \text{cm}}{\text{ft}}$ d) $\frac{1000\ \text{cm}^3}{1\ \text{L}}$
8. Which of the following does not represent a proper conversion factor?

   a) $\frac{1\ \text{g}}{100\ \text{cm}^3}$ b) $\frac{1\ \text{km}}{1000\ \text{m}}$ c) $\frac{1000\ \text{cm}^3}{1\ \text{L}}$ d) $\frac{1\ \text{cm}^3}{1\ \text{ml}}$
9. Which of the following conversion factors could you make use of, if you were converting km/h to m/s?

   a) $\frac{1\ \text{km}}{100\ \text{cm}^3}$ b) $\frac{60\ \text{s}}{1\ \text{h}}$ c) $\frac{1\ \text{km/h}}{1000\ \text{m/s}}$ d) $\frac{1000\ \text{m}}{1\ \text{km}}$
10. Which of the following conversion factors could you make use of, if you were converting L/min to ml/s?

    a) $\frac{1\ \text{L}}{100\ \text{cm}^3}$ b) $\frac{60\ \text{min}}{1\ \text{h}}$ c) $\frac{1\ \text{km}}{1000\ \text{m}}$ d) $\frac{1000\ \text{ml}}{1\ \text{L}}$

Lesson 2–6: Density Calculations

Try asking a friend why a balloon floats, and it is likely that he or she will incorrectly say, "Because it is light." Then ask the friend why a hot-air balloon with several people aboard floats, and he or she may not know what to say. Understanding density will help you understand many things that you can observe in the real world.

Density is defined as the amount of matter in a given unit of volume. A material with high density will have its particles (atoms or molecules) tightly packed together. A material with low density will have more space between its particles. There is more to density than just size. Just as you can have a large object with high density, you can have a large object with low density. The relationship between the density, volume, and mass of an object can be best understood by looking at the formula for density.

The Formula for Density

$$\textbf{Density} = \frac{\textbf{mass}}{\textbf{volume}} \quad \textbf{or} \quad \textbf{D} = \frac{\textbf{m}}{\textbf{v}}$$

We can see from the formula that the density of the object is directly proportional to the mass of the object. This means that if you could double the mass of an object without changing its volume, its density would also double. The formula also shows us that the density of an object is inversely proportional to its volume. This means that if you were to double the volume of an object, without changing its mass, its density would be divided by 2.

Whether or not something floats depends upon its density, compared to the density of the material that it is floating in. A balloon will float in the air if it is less dense than the air around it. Icebergs float in liquid water because they are less dense than the water.

Did you know that you can control your density to a certain extent? Have you ever tried to float on the surface of a pool, or any body of water? How did you do it? Did you fill your lungs up with air, before lying across the surface of the water? If so, you decreased your density!

When you fill your lungs with air, your chest expands, which increases your volume. It is true that you also increase your mass by some insignificant amount as you draw in molecules from the air around you, but your volume will increase by a much more significant amount. Look at the formula for density, and think about what happens to your density as your chest expands, and your volume increases.

$$\downarrow \textbf{Density} = \frac{\textbf{mass}}{\textbf{volume} \uparrow}$$

As your volume increases, the denominator in the formula increases and your density decreases, helping you to float.

Figure 2–6a

Have you ever tried to sink and lie down on the bottom of a pool? How did you do it? Did you try to expel as much air from your lungs as you could? If you did, your chest contracted, decreasing your volume and increasing your density.

$$\uparrow \textbf{Density} = \frac{\textbf{mass}}{\textbf{volume} \downarrow}$$

As your volume decreases, the denominator in the formula decreases and your density increases, allowing you to sink.

Figure 2–6b

In order to calculate the density of a material, you need to know its mass and volume. Both of these properties can be easily measured in the laboratory, using the methods described in Lesson 2–2. Let's imagine that you place a rock on a balance to measure its mass and find it to have a mass of 22.516 g. You then use the water displacement method to determine the volume of the rock, which you find to be 7.84 cm^3. You would then use the density formula to find the density of the rock, as shown here:

Example 1

Find the density of a rock with a mass of 22.516 g and a volume of 7.84 cm^3.

Given: mass = 22.516 g volume = 7.84 cm^3

Find: density

$$\text{Density} = \frac{\text{mass}}{\text{volume}} = \frac{22.516 \text{ g}}{7.85 \text{ cm}^3} = 2.87 \text{ g/cm}^3$$

Note: We rounded our answer to 3 significant digits, according to the rules covered in Lesson 2–4, because the number 7.84 only shows 3 significant digits.

It is likely that you will be asked, at times, to use the density formula for calculations involving a known density. You can solve a problem with only one unknown, but the unknown can be the mass, volume, or density. Let's suppose you were asked to solve the following problem.

Example 2

Gold has a density of 19.3 g/cm^3 at standard temperature and pressure (STP). What would be the mass of a 6.00 cm^3 sample of this precious metal?

The first thing that you want to do is to list what you have been given, and what you are asked to find. This will help keep you from making the mistake of solving for the incorrect unknown. Then, take the original formula for density and isolate the unknown, which in this case is the mass. You do this by multiplying both sides by "volume."

Given: volume = 6.00 cm^3 density = 19.3 g/cm^3

Find: mass

Formula: $\text{Density} = \dfrac{\text{mass}}{\text{volume}}$

$$\textit{volume} \times \text{density} = \frac{\text{mass}}{\cancel{\text{volume}}} \times \cancel{\textit{volume}}$$

After we cancel both volumes on the right-hand side, we get our new working formula, which is: mass = volume × density. Then we plug in the numbers, cross out units, solve the calculation, and round.

$$\text{mass} = \text{volume} \times \text{density}$$

$$= 6.00\ \cancel{\text{cm}^3} \times 19.3\ \text{g}/\cancel{\text{cm}^3}$$

$$= \mathbf{116\ g}$$

Once again, we rounded to three significant digits, because that is the least number of significant digits shown in the original problem.

You might also be asked to solve a density calculation for a substance when the volume is the unknown quantity. Some students have trouble isolating this unknown, and a review of algebra would be important for these students. If volume is the unknown, you isolate it by multiplying both sides of the original equation by volume, and then dividing both sides of the equation by density, as shown here.

Isolating Volume in the Density Formula

Original Fomula: $\text{Density} = \dfrac{\text{mass}}{\text{volume}}$

Step 1. Multiply both sides by volume: $\textit{volume} \times \text{density} = \dfrac{\text{mass}}{\cancel{\text{volume}}} \times \cancel{\textit{volume}}$

$\text{volume} \times \text{density} = \text{mass}$

Step 2. Divide both sides by density: $\dfrac{\text{volume} \times \cancel{\text{density}}}{\cancel{\text{density}}} = \dfrac{\text{mass}}{\text{density}}$

This gives us our working formula for solving for volume: $\text{volume} = \dfrac{\text{mass}}{\text{density}}$ or $V = \dfrac{m}{D}$

Now, let's try an example where the volume is the unknown.

Example 3

How much space would a 2.75 g sample of a substance occupy, if it has a density of 0.993 g/cm³?

Given: m = 2.75 g D = 0.993 g/cm³

Find: volume

$$v = \frac{m}{D} = \frac{2.75\ \text{g}}{0.993\ \text{g/cm}^3} = 2.7693857\ \text{cm}^3$$

which rounds to **2.77 cm³**

Try the following practice questions and check your answers at the end of the chapter before moving on to the next lesson.

Lesson 2–6 Review

1. What is the density of an object with a mass of 2.34 g and a volume of 7.3 cm³?
2. Calculate the mass of a 0.50 cm³–sized sample of a material with a density of 5.5 g/cm³.
3. What is the mass of a 1.29 cm³ sample of a material with a density of 32.7 g/cm³?

4. What would be the volume of a 1.2 g sample of a material with a density of 0.89 g/cm^3?
5. Calculate the volume of a 56.4 g sample of a material with a density of 1.12 g/cm^3.

Chapter 2 Examination

Part I—Matching

Match the following terms to the definitions that follow.

a. kilogram
b. ampere
c. centi-
d. milli-
e. mass
f. weight
g. volume
h. density
i. Kelvin

_____1. The amount of matter in an object.

_____2. The SI unit for electric current.

_____3. The attraction between two objects due to gravity.

_____4. The SI prefix that means "thousandth."

_____5. The SI base unit for mass.

_____6. The amount of space that an object occupies.

_____7. The SI base unit for temperature.

_____8. The amount of mass in a given unit of volume.

_____9. The SI prefix that means "hundredth."

Part II—Conversions and Calculations

Perform the following conversions and calculations.

10. Use the factor-label method to perform the following conversions.
    a) 2.34 g to cg
    b) 0.0335 km to cm
    c) 593 dm to m
    d) 3.45 dm^3 to cm^3
11. Use the factor-label method to perform the following conversions.
    a) 9.34 g/cm^3 to cg/L
    b) 0.0512 km/h to cm/s
    c) 1.3 L/h to ml/s
    d) 33.4 kg/dm^3 to g/cm^3

12. Identify the number of significant digits in each of the following numbers.
    a) 3009 b) 3.000 c) 0.0004 d) 4.500×10^8

13. Perform each of the following calculations and round to the correct number of significant digits.
    a) 53 + 3.59
    b) 76.32 – 34.9
    c) 3.45 x 2.9
    d) $(4.5 \times 10^8) / (3.42 \times 10^5)$

14. What would be the mass of a 4.53 cm^3 sample of a material with a density of 0.852 g/cm^3?

15. How much space would a 0.434 g sample of a material occupy if it has a density of 1.23 g/cm^3?

Part III—Multiple Choice

For each of the following questions, select the best answer.

16. Which of the following shows SI prefixes from smallest to largest?
    a) pico, milli, centi, nano
    b) pico, nano, milli, centi
    c) milli, centi, nano, pico
    d) nano, pico, milli, centi

17. Which of the following is not an SI base unit?
    a) ampere b) kilogram c) Newton d) second

18. Which of the following represents the greatest volume of water?
    a) 3.34 dm^3 b) 3412 ml c) 9213 cm^3 d) 22.9 L

Part IV—Short Essay

Answer the following questions completely.

19. Explain the difference between mass and density.

20. Explain the difference between mass and weight.

Answer Key

The actual answers will be shown in brackets, followed by the explanation. If you don't understand an explanation that is given in this section, you may want to go back and review the lesson that the question came from.

Lesson 2–1 Review

1. [meter]—As shown in Figure 2–1c, the meter measures length.
2. [electric current]—Again, see Figure 2–1c.
3. [Kelvin]—Although you will often measure temperature in Celsius degrees in your laboratory, Kelvin is the scale accepted by the International System of Measurements.
4. [100]—The prefix "centi-" means hundredth.
5. [100 000]—Each kilogram contains 1000 grams, and each gram is made up of 100 centigrams.

$$1\ \cancel{\text{kg}} \times \frac{1000\ \cancel{\text{g}}}{1\ \cancel{\text{kg}}} \times \frac{100\ \text{cg}}{1\ \cancel{\text{g}}} = 100000\ \text{cg}$$

6. [345]—Each meter is made up of 100 centimeters.

$$3.45\ \cancel{\text{m}} \times \frac{100\ \text{cm}}{1\ \cancel{\text{m}}} = 345\ \text{cm}$$

7. [0.764]—a decimeter is 1/10th of a meter, and a meter is 1/1000th of kilometer.

$$7640\ \cancel{\text{dm}} \times \frac{1\ \cancel{\text{m}}}{10\ \cancel{\text{dm}}} \times \frac{1\ \text{km}}{1000\ \cancel{\text{m}}} = 0.764\ \text{km}$$

8. [D. joule]—The joule is an SI-derived unit, made from a combination of base units.
9. [D. joule]—As shown in Figure 2–1f, the joule is a unit of energy.
10. [A. speed]—To measure the rate of motion, you need a unit of length and a unit of time.

Lesson 2–2 Review

1. [volume]—Although often confused with mass, volume is a measure of how much space an object occupies.
2. [mass]—Massive objects needn't be very large; they must contain a lot of matter.
3. [balance]
4. [1000]—Conversion factors for volume can be found on page 47
5. [3430 cm^3]— $3.43\ \cancel{\text{dm}^3} \times \dfrac{1000\ \text{cm}^3}{1\ \cancel{\text{dm}^3}} = 3430\ \text{cm}^3$
6. [0.563 L]— $563\ \cancel{\text{ml}} \times \dfrac{1\ \text{L}}{1000\ \cancel{\text{ml}}} = 0.563\ \text{L}$
7. [4300]— $4.3\ \cancel{\text{L}} \times \dfrac{1000\ \text{cm}^3}{1\ \cancel{\text{L}}} = 4300\ \text{cm}^3$
8. [graduated]

9. [B. 1 dm^3]—1 dm^3 represents 1000 ml, which is greater than answer choice A; 1 dm^3 represents 1000 cm^3, which is more than answer C; 1 dm^3 is equivalent to 1 L, which is more than answer choice D.

10. [D. 450 g]—If we convert all of the answers into common units, we find that

A. $2.6\ \cancel{cg} \times \frac{1\ g}{100\ \cancel{cg}} = 0.026\ g$; B. $124\ \cancel{mg} \times \frac{1\ g}{1000\ \cancel{mg}} = 0.124\ g$; and

C. $0.07\ \cancel{kg} \times \frac{1000\ g}{1\ \cancel{kg}} = 70\ g$. Answer D is more massive than the other choices.

Lesson 2-3 Review

1. [precision]
2. [accurate]
3. [An estimation digit is added to certain digits of each measurement in order to report the measurement with maximum precision.]
4. [1.13 cm]—The ribbon is clearly in between the lines indicating 1.1 and 1.2 cm. It is less than halfway beyond the 1.1 line, which would indicate 1.15 cm. You may judge the length of the ribbon to be 1.12 cm or 1.14, and that would be fine. What you can't do is record more or less digits, as in 1.1 cm or 1.130 cm.
5. [1.51 cm]—The ribbon is clearly just past the 1.5 cm mark. You might estimate its length at 1.51 cm or 1.52 cm, but you must include one, and only one, estimation digit.

Lesson 2-4 Review

For questions 1–6, the significant digits are shown in boldface within the parentheses.

1. [1]—(**4**0) In the number 40, the 4 is significant, because it is a nonzero digit. The 0 is only a placeholder and is not significant.
2. [3]—(**602**) In the number 602, the 6 and the 2 are significant, because they are nonzero digits. The 0 is also significant, because it is set between two significant digits.
3. [2]—(0.0000**43**) The number 0.000043 only shows two significant digits: the 4 and the 3. All of the zeros are placeholders. Remember: A zero must be to the right of **both** a decimal place **and** a significant digit to be significant.
4. [68.3]—The original answer to the calculation is 68.310 g. However, the measurement 21.0 g from the problem is considered accurate only to the tenth place. We needed to round to the tenth place.
5. [16]—The original answer to the calculation is 15.92136. The measurement of 0.21 cm from the problem only shows two significant digits, so we must round to two significant digits. The 9 in the tenth place tells us to round up to 16.

6. [B. 500.]—In this case, the decimal point wouldn't be there, unless it was to indicate that the zeros represent significant digits. This answer shows one of the three acceptable methods for showing extra significant digits.
7. [C. 3204 g]—Answer A only shows 1 significant digit, as all of the zeros are placeholders. Answers B and D show two significant digits.
8. [B. 2]—The measurement from the problem with the least number of significant digits, 4.0 m, shows two significant digits. We would need to round our answer to 2 significant digits.
9. [B. 39.1 cm]—The answer before rounding was 39.12 cm. Two of the measurements in the problem are considered accurate to the tenth place, so we must round the answer to the tenth place.
10. [A. 14 cm^2]—The original answer was 13.775 cm^2. We must round to two significant digits because the measurement 5.0 cm shows only 2 significant digits.

Lesson 2–5 Review

1. [453000 cm]—$4.53\ \cancel{\text{km}} \times \frac{1000\ \cancel{\text{m}}}{1\ \cancel{\text{km}}} \times \frac{100\ \text{cm}}{1\ \cancel{\text{m}}} = 453000\ \text{cm}$
2. [202176 s, which rounds to 202000 s]—
$$2.34\ \cancel{\text{dy}} \times \frac{24\ \cancel{\text{h}}}{1\ \cancel{\text{dy}}} \times \frac{60\ \cancel{\text{min}}}{1\ \cancel{\text{h}}} \times \frac{60\ \text{s}}{1\ \cancel{\text{min}}} = 202176\ \text{s}$$
3. [8.28 inches, which rounds to 8.3 inches]—$0.23\ \cancel{\text{yd}} \times \frac{3\ \cancel{\text{ft}}}{1\ \cancel{\text{yd}}} \times \frac{12\ \text{in}}{1\ \cancel{\text{ft}}} = 8.28\ \text{in}$
4. [1634.4 m/h, which rounds to 1630 m/h]—
$$\frac{45.4\ \cancel{\text{cm}}}{\cancel{\text{s}}} \times \frac{60\ \cancel{\text{s}}}{1\ \cancel{\text{min}}} \times \frac{60\ \cancel{\text{min}}}{1\ \text{h}} \times \frac{1\ \text{m}}{100\ \cancel{\text{cm}}} = 1634.4\ \text{m/h}$$
5. [18.144 L/wk, which rounds to 20 L/wk]—
$$\frac{0.03\ \cancel{\text{ml}}}{\cancel{\text{s}}} \times \frac{3600\ \cancel{\text{s}}}{1\ \cancel{\text{h}}} \times \frac{24\ \cancel{\text{h}}}{1\ \cancel{\text{dy}}} \times \frac{7\ \cancel{\text{dy}}}{1\ \text{wk}} \times \frac{1\ \text{L}}{1000\ \cancel{\text{ml}}} = 18.144\ \text{L/wk}$$
6. [B. $\frac{1\ \text{m}}{100\ \text{cm}}$]—Remember: In order to construct a good conversion factor, the values in the denominator and numerator must be the same. 1 m = 100 cm, but 3 yd ≠ 1 ft.
7. [D. $\frac{1000\ \text{cm}^3}{1\ \text{L}}$]—Remember: 1000 km ≠ 1 m, but 1km = 1000 m!
8. [A. $\frac{1\ \text{g}}{100\ \text{cm}^3}$]—Conversion factors must be reducible to 1.
9. [D. $\frac{1000\ \text{m}}{1\ \text{km}}$]
10. [D. $\frac{1000\ \text{ml}}{1\ \text{L}}$]—Answers B, C, and D are all proper conversion factors, but only the conversion factor from answer D will help us get where we want to go.

Lesson 2–6 Review

1. [0.32 g/cm³]—$D = \frac{m}{v} = \frac{2.34\ g}{7.3\ cm^3} = 0.3205479452\ g/cm^3$, which rounds to $0.32\ g/cm^3$
2. [2.8 g]—$m = D \times v = 5.5\ g/cm^3 \times 0.50\ cm^3 = 2.75\ g$, which rounds to 2.8 g
3. [42.2 g]—$m = D \times v = 32.7\ g/cm^3 \times 1.29\ cm^3 = 42.183\ g$, which rounds to 42.2 g
4. [1.3 cm³]—$v = \frac{m}{D} = \frac{1.2\ g}{0.89\ g/cm^3} = 1.348314607\ cm^3$, which rounds to $1.3\ cm^3$
5. [50.4 cm³]—$v = \frac{m}{D} = \frac{56.4\ g}{1.12\ g/cm^3} = 50.35714286\ cm^3$, which rounds to $50.4\ cm^3$

Chapter 2 Examination

1. [e. mass]
2. [b. ampere]
3. [f. weight]
4. [d. milli-]
5. [a. kilogram]
6. [g. volume]
7. [i. Kelvin]—Although you probably use Celsius thermometers in the lab, you need to convert the temperatures to Kelvin in many lab activities.
8. [h. density]
9. [c. centi-]—There are one hundred years in a **cent**ury, but there are 100 **centi**meters in one meter.
10. A. [234 cg]— $2.34\ g \times \frac{100\ cg}{1\ g} = 234\ cg$

    B. [33500 cm]— $0.0335\ km \times \frac{1000\ m}{1\ km} \times \frac{100\ cm}{1\ m} = 33500\ cm$

    C. [59.3 m]— $593\ dm \times \frac{1\ m}{10\ dm} = 59.3\ m$

    D. [3450 cm³]— $3.45\ dm^3 \times \frac{1000\ cm^3}{1\ dm^3} = 3450\ cm^3$
11. A. [934000 cg/L]— $\frac{9.34\ g}{cm^3} \times \frac{100\ cg}{1\ g} \times \frac{1000\ cm^3}{1\ L} = 934000\ cg/L$

    B. [1.42 cm/s]— $\frac{0.0512\ km}{h} \times \frac{1000\ m}{1\ km} \times \frac{100\ cm}{1\ m} \times \frac{1\ h}{3600\ s} = 1.42\ cm/s$

    C. [0.36 ml/s]— $\frac{1.3\ L}{h} \times \frac{1000\ ml}{L} \times \frac{1\ h}{3600\ s} = 0.36\ ml/s$

    D. [33.4 g/cm³]— $\frac{33.4\ kg}{dm^3} \times \frac{1000\ g}{1\ kg} \times \frac{1\ dm^3}{1000\ cm^3} = 33.4\ g/cm^3$
12. A. [4]—The 3 and the 9 are significant because they are nonzero digits. The two zeros are significant because they are found between two significant digits.

B. [4]—The three is significant because it is a nonzero digit. All three zeros are significant because they are found to the right of both a significant digit **and** a decimal.

C. [1]—Only the 4 is significant, being a nonzero digit. All of the zeros are placeholders. Some of them are to the right of a decimal, but in order to be significant, they need to be to the right of a significant digit as well.

D. [4]—Remember, we only focus on the coefficient. The 4 and 5 are nonzero digits, and both zeros are to the right of a decimal and a significant digit.

13. A. [57]—(53 + 3.59 = 56.59) We need to round to the "ones" place, according to the rule for addition and subtraction.

B. [41.4]—(76.32 – 34.9 = 41.42) We need to round to the "tenth" place, because 34.9 is considered accurate to the tenth place.

C. [$1\bar{0}$]—(3.45 x 2.9 = 10.005) We want our answer to show two significant digits, so I put a line over the zero to make it significant. I could have used scientific notation instead, and written 1.0×10^1.

D. [1.3×10^3]—($\frac{4.5 \times 10^8}{3.42 \times 10^5} = 1.315789474 \times 10^3$) According to the rule for multiplying and dividing, we want our answer to show only two significant digits.

14. [3.86 g]—$m = D \times v = 0.852\ g/\cancel{cm^3} \times 4.53\ \cancel{cm^3} = 3.85956\ g$

15. [0.353 cm³]—$v = \frac{m}{D} = \frac{0.434\ \cancel{g}}{1.23\ \cancel{g}/cm^3} = 0.3528455285\ g/cm^3$

16. [B. pico, nano, milli, centi]—As shown in Figure 2–1d, answer B has the prefixes arranged from smallest to largest.

17. [C. Newton]—Although used very often in physics, the Newton is an SI-derived unit, rather than a base unit.

18. [D. 22.9 L]—By converting all of the units to liters, we can see that none of the other answers are nearly as large as answer choice D.

19. [Mass is a measure of the amount of matter in an object. Density is a measure of how much matter is in a given unit of volume.]—The density of an object is directly proportional to its mass, and inversely proportional to its volume. The formula for density is $\text{Density} = \frac{\text{mass}}{\text{volume}}$. The density of an object might be expressed in g/cm³, Kg/L, or any other unit of mass over a unit of volume.

20. [Mass is a measure of how much matter an object has, whereas weight is a measure of the force of attraction between two objects (one of these is usually the Earth).]—More massive objects do weigh more than less massive objects, under the same conditions, but there is more to weight than just mass. The weight of an object also has to do with the distance between it and the object (usually the Earth) that is attracting it.

Atomic Structure

Lesson 3–1: Modern Atomic Theory and Model

The concept of "atoms" is believed to have originated in ancient Greece. Democritus and his teacher Leppicutius were supporters of what has been called a ***discontinuous theory*** of matter, which means that all matter is made up of tiny particles, which are separated by spaces. They argued that if you kept cutting a sample of a substance up into smaller and smaller pieces, you would eventually get down to the smallest complete piece of that substance. Democritus called this smallest particle ***atomos***, which means indivisible.

This atomic theory didn't attract too many followers over the next 2,000 years or so. However, in the 1600s and 1700s, early chemists began publishing the results of experiments that they were carrying out. Evangelista Torricelli (1608–1647) experimented with air pressure. Robert Boyle (1627–1691) discovered and published his gas law, which we will study in Chapter 8. In 1661, Robert Boyle published *The Sceptical Chymist*, which argued the virtues of an atomic theory.

Despite Robert Boyle's best efforts, the atomic theory did not become widely accepted during his lifetime. Most scientists agree that the "birth" of modern chemistry had to wait almost another 100 years after Boyle's death, when Antoine Lavoisier (1743–1794) would publish his great work, *Traite Elementaire de Chimie*, in 1789. Considered by many to be the founder of modern chemistry, Lavoisier carried out carefully controlled experiments, which provided real evidence for the Law of Conservation of Mass, which we covered in Lesson 1–4.

John Dalton (1766–1844) summarized the works of Lavoisier and many other early chemists in a "modern" atomic theory, which stated:

1. All matter is made up of tiny indivisible particles called atoms.
2. Atoms of the same element have identical properties, including identical mass.
3. Atoms of different elements have different properties, including different mass.
4. Atoms combine in fixed whole number ratios to form compounds.
5. A specific compound is always made up of atoms in a specific proportion.

For his work, John Dalton is often considered the "father" of modern atomic theory.

As has the atomic theory, the atomic model has changed and developed over the years since Democritus. Democritus pictured atoms as being solid, indestructible, and completely uniform. Two thousand years later, J.J. Thomson proposed the "plum-pudding" model of the atom, with negatively charged electrons embedded in the positively charged bulk of the atom. Ernest Rutherford's famous "gold-foil" experiment showed that most of the atom was empty space, leading to Niels Bohr's "planetary" model of the atom, where the electrons orbited around the nucleus the way planets do a star. Werner Heisenberg's "Uncertainty Principle" showed that the planetary model of the atom was too specific. Erwin Schrodinger introduced the Wave-Mechanical model of the atom, which is still the way that most people think of atoms today. More recently, however, the idea that even the subatomic particles are made up of smaller particles called quarks suggests that our model of the atom is still far from complete.

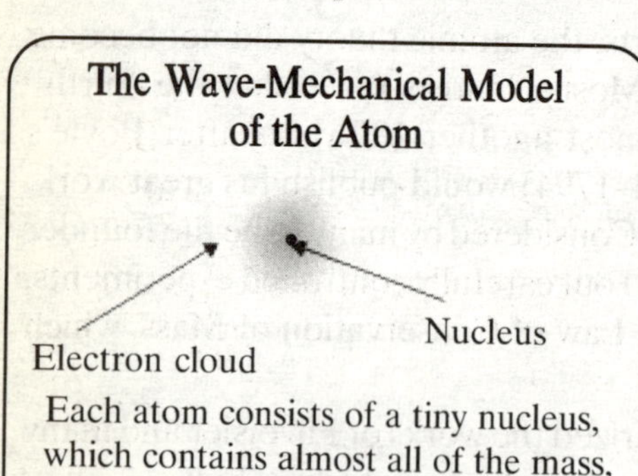

Figure 3–1a

In the current model of the atom we imagine a tiny nucleus, which contains the vast majority of the mass of the atom. It is in this nucleus that we find the protons (p^+) and neutrons (n^0) of the atom. The electrons (e^-) represent the third type of subatomic particle, and they are found outside of the nucleus, occupying an area called the electron cloud.

A ***proton*** is a positively charged subatomic particle with a mass of approximately 1 atomic mass unit (amu) and a charge of +1. It is the number of protons in the nucleus of an atom, or the ***nuclear charge***, that gives the element its identity. For example, any atom containing only one proton in its nucleus is considered an atom of the element hydrogen, whereas any atom containing eight protons in its nucleus is considered an oxygen atom. The number of protons in an atom is called the ***atomic number*** of that element. So, we can say that the atomic number of hydrogen is 1, and the atomic number of oxygen is 8. A single proton is often represented by the symbol (p^+) or ($^{1}_{1}p^+$). Because a single proton is identical to the nucleus of a form of hydrogen, the symbol (H^+), which will we see often when we study acids, also represents a proton.

A ***neutron*** ($^{1}_{0}n$) is a neutrally charged subatomic particle, which, as does the proton, has a mass of approximately 1 atomic mass unit (amu). When we add the total number of protons and neutrons in the nucleus of an atom, we get the atom's ***mass number***. Because the neutron has no charge, it does not affect the atomic number and does not alter the identity of the element. For this reason, it is possible to have two atoms of the same element with differing mass numbers, because they have different numbers of neutrons. Atoms of the same element with different masses are called ***isotopes***. For example, there are three different isotopes of hydrogen, as shown in Figure 3–1b. Please note that the images in this figure are not drawn anywhere near to scale.

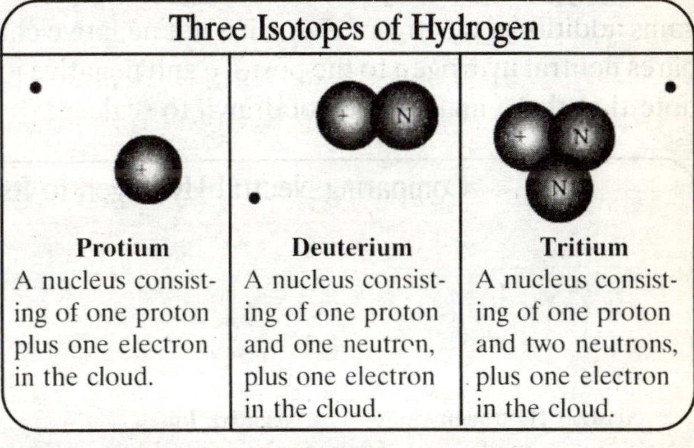

Figure 3–1b

The atomic mass unit (amu), which is represented with the symbol "u," is based on a particular isotope of carbon, called carbon-12. Carbon-12 is considered to have a mass of exactly 12 u, and all of the other elemental isotopes are measured relative to that isotope. The ***atomic masses***, shown on the periodic table, represent a weighted average of the masses of the naturally occurring isotopes of each element. For example, some periodic tables show an atomic mass of 1.00794 u for hydrogen, despite the fact that no particular isotope of hydrogen has a mass number equal to that value.

Chemists come up with an average, based on the mass numbers of the isotopes and the relative abundance by which they appear. The fact that the atomic number of hydrogen is so close to the mass number of the isotope of hydrogen known as ***protium*** indicates that the vast majority of the hydrogen atoms found in nature (approximately 99%) are of this type.

Electrons (e^-), which are located outside of the nucleus, have essentially no mass and a charge of -1. Each electron has a charge that is the equal and opposite to the charge of a proton but contains only about 1/1836th of the mass of a proton. Because atoms typically have the same number of electrons as protons, they normally have a net charge of zero and are said to be neutral. For example, if an atom of oxygen has 8 protons ($8 \times +1 = +8$) and 8 electrons ($8 \times -1 = -8$), then ($(+8) + (-8) = 0$) the atom of oxygen has a net charge of zero.

Atoms will often gain or lose one or more electrons to become charged atoms called ***ions***. An atom that has lost some of its electrons would have more protons than electrons and therefore have a net positive charge. We call this type of atom a ***positive ion***. ***Negative ions*** are formed when an atom gains additional electrons, giving it a net negative charge. Figure 3–1c compares neutral hydrogen to the positive and negative ions of hydrogen. Please note that these images are not drawn to scale.

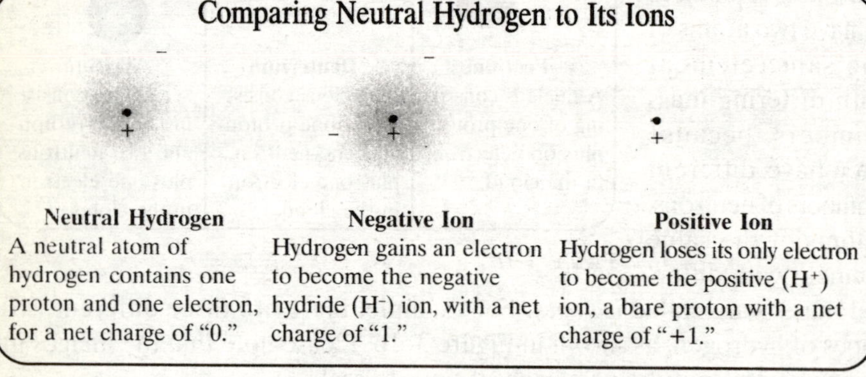

Figure 3–1c

Lesson 3–1 Review

1. A(n) _______________ is an atom that has gained or lost one or more electrons.
2. A(n) _______________ is a neutrally charged subatomic particle found in the nucleus.

3. The atomic mass unit (u) is based on an atom of the isotope ________________.

4. The ________________ number of an atom is equal to the number of protons it has.

5. The ________________ number of an atom is equal to the total number of protons and neutrons in its nucleus.

6. Atoms of the same element with differing masses are called ________________.

Lesson 3–2: Elemental Symbols

In Chapter 5 we will go over chemical formulas, which are really the "words" that make up the language of chemistry. The elemental symbols, which are shown on the Periodic Table of Elements, are the letters that make up the (words) chemical formulas. Chemical formulas combine in chemical equations to form the "sentences" in the language of chemistry. Before you can be expected to correctly write the chemical equations (sentences) or the chemical formulas (words), you must make sure that you are using the elemental symbols (letters) correctly.

One chemical equation that you may be familiar with, from your biology studies, is the equation for photosynthesis, shown here:

$$\mathbf{6CO_2 + 6H_2O \rightarrow C_6H_{12}O_6 + 6O_2}$$

The entire statement represents the chemical equation. The chemical equation is made up of four chemical formulas: CO_2, H_2O, $C_6H_{12}O_6$, and O_2. These chemical formulas are made up of either one, two, or three elemental symbols. For example, $C_6H_{12}O_6$ is made up of three elemental symbols (C, H, and O).

Really, the only "trick" to working with elemental symbols is to be careful to distinguish between uppercase and lowercase letters. There is actually a huge difference between the substances shown in the formulas "CO" and "Co." **CO** is the chemical formula for the compound known as carbon monoxide, which is a colorless, poisonous gas. **Co** is the elemental symbol for the element cobalt, which is a metal. Remember: Regardless of how many letters are used to represent an element, each elemental symbol can only contain a single capital letter. The elemental symbols for some of the elements that your instructor will likely discuss most often are shown in Figure 3–2a.

Elemental Symbols of Sample Elements

Element	Symbol	Element	Symbol	Element	Symbol
Aluminum	Al	Fluorine	F	Nickel	Ni
Argon	Ar	Gold	Au	Nitrogen	N
Barium	Ba	Hydrogen	H	Oxygen	O
Beryllium	Be	Iodine	I	Radon	Rn
Boron	B	Iron	Fe	Selenium	Se
Bromine	Br	Krypton	Kr	Silicon	Si
Calcium	Ca	Lead	Pb	Sodium	Na
Carbon	C	Magnesium	Mg	Sulfur	S
Chlorine	Cl	Mercury	Hg	Uranium	U
Copper	Cu	Neon	Ne	Zinc	Zr

Figure 3–2a

When the elemental symbol of an element is combined with additional information, such as the atomic number, mass number, or charge, it is usually called ***elemental notation***. Elemental notation is a simple way of summarizing a good deal of information in a small space.

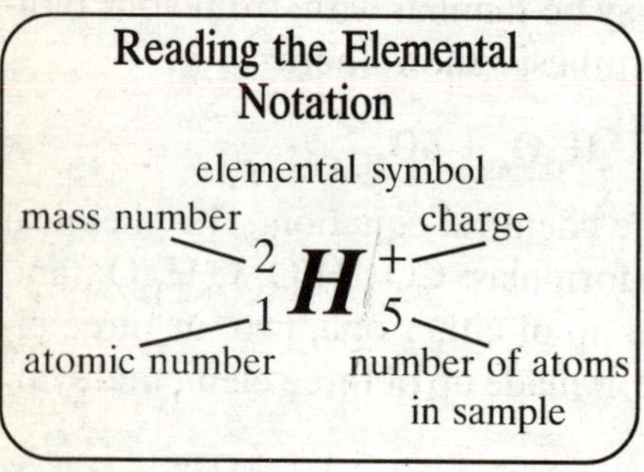

Figure 3–2b

Some other examples of elemental notation are shown here:

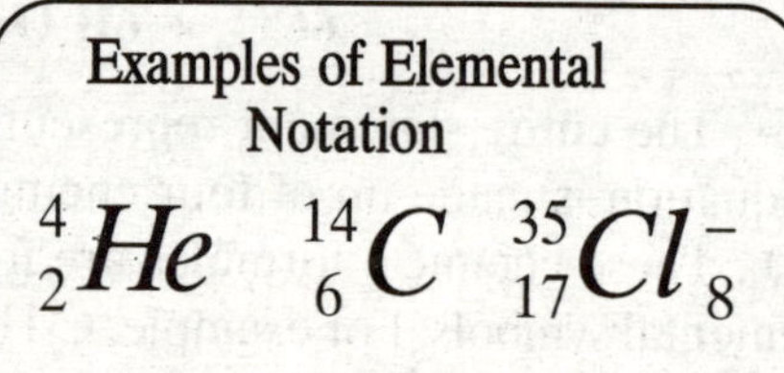

Figure 3–2c

Not all examples of elemental notation will be written with the same amount of information. Notice that the first two examples show only two numbers. The number in the upper-left corner is the mass number, so, for example, the carbon sample shown is of the isotope called carbon-14, which, as you may know, is used for radioactive dating. The number in the lower-left corner is the atomic number. This number is not always given, because the elemental symbol already tells you the identity of the element, so the atomic number represents redundant information. If you needed to know the atomic number you could simply look it up on the periodic table. The number in the lower-right corner of the hydrogen example indicates the number of atoms in the sample. The plus in the upper-right corner of the hydrogen example represents the net charge on each of the atoms, so we are dealing with positive ions in this example.

We can use the information in the elemental notation in order to determine the number of each of the subatomic particles in the atom. For example, look at the following elemental notation: ${}^{56}_{26}Fe^{2+}$

We know that the number in the lower left-hand corner is the atomic number, which gives us the number of protons, so this atom of iron (Fe) has 26 protons. We know that the number in the upper left-hand corner is the mass number, which is the total number of protons and neutrons. If we subtract the atomic number from the mass number, we find the number of neutrons to be 30 (56 – 26 = 30). This particular atom of iron has a +2 charge, which means that it has 2 more protons than electrons. By subtracting the charge number from the number of protons, we find that there are 24 electrons in the atom (26 – (+2) = 24).

Let's try a few example problems.

Example 1

Use the following elemental notations to fill in the missing information in the table below.

	${}^{20}_{10}Ne^{0}$	${}^{40}_{18}Ag^{+}$	${}^{40}Ca^{2+}$	${}^{32}_{16}S^{2-}$
Number of protons				
Number of electrons				
Number of neutrons				

The first thing asked for is the number of protons in each element, which is very easy to determine. The atomic number, which is identical to the number of protons, is located in the lower left-hand corner of each of the elemental notations. But wait! Where is the atomic number for calcium? It appears to be missing, but does that mean that we can't solve it? Not at all. As I mentioned earlier, the atomic number is redundant; if you know the identity of the element, you can look up the atomic number on the periodic table. Find calcium on the periodic table, and you will see that it has an atomic number of 20. Now you can fill in the number of protons for each of the elements.

	$^{20}_{10}Ne^{0}$	$^{40}_{18}Ag^{+}$	$^{40}Ca^{2+}$	$^{32}_{16}S^{2-}$
Number of protons	10	18	20	16
Number of electrons				
Number of neutrons				

Next, you'll need to determine the number of electrons in each of the samples. Remember: If we subtract the charge number from the upper right-hand side of the elemental notation from the atomic number, found on the lower left-hand side of the elemental notation, we get the number of electrons in the atom. Following are the calculations for this step.

- Number of electrons shown for Ne^0 = 10 – 0 = 10. This neon atom is neutral, so it has an equal number of protons and electrons.
- Number of electrons shown for Ag^+ = 18 – 1 = 17. This silver atom has a charge of +1, because it lost one of its electrons (negative charges).
- Number of electrons shown for Ca^{2+} = 20 – 2 = 18. This calcium atom has a charge of +2, because it lost two of its electrons.
- Number of electrons shown for S^{2-} = 16 – (-2) = 18. This sulfur atom has a charge of -2 because it has gained two extra electrons.

Let's add this information to our table.

	$^{20}_{10}Ne^{0}$	$^{40}_{18}Ag^{+}$	$^{40}Ca^{2+}$	$^{32}_{16}S^{2-}$
Number of protons	10	18	20	16
Number of electrons	10	17	18	18
Number of neutrons				

Lastly, we need to fill in the information for the number of neutrons shown in each sample. We simply subtract the atomic number, found in the lower left-hand corner of the elemental notation, from the mass number, found in the upper left-hand corner of the elemental notation. For calcium, remember that the atomic number is 20.

- The neon atom has (20 – 10) 10 neutrons.
- The argon atom has (40 – 18) 22 neutrons.
- The calcium atom has (40 – 20) 20 neutrons.
- The sulfur atom has (32 – 16) 16 neutrons.

	$^{20}_{10}Ne^0$	$^{40}_{18}Ag^+$	$^{40}Ca^{2+}$	$^{32}_{16}S^{2-}$
Number of protons	10	18	20	16
Number of electrons	10	17	18	18
Number of neutrons	10	22	20	16

Notice that it is possible for atoms of two different elements to have the same number of electrons. They can also have the same number of neutrons, but two atoms of different elements cannot have the same number of protons. As I said earlier, it is the number of protons that determines the identity of an element.

Now, let's look at elemental notations for the three isotopes of hydrogen.

Elemental Notation for the Three Isotopes of Hydrogen

^1_1H	^2_1H	^3_1H
Protium	Deuterium	Tritium

Figure 3–2d

Notice that the three isotopes of hydrogen have the same atomic number but different mass numbers. This is because they have the same number of protons but a different number of neutrons. If you change the number of protons, you change the identity of the element.

Let's try some practice questions.

Lesson 3–2 Review

1. Determine the number of protons shown in each of the following elemental notations.
   a) $^{39}_{19}K$ b) $^{56}_{26}Fe^{2+}$ c) $^{22}_{11}Na$ d) $^{88}_{38}Sr^{2+}_{4}$
2. Determine the number of neutrons shown in each of the following elemental notations.
   a) $^{27}_{13}Al$ b) $^{24}_{12}Mg^{2+}_{5}$ c) $^{86}_{37}Rb$ d) $^{31}_{16}S^{2-}$
3. Determine the number of electrons shown in each of the following elemental notations.
   a) $^{137}_{56}Ba$ b) $^{55}_{25}Mn$ c) $^{14}_{7}N^{3-}$ d) $^{7}_{3}Li^{+}$
4. Which of the following atoms has an equal number of protons and neutrons?
   a) $^{27}_{13}Al$ b) $^{55}_{25}Mn$ c) $^{39}_{19}K$ d) $^{22}_{11}Na$
5. Which of the following would represent different isotopes of the same element?
   a) $^{12}_{6}X$ and $^{12}_{7}X$ b) $^{12}_{6}X$ and $^{13}_{7}X$ c) $^{12}_{6}X$ and $^{12}_{6}X$ d) $^{12}_{6}X$ and $^{14}_{6}X$

6. The atomic mass unit (amu) is defined as exactly 1/12 of an atom of ___.

   a) ${}^{88}_{38}Sr^{2+}_{4}$ b) ${}^{12}_{6}C$ c) ${}^{14}_{6}C$ d) ${}^{137}_{56}Ba$

Lesson 3–3: Quantum Numbers

Compared to the old planetary model of the atom, the quantum-mechanic model seems much less certain. Instead of thinking of electrons as occupying fixed orbits with predictable paths, we now think in terms of probability. Although you shouldn't think of the electrons as being locked into fixed orbits, the way planets are, we are able to predict the areas where electrons are most likely to be found. Imagine that you needed to find your chemistry teacher outside of class, perhaps to give him or her a lab report. You might start by going to the floor of the building where the science classes are found. You might then narrow your search to the specific hallway in which the classroom is located. You might then check in the specific classroom. Your teacher might not be in that classroom, but this may be the place where he or she probably is.

Our model of the electron cloud is broken up into areas, as in the example of finding the chemistry teacher. Each electron is located in a specific energy level (in our example, the floors of a building). Each energy level is broken up into sublevels (the different hallways on the floor of a building). Each sublevel contains a certain number of orbitals (just as a hallway can contain some number of rooms).

Just as levels of a building are given numbers, energy levels are given numbers as well. As a floor, or level of a building, can contain several hallways, each energy level is broken up into sublevels, which are designated by specific letters. Each sublevel can contain a certain number of orbitals, which are analogous to the rooms in the hallway. Each orbital can contain up to two electrons.

The rooms in many buildings contain numbers on their doors, so that they can be found easily. So to, numbers are given to electrons, so that we can picture where they are likely to be found. The numbers that are given to electrons are called quantum numbers, and each electron is given four.

n The principal, or first, quantum number is used to indicate the ***energy level*** that the electron is found in. The value for ***n*** will always be a whole number, and the higher the number, the further away from the nucleus the electron described by ***n*** tends to be. For example, an electron with a value of 3 for ***n*** is in the third energy level, so it is likely to be located further away from the nucleus than an electron with a value of 1 for ***n***.

Each energy level is divided into sublevels, the number of which is equal to the value of ***n***. So, for example, the third energy level (***n*** = 3) contains 3 sublevels, whereas the fifth energy level (***n*** = 5) would contain 5 sublevels.

ℓ The second quantum number, called the angular momentum quantum number, is used to indicate the type of ***sublevel*** that the electron occupies. The possible values for ℓ refer to the types of sublevels, each of which contain a certain number of orbitals. An ***orbital*** is a space that can be occupied by up to two electrons, which occupy a specific three-dimensional area. Depending on the course that you are in, you may or may not need to know the shapes of the sublevels, but you are very likely to need to know the number of orbitals and electrons that each one holds. Values of ℓ of 0, 1, 2, and 3 refer to sublevels, which are, in turn, designated by the letters s, p, d, and f respectively. So, an electron that has a value 3 for ***n*** (the first quantum number) and a value of 1 for ℓ (the second quantum number) would be found in the third energy level, in a "p" sublevel.

$\mathbf{M}_\ell$ The third quantum number, called the magnetic quantum number, indicates the spatial orientation of the orbital that the electron occupies. The possible values of $\mathbf{M}_\ell$ for an electron depend upon the value of the electron's second quantum number (ℓ). For any given electron, the possible values of $\mathbf{M}_\ell$ range from $+\ell$ to $-\ell$, including the value 0. So, in other words, an electron with a value of 2 for ℓ would occupy a "d" sublevel, and could have values of +2, +1, 0, -1, or -2 for $\mathbf{M}_\ell$.

$\mathbf{M}_s$ The fourth quantum number, called the spin quantum number, indicates the "spin" direction of the electron in a particular orbital. Pauli's exclusion principle states that no two electrons in an atom can have the exact same set of four quantum numbers. Electrons that are in the same orbital have the same value for the first three quantum numbers (***n***, ℓ, and $\mathbf{M}_\ell$), which means that they occupy the same energy level, sublevel, and orbital. In order to satisfy the exclusion principle, electrons in the same orbital must have a different value for $\mathbf{M}_s$. Each orbital can hold up to two electrons, so only two values for $\mathbf{M}_s$ are required. Those values have been designated as +1/2 and -1/2. Combined with the other three quantum numbers, the fourth quantum number indicates the actual electron that you are studying.

The four quantum numbers are summarized in Figure 3–3a on page 84.

The Four Quantum Numbers

	Called	Symbol	Indicates	Values*
First Quantum Number	Principal	n	Energy Level	1–7
Second Quantum Number	Angular momentum	ℓ	Sublevel	0, 1, 2, 3 (or s, p, d, f)
Third Quantum Number	Magnetic	M_ℓ	Orbital	+3, +2, +1, 0, -1, -2, -3
Fourth Quantum Number	Spin	M_s	Electron	+1/2 and -1/2

* There are other possible values, but the listed values will cover all that you need to use for the elements that are currently known.

Figure 3–3a

The information that we have covered on quantum numbers will be used to construct a type of notation called the electron configurations of elements in the next chapter. You will want to be able to remember how many sublevels, and which types, are found in each energy level. I ask my students to remember a simple sentence such as, "*s*ome ***p***eople ***d***on't ***f***orget" to help them recall the order of the sublevels. To recall which sublevels the third energy level has, for example, they might say to themselves, "*s*ome ***p***eople ***d***on't."

Also remember that each orbital can hold up to two electrons and that the sublevels contain different numbers of orbitals. The number of sublevels, orbitals, and electrons that the first four energy levels hold is summarized in Figure 3–3b.

Remember: Each orbital can hold up to two electrons, so a p sublevel, which contains 3 orbitals, can hold up to 6 electrons (2 electrons × 3 orbitals). To find out how many total electrons a certain energy level can hold, you can do one of two things. First, you could add up the total number of electrons that each of its sublevels can hold. For example, the fourth energy level has four sublevels (s, p, d, and f), which

Summary of the First Four Energy Levels

Principal Energy Level (n)	Type(s) of Sublevel	Number of Orbitals	Maximum Number of Electrons
1	s	1	2
2	s	1	2
	p	3	6
3	s	1	2
	p	3	6
	d	5	10
4	s	1	2
	p	3	6
	d	5	10
	f	7	14

Figure 3–3b

hold 1, 6, 10, and 14 electrons respectively. We can add these numbers together to find that the fourth energy level can hold 32 e^- (2 + 6 + 10 + 14 = 32). Another way to calculate the total number of electrons that an energy level can hold is to use the formula $2n^2$, where n is the energy level number. Solving for $n = 4$, we find that $2(4)^2 = 32$ e^-.

Try the following practice questions and check your answers.

Lesson 3–3 Review

1. How many quantum numbers are used to describe each electron?
   a) 1 b) 2 c) 3 d) 4
2. Which quantum number indicates the sublevel that an electron is found in?
   a) first b) second c) third d) fourth
3. How many sublevels would the energy level represented by **n** = 4 be broken up into?
   a) 1 b) 2 c) 3 d) 4
4. A value for the second quantum number of 0 ($\ell = 0$) indicates which type of sublevel?
   a) s b) p c) d d) f
5. How many orbitals does a p sublevel contain?
   a) 1 b) 3 c) 5 d) 7
6. How many total electrons can the second energy level hold?
   a) 2 b) 6 c) 8 d) 16

Lesson 3–4: Electron Configuration

To a chemist, the electrons in the highest energy level, which are called ***valence electrons***, are probably the most important part of the atom. It is these electrons that determine the reactivity of an element. When we talk about ***electron configuration***, we are speaking of the arrangement of the electrons in a particular atom or element. By being able to determine these arrangements, you will be able to predict how an element reacts with other elements and what types of compounds it will form. We

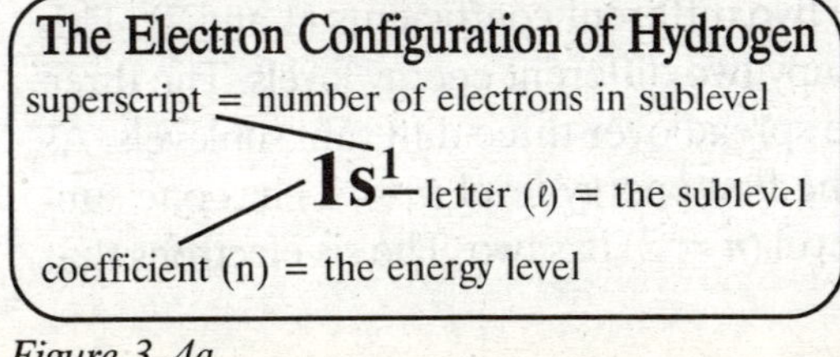

Figure 3–4a

will start by learning how to read an electron configuration. Look at the electron configuration for hydrogen, shown in Figure 3–4a.

The superscript shows us that hydrogen has only one electron. The large (coefficient) number represents the principal quantum number for the electron. In this case, because the value for n = 1, it means that this electron is located in the first energy level. The letter indicates the sublevel that the electron is located in. Because there is only one energy level in this particular atom, it represents the ***valence shell***, or outer energy level. Hydrogen is said to have one electron in its valence shell.

The Electron Configuration of Helium

superscript = number of electrons in sublevel

$1s^2$ — letter (ℓ) = the sublevel

coefficient (n) = the energy level

Figure 3–4b

Now, let's compare that electron configuration of hydrogen to the electron configuration of helium, shown in Figure 3–4b.

Now, we see that the superscript for helium is 2, which means that helium contains two electrons. This shouldn't surprise you, because a quick check of the periodic table shows that the atomic number for helium is 2, indicating that a neutral atom of helium contains two protons and two electrons. The letter "s" tells us that both of helium's electrons occupy an "s" sublevel. The value of n (coefficient) is 1, which tells us that both of these electrons are in the first energy level. Again, because the electrons only occupy one energy level, the first energy level is also helium's valence shell. It has two electrons in its valence shell.

For clarity's sake, let's look at the electron configuration of an atom that shows electrons in more than one energy level. Following is the electron configuration for oxygen.

The Electron Configuration of Oxygen

$$1s^2 2s^2 2p^4$$

What are we looking at here? Well, first add up the superscripts and you will see that there are a total of 8 electrons (2 + 2 + 4 = 8). This makes sense, because oxygen, with an atomic number of 8, has a total of 8 electrons. Now, you see that there are two different coefficients (1 and 2). This tells us that oxygen's electrons occupy two different energy levels. The three letters tell us that the electrons are spread over three different sublevels. As you will recall from Lesson 3–3, the first energy level (n = 1) has one sublevel, whereas the second energy level (n = 2) has two. The six electrons that occupy the second energy level ($2s^2$ $2p^4$) represent oxygen's valence electrons. The nucleus of the atom and the 2 electrons in the first energy level represent the "***kernel,***" or core, of the atom.

Did you notice that the kernel configuration of the oxygen atom looks just like the electron configuration of the helium atom? This leads to an interesting shorthand notation that is often used for electron configurations. We can replace the kernel configuration with the noble gas symbol that it matches. See the example here:

The Shorthand Electron Configuration of Oxygen

[He] $2s^2$ $2p^4$

This may not seem to be such a space-saver now, but it will when we see electron configurations for the larger atoms. Let's compare the full electron configuration of potassium to the shorthand notation. A neutral atom of potassium, with an atomic number of 19, contains 19 electrons.

The Electron Configuration of Potassium

$1s^2$ $2s^2$ $2p^6$ $3s^2$ $3p^6$ $4s^1$

As you can see, the 19 electrons of potassium are spread over four different energy levels and six sublevels. Only the electron(s) found in the energy level with the highest value for n are considered valence electrons, so potassium only has 1, designated by the notation $4s^1$. The other 18 electrons, which are part of the kernel, show the exact same configuration as the noble gas called argon. Replacing the kernel configuration, we get the shorthand notation shown here:

The Shorthand Electron Configuration of Potassium

[Ar] $4s^1$

This notation tells us that the kernel configuration of potassium is identical to the full configuration of argon. In addition, the valence shell of the potassium atom consists of 1 electron in the "s" sublevel of the fourth energy level.

Now that we have gone over how to read the electron configuration of the atom, we will learn how to write the electron configuration for elements. The first thing that you will want to have in front of you is something called the "arrow diagram." Depending upon your teacher, you may need to be able to construct such a diagram by yourself.

If you need to construct such a diagram, recall the sentence "***s**ome **p**eople **d**on't **f**orget,*" which will help you remember the order for the sublevels, and follow the following steps.

1. Start by numbering the lines down a page, from one to seven.
2. On the first line, add only one letter, "s," which begins our sentence.
3. Now, start adding one more letter each line, so on the second line, write two letters, "s" and "p," leaving some space between them.
4. On the third line, write three letters, "s," "p," and "d," again leaving some space between them.
5. On the fourth line, write all four letters, "s," "p," "d," and "f." By now, your diagram should look like Figure 3–4c.
6. On the fifth line, write all four letters again, "s," "p," "d," and "f."
7. Now, start taking away one letter each time, so write "s," "p," and "d" on the sixth line and "s" and "p" on the seventh.
8. Next, fill in the line numbers straight across, in front of each letter. So, for example, on the fifth line you will add the number "5" in front of each of the other three letters in the line.

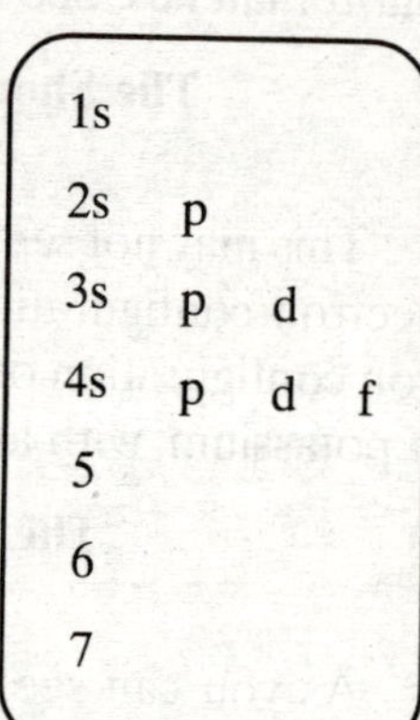

Figure 3–4c

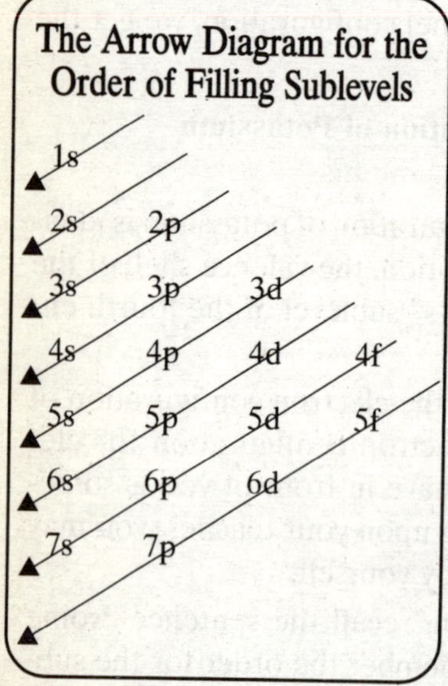

Figure 3–4d

9. Finally, draw arrows through the diagram at such an angle that the first two arrows only cut through 1s and 2s, respectively. Continue at the same angle, and your final diagram should look like Figure 3–4d.

The next thing that you will want to be sure of is how many electrons fit in each of the types of sublevels. The trick to remembering this is to remember both your sentence for the order of the sublevels (*s*ome *p*eople *d*on't *f*orget!) and the first four odd numbers (1, 3, 5, and 7). You will also want to recall that each orbital can hold up to 2 electrons. Figure 3–4e on page 89 summarizes the information that you will need.

Sublevel Summary		
Sublevel	**Number of orbitals it contains (first 4 odd numbers)**	**Total number of electrons it can hold (2 × the number of orbitals)**
s	1	2
p	3	6
d	5	10
f	7	14

Figure 3-4e

Now, you should have all of the tools that you need to do the electron configuration. Become familiar with the steps shown here for writing the electron configuration of an element, and then we will review them.

1. Look up the atomic number of the element.
2. Determine the number of electrons for the specific element.
   A. If the atom is neutral, then the number of electrons is equal to the number of protons.
   B. If the atom is charged, then algebraically subtract the charge from the atomic number of the element. (*Example 1:* The atomic number of sodium is 11. An ion of Na+ would have a total of 11 – 1 = **10** electrons. *Example 2:* The atomic number of sulfur is 16. An ion of S^{2-} would have 16 – (-2) = **18** total electrons.)
3. Consult the arrow diagram to determine the order in which the sublevels should be filled.
4. Write the configuration, filling in up to 2 electrons in each "s" sublevel, up to 6 electrons in each "p" sublevel, up to 10 electrons in each "d" sublevel, and up to 14 electrons in each "f" sublevel.
5. When you think that you are finished, add up the exponents (superscripts) to see if you have the correct number of electrons.

Let's do an example.

Example 1

Write the full electron configuration for the element neon (Ne).

Step 1. Look up the atomic number of the element. The periodic table shows us that the atomic number of neon is 10.

Step 2. Determine the number of electrons for the specific element.

A. ***If the atom is neutral, then the number of electrons is equal to the number of protons.*** Because no charge was mentioned, we know that this is a neutral atom. So, the number of electrons = 10.

Step 3. Consult the arrow diagram to determine the order in which the sublevels should be filled. We look at the arrow diagram (Figure 3–4d) and we see the order in which the sublevels are cut by the arrows. You start by tracing the top arrow from tail to tip, as it passes through "1s," which becomes the first sublevel that we will fill in. When you get to the tip of the first arrow you proceed to the next and trace it from tail to tip, as it passes through "2s," which becomes the second sublevel we will fill. When we get to the tip of the second arrow, we move to the tail of the third arrow and trace it through, and the order continues with 2p, 3s, 3p, 4s, 3d, and so forth. We won't need to fill all of these sublevels, but we need to be able to read the order.

Step 4. Write the configuration, filling in up to 2 electrons in each "s" sublevel, up to 6 electrons in each "p" sublevel, up to 10 electrons in each "d" sublevel, and up to 14 electrons in each "f" sublevel. We have $1s^2\ 2s^2\ 2p^6$, and that brings us up to 10 electrons.

Step 5. When you think that you are finished, add up the exponents (superscripts) to see if you have the correct number of electrons. Adding the exponents, $1s^2\ 2s^2\ 2p^6$ or $(2 + 2 + 6)$ we get 10 electrons, so our configuration is probably correct.

The full electron configuration for the element neon (Ne):
$1s^2\ 2s^2\ 2p^6$

Neon has a total of 8 electrons in its valence shell, because both the 2s and 2p sublevels are in the highest energy level. As you will learn, elements with 8 valence electrons are said to have a complete ***octet*** and are considered chemically stable—that is, they are less likely to react.

For our next example, let's try an electron configuration for an element that does not have a complete valence shell.

Example 2

Write the full electron configuration for the element aluminum (Al).

Step 1. Look up the atomic number of the element. The periodic table shows us that the atomic number of aluminum is 13.

Step 2. Determine the number of electrons for the specific element.

A. ***If the atom is neutral, then the number of electrons is equal to the number of protons.*** Because no charge was mentioned, we know that this is a neutral atom. So, the number of electrons = 13.

Step 3. Consult the arrow diagram to determine the order in which the sublevels should be filled. We look at the arrow diagram, and we see the order in which the sublevels are cut by the arrows. We start with the tail of the top arrow, and trace it to the tip. When you get to the tip of an arrow, you proceed to the tail of the next arrow down. It begins with 1s, then 2s, then 2p, 3s, 3p, 4s, 3d, and so forth. We won't need to fill all of these sublevels, but we need to be able to read the order.

Step 4. Write the configuration, filling in up to 2 electrons in each "s" sublevel, up to 6 electrons in each "p" sublevel, up to 10 electrons in each "d" sublevel, and up to 14 electrons in each "f" sublevel. We have $1s^2\ 2s^2\ 2p^6\ 3s^2\ 3p^1$, and that brings us up to 13 electrons.

Step 5. When you think that you are finished, add up the exponents (superscripts) to see if you have the correct number of electrons. Adding the exponents of $1s^2\ 2s^2\ 2p^6\ 3s^2\ 3p^1$ $(2 + 2 + 6 + 2 + 1)$, we get 13 electrons, so our configuration is probably correct.

The full electron configuration for the element aluminum (Al):

$$1s^2\ 2s^2\ 2p^6\ 3s^2\ 3p^1$$

Were you surprised that we only placed 1 electron in the 3p sublevel? Remember that when we say that a "p" sublevel can hold up to 6 electrons, we mean that 6 is the maximum that it can hold. In this case, we only had to place 1 electron in the final "p" sublevel to get up to 13 total electrons.

Can you figure out what the shorthand notation would be for the element aluminum? Notice that its kernel configuration is identical to the configuration we did for neon, in Example 1. The shorthand notation for aluminum is shown here:

The Shorthand Electron Configuration of Aluminum

$$[Ne]\ 3s^2\ 3p^1$$

For our last example, let's try an example of an atom that has obtained a charge, which, as you know, is called an ion.

Example 3

Write the full electron configuration for a magnesium (Mg^{2+}) ion.

Notice how working with an ion changes Step #2.

Step 1. Look up the atomic number of the element. The periodic table shows us that the atomic number of magnesium is 12.

Step 2. Determine the number of electrons for the specific element.

B. If the atom is charged, then algebraically subtract the charge from the atomic number of the element. (**Example 1:** *The atomic number of sodium is 11. An ion of Na+ would have a total of 11 – 1 = 10 electrons.* **Example 2:** *The atomic number of sulfur is 16. An ion of S^{2-} would have 16 – (-2) = 18 total electrons.)*

Because this atom of magnesium has a +2 charge, we must subtract 2 from the number of electrons (12 – 2 = 10). So, the number of electrons = 10.

Step 3. Consult the arrow diagram to determine the order in which the sublevels should be filled. Once again, it begins with 1s, then 2s, then 2p, 3s, 3p, 4s, 3d, and so on. We won't need to fill all of these sublevels, but we need to be able to read the order.

Step 4. Write the configuration, filling in up to 2 electrons in each "s" sublevel, up to 6 electrons in each "p" sublevel, up to 10 electrons in each "d" sublevel, and up to 14 electrons in each "f" sublevel. We have $1s^2\ 2s^2\ 2p^6$, and that brings us up to 10 electrons.

Step 5. When you think that you are finished, add up the exponents (superscripts) to see if you have the correct number of electrons. Adding the exponents of $1s^2 2s^2 2p^6$ (2 + 2 + 6), we get 10 electrons, so our configuration is probably correct.

The full electron configuration for a magnesium (Mg^{2+}) ion:

$1s^2\ 2s^2\ 2p^6$

Do you notice anything strange about this configuration? It is exactly the same electron configuration as the neutral atom of neon! Does that mean that the Mg^{2+} ion is the same as neon? The answer is yes and no. The nucleus of the magnesium ion contains 2 more protons than the neon atom, so they have different masses and different atomic numbers and are certainly not the same element. However, having the same configuration as neon, with 8 valence electrons, is a more stable state for the magnesium atom to be in. Pieces of the puzzle continue to fall into place.

Up to this point, we have been discussing the "ground state" or normal state of the various elements for which we did the electron configuration. When an electron gains additional energy (from a source of heat, light, electricity, and so forth), it can temporarily move up to a higher energy level. This "excited state" is unstable, and the electron quickly releases the additional energy in the form of visible light and falls back to its ground state. Most instructors will want their students to be able to recognize the electron configuration of an atom in this excited state. You probably won't be asked to write the electron configuration of an atom in the excited state, because you wouldn't know which electron moves up to a higher energy level, but you should be able to recognize when a configuration for an excited state is being shown.

Look at some examples of the normal (ground state) electron configurations, and compare them to their configurations in examples of excited state.

Comparing the Ground State Configuration to the Excited State

Element	Ground State Configuration	Excited State Configuration
Fluorine	$1s^2\,2s^2\,2p^5$	$1s^2\,2s^2\,2p^4\,3s^1$
Neon	$1s^2\,2s^2\,2p^6$	$1s^2\,2s^2\,2p^5\,3s^1$
Sodium	$1s^2\,2s^2\,2p^6\,3s^1$	$1s^2\,2s^2\,2p^6\,3p^1$

Figure 3–4f

Can you recognize the "look" of an atom in the excited state? In each case, one or more electron has been "promoted" to a higher energy level. For example, when the "$3s^1$" electron in the sodium atom gets excited, it jumps up to the "$3p^1$" position. Take note of the fact that the atomic number, which we determine by adding all of the exponents, does not change. We still have the same number of electrons in the excited state.

Let's review some questions from this somewhat difficult, yet important, lesson. Be sure to check your answers at the end of the chapter.

Lesson 3–4 Review

1. How many sublevels does the second energy level contain?
2. How many orbitals does a "p" sublevel contain?
3. How many total electrons can the fourth energy level hold?
4. How many electrons can an "f" orbital hold?

Determine the proper ground state electron configuration for the following elements.

5. Potassium

6. Krypton

Lesson 3–5: Orbital Notation

The orbital notation of an element is related to the electron configuration, and both types of notation are often shown at the same time. The real difference in the two types of notation has to do with the level of detail shown. In electron configuration, no details are shown about the actual orbitals. When we see "$3p^4$," we know that there are 4 electrons in the p sublevel, but we don't know in which orbitals the electrons are located. When we do the orbital notation for the element, we get that level of detail for the orbitals.

The first step for doing the orbital notation is to do the electron configuration. You will then want to remember ***Hund's Rule***, which states that the most stable arrangement of electrons in sublevels is the one with the greatest number of parallel spins. The way that we make use of this rule is to put one electron in each of the orbitals in a particular sublevel, before we double up the electrons in any particular orbital.

When we write an orbital notation, the orbitals are represented by circles or squares, and the electrons are represented by arrows or slashes. For the sake of clarity, the orbital notation is usually accompanied by the proper electron configuration.

One last thing: In order to do orbital notation, you must be able to do electron configuration. So, if you haven't mastered Lesson 3–4, you should go back and do more practice.

Let's try an example.

Example 1

Write the proper orbital notation for an atom of aluminum.

The first thing that we need to do is find the electron configuration for aluminum. Because we did this example in the last lesson, I won't go through the steps. Refer back to the last lesson if you don't remember how to do this. We will space the configuration out a bit more this time, so that we have room to write the orbital notation.

$1s^2$ $2s^2$ $2p^6$ $3s^2$ $3p^1$

Next, we will draw one circle for each orbital. Remember from the last lesson that "s" sublevels contain 1 orbital (as shown in Figure 3–4e on page 89), "p" sublevels contain 3, and the "d" and "f" sublevels hold 5 and 7 orbitals, respectively.

○ ○ ○○○ ○ ○○○

$1s^2$ $2s^2$ $2p^6$ $3s^2$ $3p^1$

Now, if you choose to represent the electrons by arrows, you put one arrow in the orbitals with only one electron and two arrows in the orbitals with two electrons, as shown below. Remember: An electron can only hold up to two electrons, so there will never be more than two arrows in a circle.

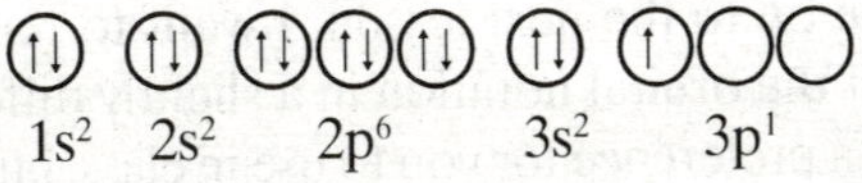

Notice that you will not always fill in all of the circles that you draw. Aluminum only has one electron in the 3p sublevel, so only a single 3p orbital has an electron. For our next example, let's try an orbital notation in which Hund's Rule becomes more important.

Example 2

Write the proper orbital notation for an atom of phosphorus.

We start be looking up the atomic number for phosphorus, which is 15. We then do the electron configuration for the element, based on the rules that we learned early in this chapter, and find it to be $1s^2\ 2s^2\ 2p^6\ 3s^2\ 3p^3$. We space our electron configuration out, to make room for the circles representing our orbitals, and fill in the arrows to represent the electrons, as shown here.

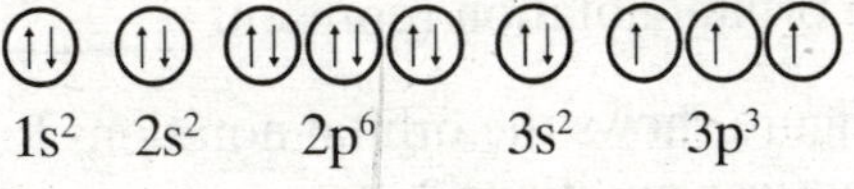

Do you see how we apply Hund's Rule? The atom is more stable if it has more electrons with parallel spin, so we place one electron in each of the 3p orbitals, rather than putting, say, two electrons in the first orbital and one in the second. It is easy to remember what to do: When you are filling the orbitals within a sublevel, each orbital gets one electron before any orbital gets two. When you have more electrons than orbitals, you need to double up, as shown by the next example.

Example 3

Write the proper orbital notation for an atom of sulfur.

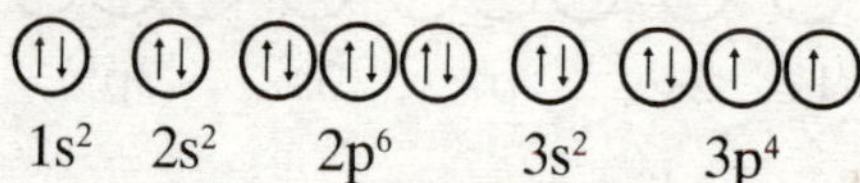

So, if you can do the electron configuration for an element, you should be able to do the orbital notation. Just study the examples from this lesson and make sure you understand them.

Before we move on to the next lesson, I want to remind you that some books will represent the orbital notation in a slightly different fashion. Your instructor may have a preference for you to use in class, but you should be able to recognize any of the alternate versions that you might see. Figure 3–5a shows the three most common versions for the orbital notation of carbon.

Three Versions of the Orbital Notation of Carbon

	$1s^2$	$2s^2$	$2p^2$
Circles for orbitals, arrows for electrons:	(↑↓)	(↑↓)	(↑)(↑)( )
Squares for orbitals, arrows for electrons:	[↑↓]	[↑↓]	[↑][↑][ ]
Circles for orbitals, slashes for electrons:	(⊗)	(⊗)	(⊘)(⊘)( )

Figure 3–5a

Lesson 3–5 Review

1. The circles in an orbital notation represent ____________.
2. The arrows in an orbital notation represent ____________.

The following figure shows the orbital notations for six elements. Use these notations to answer questions 3–8.

A. (↑↓) (↑↓) (↑↓)(↑↓)(↑)

B. (↑↓) (↑↓) (↑)( )( )

C. (↑↓) (↑↓) (↑↓)(↑↓)(↑↓) (↑↓) (↑↓)(↑↓)(↑)

D. (↑↓)

E. (↑↓) (↑↓) (↑↓)(↑↓)(↑↓) (↑↓)

F. (↑↓) (↑)

Which orbital notation matches up to each of the following elements?

3. Chlorine
4. Helium
5. Fluorine
6. Lithium
7. Magnesium
8. Boron

Lesson 3–6: Lewis Dot Notation

Another type of notation that is used in chemistry is called Lewis dot notation. In Lewis dot notation, the ***kernel*** of the atom—that is, the nucleus and all of the inner electrons—is represented by the elemental symbol. The valence electrons are represented by dots, and each of the four sides around the elemental symbol represents one of the orbitals in the valence shell. The rules for orbitals still apply, so no side can have more than two dots, and each of the "p" orbital sides gets one dot, before you double up.

Figure 3–6a shows the general configuration for the Lewis dot notation.

You can really place the "s" orbital on any side of the elemental symbol, but in this book, I will usually begin with the "s" orbital on the left-hand side. Let's start off with an easy example.

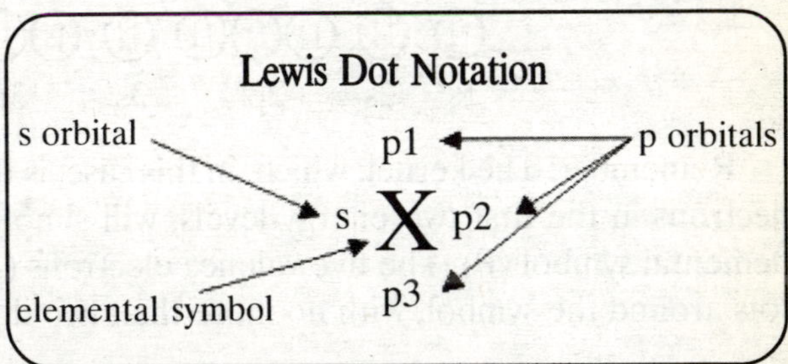

Figure 3–6a

Example 1

Show the correct Lewis dot notation for the element beryllium.

The key to solving this type of problem is to look at the electron configuration of the element. Beryllium, with an atomic number of 4, has 4 electrons, giving it an electron configuration of $1s^2\ 2s^2$. The nucleus and the two electrons in the first energy level ($1s^2$) make up the kernel of the atom, and they are represented by the elemental symbol in our Lewis dot notation. The two electrons in the second energy level ($2s^2$) represent the valence electrons, which are represented by dots in the Lewis dot notation. So, to construct the proper notation, we write the elemental symbol and two dots to the left of the symbol, as shown here.

:Be

Next, we will look at the Lewis dot diagram for carbon, which has an electron configuration of $1s^2\ 2s^2\ 2p^2$. The valence shell of carbon contains a total of four electrons ($2s^2\ 2p^2$), so our dot diagram will show four dots. The two s electrons will go on the left-hand side of the elemental notation, but the two p electrons will split up, one per side, according to Hund's Rule.

Example 2

Show the correct Lewis dot notation for the element carbon.

Let's look at the Lewis dot notation for an atom of phosphorus. As you recall, the following is the electron configuration and the orbital notation that we found in the last lesson.

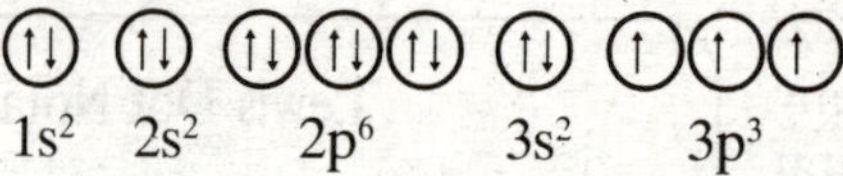

Remember: The kernel, which, in this case, is the nucleus and all of the electrons in the first two energy levels, will simply be represented by the elemental symbol (P). The five valence electrons ($3s^2\ 3p^3$) will be shown as dots around the symbol, with no more than two dots per side.

Lewis Dot Notation for Phosphorus

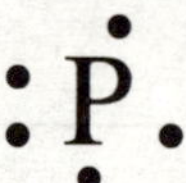

Note that the sides, which represent the "p" orbitals, received only one dot each, just as the circles representing those same orbitals in the orbital notation for phosphorus received only one dot each. Hund's Rule instructs us to find an empty orbital for each p electron, until we need to double them up. Why, then, do the s electrons double up? Because there is only one s orbital, the electrons in the s sublevel have no choice; they must share the only s orbital.

Next, let's compare the electron configuration and orbital notation of the element sulfur to its Lewis dot structure. Sulfur has six electrons in its valence shell, so its dot diagram will show six dots. Does the diagram below make it easier to see how the valence electrons are represented in the Lewis dot structure?

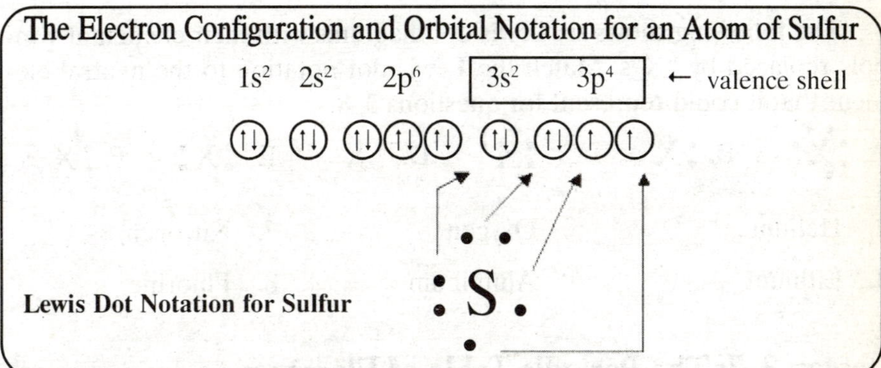

Figure 3–6b

For our final example, let's try a chorine (Cl^-) ion. We didn't do this example in Lesson 3–4, try it on your own and see if you get the same answer that I show in Figure 3–6c. Remember to go back and do the electron configuration and the orbital notation first, paying careful attention to the charge on the ion. Then construct the Lewis dot notation for the ion, and check it against the answer.

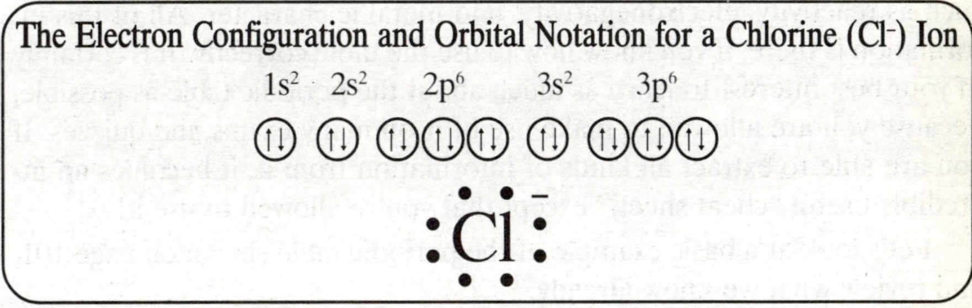

Figure 3–6c

Notice that we include the charge on the ion in our Lewis dot diagram, in order to differentiate it from the neutral atom. Did you get 8 valence electrons? If not, go back and review what you do to the atomic number in order to get the proper number of electrons for an ion.

Lesson 3–6 Review

1. What do the dots in a Lewis dot diagram represent?
2. Which electrons are shown as dots in a Lewis dot diagram?

The following shows six Lewis dot diagrams with their elemental symbols replaced by "X"s. Match the Lewis dot notation to the neutral element that it could represent for questions 3–8.

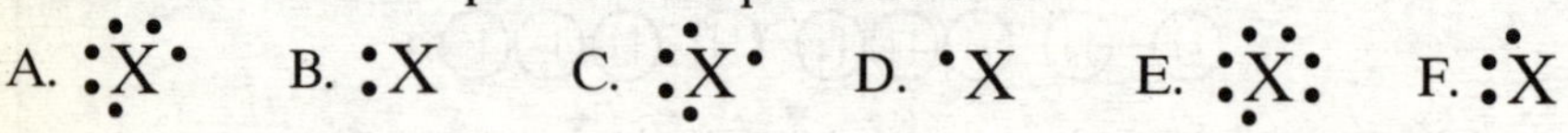

A. :Ẍ· (with one dot below) B. :X C. :Ẋ· (with one dot below) D. ·X E. :Ẍ: (with one dot below) F. :Ẋ

3. Helium
4. Lithium
5. Oxygen
6. Aluminum
7. Nitrogen
8. Fluorine

Lesson 3–7: The Periodic Table of Elements

Perhaps the most interesting thing about the Periodic Table of Elements is how much information is hidden in it. To the untrained eye, the periodic table appears to only show the elemental symbols, elemental names, atomic numbers, and atomic masses. However, someone who knows a bit more chemistry can "squeeze" much more information out of the same table. You will learn to use the periodic table to check your electron configurations, orbital notations, and Lewis dot notations. You will also learn to use the periodic table to check many relative properties of the elements, such as reactivity, electronegativity, and metallic character. All of this information is there, if you know how to use the table correctly. It is certainly in your best interest to learn as much about the periodic table as possible, because you are allowed to make use of it on many exams and quizzes. If you are able to extract all kinds of information from it, it becomes an incredibly useful "cheat sheet," except that you're allowed to use it!

Let's look at a basic example of the periodic table shown on page 101, and review what we know already.

Each of the boxes in Figure 3–7a contains two numbers and one or more letters. The top number in each box is called the atomic mass. The letter or letters in each box represent the elemental symbol. The bottom number in each box is called the atomic number.

The ***atomic number*** is the number of protons found in the nucleus of an atom of that particular element. Oxygen, for example, with its atomic number of eight, is made up of atoms with eight protons in each nucleus. In a neutral atom, the number of electrons in the electron cloud is equal to the number of protons in its nucleus.

The ***elemental symbol*** for each element is made up of only one capital letter and, in some cases, one or more lowercase letters. With only twenty-six letters and more than 100 elements, it is necessary to combine the letters to

The Periodic Table of Elements

	1	2	3	4	5	6	7	8	9	10	11	12	13	14	15	16	17	18
1	Hydrogen 1.0079 H 1																	Helium 4.003 He 2
2	Lithium 6.941 Li 3	Beryllium 9.012 Be 4											Boron 10.811 B 5	Carbon 12.011 C 6	Nitrogen 14.007 N 7	Oxygen 15.999 O 8	Fluorine 18.998 F 9	Neon 20.180 Ne 10
3	Sodium 22.990 Na 11	Magnesium 24.305 Mg 12											Aluminum 26.982 Al 13	Silicon 28.086 Si 14	Phosphorus 30.974 P 15	Sulfur 32.065 S 16	Chlorine 35.453 Cl 17	Argon 39.948 Ar 18
4	Potassium 39.098 K 19	Calcium 40.078 Ca 20	Scandium 44.956 Sc 21	Titanium 47.867 Ti 22	Vanadium 50.942 V 23	Chromium 51.996 Cr 24	Manganese 54.938 Mn 25	Iron 55.845 Fe 26	Cobalt 58.933 Co 27	Nickel 58.693 Ni 28	Copper 63.546 Cu 29	Zinc 65.39 Zn 30	Gallium 69.723 Ga 31	Germanium 72.64 Ge 32	Arsenic 74.922 As 33	Selenium 78.96 Se 34	Bromine 79.904 Br 35	Krypton 83.80 Kr 36
5	Rubidium 85.468 Rb 37	Strontium 87.62 Sr 38	Yttrium 88.906 Y 39	Zirconium 91.224 Zr 40	Niobium 92.906 Nb 41	Molybdenum 95.94 Mo 42	Technetium {98} Tc 43	Ruthenium 101.07 Ru 44	Rhodium 102.906 Rh 45	Palladium 106.42 Pd 46	Silver 107.868 Ag 47	Cadmium 112.411 Cd 48	Indium 114.818 In 49	Tin 118.710 Sn 50	Antimony 121.760 Sb 51	Tellurium 127.60 Te 52	Iodine 126.904 I 53	Xenon 131.293 Xe 54
6	Cesium 132.905 Cs 55	Barium 137.327 Ba 56	Lutetium 174.967 Lu 71	Hafnium 178.49 Hf 72	Tantalum 180.948 Ta 73	Tungsten 183.84 W 74	Rhenium 186.207 Re 75	Osmium 190.23 Os 76	Iridium 192.217 Ir 77	Platinum 195.078 Pt 78	Gold 196.967 Au 79	Mercury 200.59 Hg 80	Thallium 204.383 Tl 81	Lead 207.2 Pb 82	Bismuth 208.980 Bi 83	Polonium {209} Po 84	Astatine {210} At 85	Radon {222} Rn 86
7	Francium {223} Fr 87	Radium {226} Ra 88	Lawrencium {262} Lr 103	Rutherfordium {261} Rf 104	Dubnium {262} Db 105	Seaborgium {266} Sg 106	Bohrium {264} Bh 107	Hassium {277} Hs 108	Meitnerium {268} Mt 109	Darmstadtium {281} Ds 110	Unununium {272} Uuu 111	Ununbium {285} Uub 112	113	Ununquadium {289} Uuq 114	115	Ununhexium {292} Uuh 116	117	118

Lanthanum 138.906 La 57	Cerium 140.116 Ce 58	Praseodymium 140.908 Pr 59	Neodymium 144.24 Nd 60	Promethium {145} Pm 61	Samarium 150.36 Sm 62	Europium 151.964 Eu 63	Gadolinium 157.25 Gd 64	Terbium 158.925 Tb 65	Dysprosium 162.50 Dy 66	Holmium 164.930 Ho 67	Erbium 167.259 Er 68	Thulium 168.934 Tm 69	Ytterbium 173.04 Yb 70
Actinium {227} Ac 89	Thorium 232.038 Th 90	Protactinium 231.036 Pa 91	Uranium 238.029 U 92	Neptunium {237} Np 93	Plutonium {244} Pu 94	Americium {243} Am 95	Curium {247} Cm 96	Berkelium {247} Bk 97	Californium {251} Cf 98	Einsteinium {252} Es 99	Fermium {257} Fm 100	Mendelevium {258} Md 101	Nobelium {259} No 102

Figure 3–7a

represent some elements. Some elements, such as carbon (C), have the first letter in their name as the symbol. Other elements, such as aluminum (Al), use the first two letters in their name as the symbol. Gold (Au) derives its symbol from that Latin word *aurum*, which means gold. Some elements, such as ununbium (Uub), have temporary systematic names and symbols given to them, until a common name can be agreed upon.

In Figure 3-7b the bottom number in each box represents the ***atomic mass*** of the element. The atomic mass, which is still misleadingly called "atomic weight" by some people, actually represents a weighted average of the atomic masses of the various isotopes of an element.

All of the elements on the periodic table break up into three major categories: metals, nonmetals, and semimetals. ***Metals*** are elements with relatively few valence electrons, which tend to form positive ions by losing one or more electrons. Metals tend to be good conductors of heat and electricity. ***Nonmetals*** are elements that have more valence electrons and tend to form negative ions by gaining one or more electrons. Nonmetals tend to be poor conductors of heat and electricity. ***Semimetals***, which are also called ***metalloids*** or ***semiconductors,*** tend to have some characteristics of metals and some of nonmetals. Figure 3–7b shows a periodic table that shows the locations of the three basic types of elements.

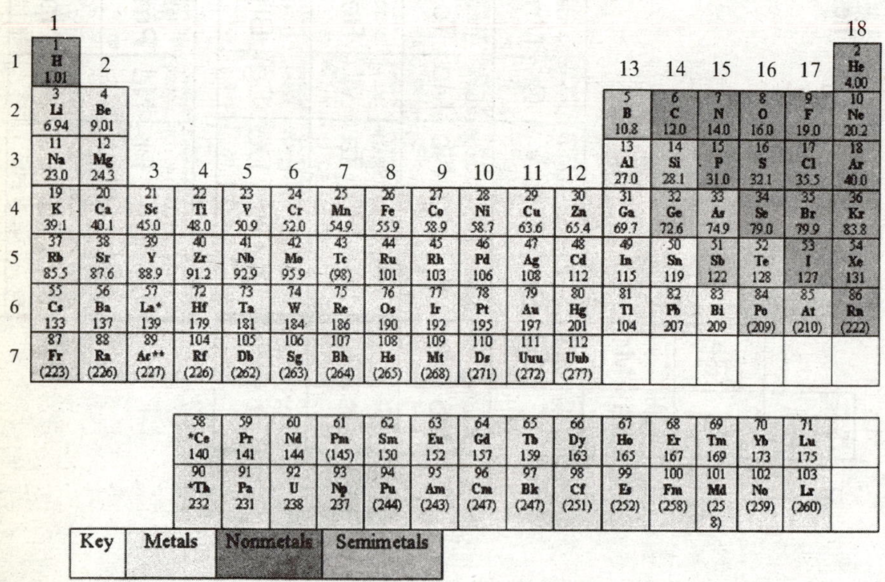

	1	2	3	4	5	6	7	8	9	10	11	12	13	14	15	16	17	18
1	1 **H** 1.01																	2 **He** 4.00
2	3 **Li** 6.94	4 **Be** 9.01											5 **B** 10.8	6 **C** 12.0	7 **N** 14.0	8 **O** 16.0	9 **F** 19.0	10 **Ne** 20.2
3	11 **Na** 23.0	12 **Mg** 24.3											13 **Al** 27.0	14 **Si** 28.1	15 **P** 31.0	16 **S** 32.1	17 **Cl** 35.5	18 **Ar** 40.0
4	19 **K** 39.1	20 **Ca** 40.1	21 **Sc** 45.0	22 **Ti** 48.0	23 **V** 50.9	24 **Cr** 52.0	25 **Mn** 54.9	26 **Fe** 55.9	27 **Co** 58.9	28 **Ni** 58.7	29 **Cu** 63.6	30 **Zn** 65.4	31 **Ga** 69.7	32 **Ge** 72.6	33 **As** 74.9	34 **Se** 79.0	35 **Br** 79.9	36 **Kr** 83.8
5	37 **Rb** 85.5	38 **Sr** 87.6	39 **Y** 88.9	40 **Zr** 91.2	41 **Nb** 92.9	42 **Mo** 95.9	43 **Tc** (98)	44 **Ru** 101	45 **Rh** 103	46 **Pd** 106	47 **Ag** 108	48 **Cd** 112	49 **In** 115	50 **Sn** 119	51 **Sb** 122	52 **Te** 128	53 **I** 127	54 **Xe** 131
6	55 **Cs** 133	56 **Ba** 137	57 **La*** 139	72 **Hf** 179	73 **Ta** 181	74 **W** 184	75 **Re** 186	76 **Os** 190	77 **Ir** 192	78 **Pt** 195	79 **Au** 197	80 **Hg** 201	81 **Tl** 104	82 **Pb** 207	83 **Bi** 209	84 **Po** (209)	85 **At** (210)	86 **Rn** (222)
7	87 **Fr** (223)	88 **Ra** (226)	89 **Ac**** (227)	104 **Rf** (226)	105 **Db** (262)	106 **Sg** (263)	107 **Bh** (264)	108 **Hs** (265)	109 **Mt** (268)	110 **Ds** (271)	111 **Uuu** (272)	112 **Uub** (277)						
				58 ***Ce** 140	59 **Pr** 141	60 **Nd** 144	61 **Pm** (145)	62 **Sm** 150	63 **Eu** 152	64 **Gd** 157	65 **Tb** 159	66 **Dy** 163	67 **Ho** 165	68 **Er** 167	69 **Tm** 169	70 **Yb** 173	71 **Lu** 175	
				90 ***Th** 232	91 **Pa** 231	92 **U** 238	93 **Np** 237	94 **Pu** (244)	95 **Am** (243)	96 **Cm** (247)	97 **Bk** (247)	98 **Cf** (251)	99 **Es** (252)	100 **Fm** (258)	101 **Md** (25 8)	102 **No** (259)	103 **Lr** (260)	

Key	Metals	Nonmetals	Semimetals

Figure 3–7b

As you can see, most of the elements are considered metals. Even elements such as calcium, which you might not think of as a metal, contain the characteristics, which result in its classification as a metal.

Did you ever wonder where the periodic table gets it name? When the elements on the periodic table are arranged according to their atomic numbers—that is, by the number of protons in their nuclei—certain properties repeat themselves *periodically*. The table was designed in such a way that the elements with similar properties all end up in the same columns.

Look at column 17 in the periodic table in Figure 3–7b. Ions of the elements in column 17 have very similar uses. Fluorides, chlorides, and bromides are all used to treat drinking water. With the exception of hydrogen, all of the elements in column 1 are metals that are so reactive that they can't be found in nature in their pure elemental form. Hydrogen is actually a strange exception, being a nonmetallic metal stuck in a column of metallic elements.

The columns of elements with similar properties are called ***families*** or ***groups***. Four of the groups, those found in columns 1, 2, 17, and 18 specifically, have special family names. Your instructor may require you to know the names of some of the families, shown in the periodic table that follows in Figure 3–7c. In addition to these family names, other sections of elements are also given special group names, as seen in the examples of the transition metals, lanthanides, and actinides shown here.

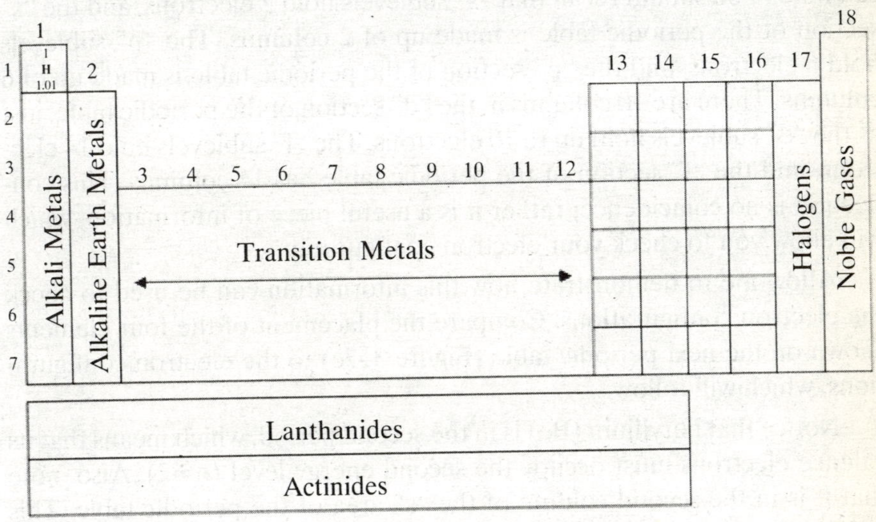

Figure 3–7c

The horizontal rows of the periodic table are called ***periods***. The period that an element is in will indicate the energy level in which its valence electrons reside. For example, sulfur is found in the third period. Using the steps in Lesson 3–4, we can conclude that it has the electron configuration of $1s^2\ 2s^2\ 2p^6\ 3s^2\ 3p^4$. As you can see, its valence electrons are in the third energy level.

We can use the periodic table to check the electron configurations that we write. The periodic table can be thought of as being made up of "s," "p," "d," and "f" sections, as shown in Figure 3–7d.

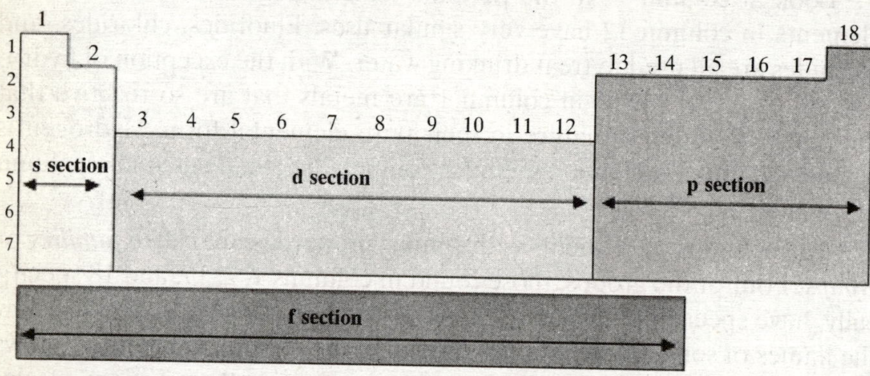

Figure 3–7d

Compare Figure 3–7d to the number of electrons that each sublevel can hold. You should recall that "s" sublevels hold 2 electrons, and the "s" section of the periodic table is made up of 2 columns. The "p" sublevels hold 6 electrons, and the "p" section of the periodic table is made up of 6 columns. There are 10 columns in the "d" section of the periodic table, just as the "d" sublevels hold up to 10 electrons. The "f" sublevels hold 14 electrons, and the "f" section of the periodic table has 14 columns. This connection is no coincidence; rather it is a useful piece of information, which will allow you to check your electron configurations.

Allow me to demonstrate how this information can be used to check the electron configurations. Compare the placement of the four elements shown on the next periodic table (Figure 3–7e) to the electron configurations, which will follow.

Notice that beryllium (Be) is in the second period, which means that its valence electrons must occupy the second energy level (n=2). Also, note that it is in the second column of the "s" area of the periodic table. This tells us that its valence configuration must end in "s^2." The combination of these two pieces of information tells us that the valence configuration of

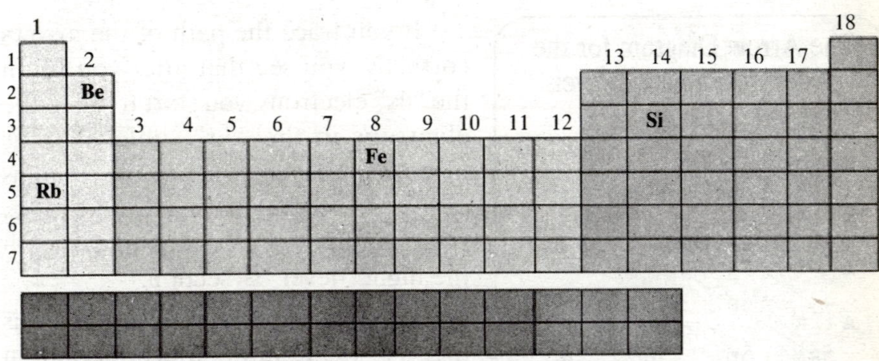

Figure 3–7e

beryllium is "$2s^2$." When we construct the full electron configuration as described in Lesson 3–4, we find that beryllium, which has an atomic number (often represented by the symbol "Z") of 4, shows the configuration: $1s^2\ 2s^2$. Do you see how we can check the configuration using the periodic table?

Let's look at the element rubidium (Rb) next. What can we tell from its position on the periodic table? It is in the fifth period (horizontal row), and in the first column of the "s" section of the periodic table. This tells us that the valence configuration of rubidium must end in $5s^1$. When we construct the full electron configuration of rubidium, with its 37 electrons, we get $1s^2\ 2s^2\ 2p^6\ 3s^2\ 3p^6\ 4s^2\ 3d^{10}\ 4p^6\ 5s^1$. Again, our resultant configuration matches our predictions.

Iron (Fe) is slightly trickier, because it is located in the "d" section of the periodic table. The trick to the "d" section of the periodic table is that the valence shell of these elements is not made up of "d" electrons. Remember that there are only up to 8 electrons in the valence shell, which is why only 8 electrons can fit around the Lewis dot notation of any given element. Refer to Figure 3–7f to refresh your memory.

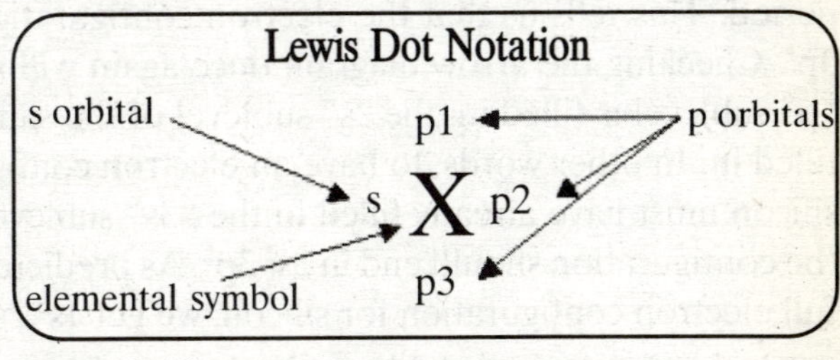

Figure 3–7f

You see, the "d" sublevels can hold up to 10 electrons, and there is no room for them on the Lewis dot notation. This means that when you are filling up a "d" sublevel, you are not filling up the valence shell. How is that possible? The answer can be found by looking once more at the arrow diagram.

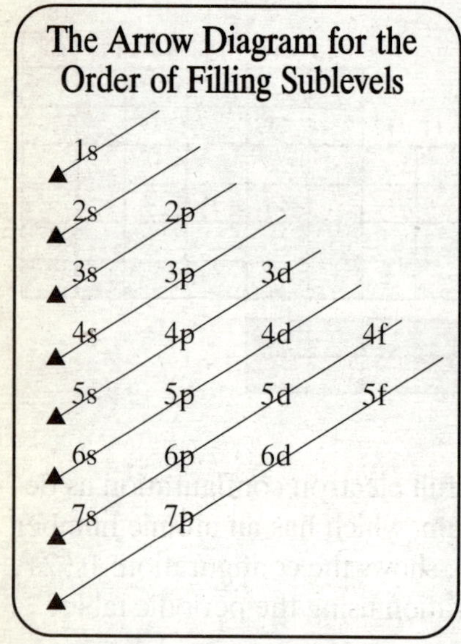

Figure 3–7g

If you trace the path of the arrows correctly, you see that after you fill in the "4s" electrons, you start to fill in the electrons in the "3d" sublevel. This means that, as you go across the columns of the "d" section of the periodic table, you are filling in electrons underneath the higher-level "s" section.

The way this applies to iron (Fe) is that it is found in the fourth period, in the sixth column of the "d" section. The fact that it is in the fourth period tells us that its valence electrons must be in the fourth energy level. However, the "d" section that is being filled in last is in the next lowest energy level. I combine this information to predict that the configuration of iron should end in "$4s^2$ $3d^6$," where the "$4s^2$" is actually the valence shell, because it has the highest value for the principal quantum number (***n***).

Constructing the full electron configuration, as I reviewed in Lesson 3–4, I get $1s^2 2s^2$ $2p^6$ $3s^2$ $3p^6$ $4s^2$ $3d^6$. We filled in the "3d" section last, but the "$4s^2$" actually represents the valence shell, so it is not uncommon to rewrite this configuration as $1s^2$ $2s^2$ $2p^6$ $3s^2$ $3p^6$ $3d^6$ $4s^2$, or with the shorthand notation: [Ar] $3d^6$ $4s^2$.

Finally, let's try the element silicon (Si), which has 14 electrons in its cloud. Silicon is located in the second column of the "p" section of the third period. This tells us that the electron configuration of silicon must end in $3p^2$. Checking the arrow diagram once again will remind you that before a "p" sublevel is filled in, the "s" sublevel of the same energy level is already filled in. In other words, to have an electron configuration that ends in $3p^2$, silicon must have already filled in the "3s" sublevel, so we can predict that the configuration should end in $3s^2$ $3p^2$. As predicted, when we construct the full electron configuration for silicon, we get $1s^2$ $2s^2$ $2p^6$ $3s^2$ $3p^2$.

Using the periodic table to check your electron configurations will give you a huge advantage. Anytime you are asked to construct an electron configuration, find the period and column that the element is in, and predict the end of the configuration. If your complete configuration doesn't end as you predicted, you know to go back and check your work. There are some elemental exceptions to the method described here, but your instructor may not expect you to know about them. If he or she does, consult a reference text to see the complete configurations of all elements.

I hope that you now see that the periodic table contains more information than you first suspected. Let's review what you just learned.

Lesson 3–7 Review

1. The ______________ number is the number of protons found in the nucleus of an atom.
2. Semimetals are also called ______________.
3. The columns of the periodic table are called ______________ of families.
4. The horizontal rows of the periodic table are called ______________.
5. The elements in the first column of the periodic table are called ______________.
6. ______________ are elements that tend to form positive electrons by losing electrons.

Chapter 3 Examination

Part I—Matching

Match the following terms to the definitions that follow. Not all answers will be used.

a. proton
b. neutron
c. electron
d. mass number
e. atomic number
f. kernel
g. negative
h. elemental notation
i. positive
j. quantum numbers
k. valence
l. orbital
m. Hund's Rule

_____1. A negatively charged particle found in the cloud region of the atom.

_____2. The nucleus and all of the electrons, except the valence electrons.

_____3. The total number of protons and neutrons in an atom.

_____4. A space that can be occupied by up to two electrons.

_____5. This type of ion is formed when an atom loses some electrons.

_____6. Each electron is described by a set of four of these.

_____7. This number is equal to the number of protons in an atom.

_____8. This type of ion is formed when an atom gains additional electrons.

_____9. A positively charged particle found in the nucleus of an atom.

Part II—Elemental Notation

Base your answers to questions 10–15 on the following elemental notation.

$$^{3}_{1}H^{+}$$

10. Which element is represented by this elemental notation?
11. What is the atomic number of the element?
12. What is the mass number of the element?
13. How many protons does the element have?
14. How many neutrons does the element have?
15. How many electrons does the element have?

Part III—Electron Configuration

Base your answers to questions 16–20 on the following electron configurations of neutral atoms.

a. $1s^2\ 2s^2\ 2p^6\ 3s^1$	c. $1s^2$	e. $1s^2\ 2s^2\ 2p^5$
b. $1s^2\ 2s^2\ 2p^2$	d. $1s^2\ 2s^1\ 2p^3$	f. $1s^2\ 2s^2\ 2p^4$

16. Which electron configuration represents a noble gas?
17. Which electron configuration shows an atom in the excited state?
18. Which electron configuration shows an alkali metal?
19. Which electron configuration represents a halogen?
20. Which ground state configuration shows a total of 4 valence electrons?

Answer Key

The actual answers will be shown in brackets, followed by the explanation. If you don't understand an explanation that is given in this section, you may want to go back and review the lesson that the question came from.

Lesson 3–1 Review

1. [ion]—Remember: A negative ion has extra electrons, meaning extra negative charges. A positive ion has lost some of its electrons or negative charges, leaving it with a net positive charge.
2. [neutron]—*N*eutral ***n***eutrons are found in the ***n***ucleus.
3. [carbon-12]
4. [atomic]—The nuclear charge is also equal to the number of protons in the nucleus of the atom.
5. [mass]—Because the electrons have so little mass, they are not taken into consideration when calculating the mass number of an atom.
6. [isotopes]—Carbon-14 is a radioactive isotope of carbon.

Lesson 3–2 Review

1. A. [19] B. [26] C. [11] D. [38]
   The number of protons is equal to the atomic number, so look in the lower left-hand corner of the elemental notation.
2. A. [14]—(27 – 13 = 14) B. [12]—(24 – 12 = 12)
   C. [49]—(86 – 37 = 49) D. [15]—(31 – 16 = 15)
   The number of neutrons is equal to the mass number—the atomic number. The mass number is found in the upper left-hand corner of the elemental notation; the atomic number is found in the lower left-hand corner.
3. A. [56]—(56 – 0 = 56) B. [25]—(25 – 0 = 25)
   C. [10]—(7 – (-3) = 10) D. [2]—(3 – 1 = 2)
   The number of electrons can be determined by algebraically subtracting the charge from the atomic number.
4. [D. $^{22}_{11}Na$]—We must subtract the atomic number from the mass number to get the number of neutrons. When the atomic number is exactly half the mass number, then the number of neutrons matches the number of protons.
5. [D. $^{12}_{6}X$ and $^{14}_{6}X$]—In order for two elemental notations to represent different isotopes of the same element, they must have the same atomic number (number of protons) but a different mass number. Answers A and B show different elements, because they have different atomic numbers.

Answer C shows the same element but also the same isotope, because the mass number is also the same.

6. [B. $^{12}_{6}C$]—Carbon-12 has a mass of exactly 12 atomic mass units.

Lesson 3–3 Review

1. [D. four]—Each electron is described by a set of four quantum numbers.
2. [B. second]—The second quantum number indicates the sublevel that the electron is found in.
3. [D. four]—The number of sublevels that an energy level contains is equal to the value for n. So, the third energy level (n = 3) contains 3 sublevels and the fourth energy level (n = 4) contains 4 sublevels.
4. [A. s]—There are four possible values for ℓ: 0, 1, 2 and 3, which correspond to sublevels of s, p, d, and f respectively.
5. [B. 3]—I remember the number of orbitals that the sublevels contain by remembering that the values represent the first 4 odd numbers, or 1, 3, 5, and 7. Because "p" comes from the second word in my sentence, "*s*ome *p*eople *d*on't *f*orget," I know that I use the second odd number for the number of orbitals it holds.
6. [C. 8]—The total number of electrons that an energy level can hold is given by the formula $2n^2$. Filling in 2 for n, we get $2(2^2) = 8$. So the second energy level (n = 2) holds 8 electrons. Alternatively, we could have added up the electrons that can be held by the sublevels. The second energy level has two sublevels, an "s" and a "p." The "s" sublevel can hold up to 2 electrons and the "p" sublevel holds up to 6, for a total of 8 electrons.

Lesson 3–4 Review

1. [2]—Each level has a number of sublevels that is equal to its energy level (n) number. The first energy level (n = 1) has one sublevel, the second (n = 2) has two, the third (n = 3) has three, and so on.
2. [3]—The four sublevels (s, p, d, f) contain 1, 3, 5, and 7 orbitals, in that order. The p sublevel contains 3 orbitals, which can hold up to a total of 6 electrons.
3. [32]—The total number of electrons that an energy level can hold is given by the formula $2n^2$. So, when n = 4, $2(4^2) = 32$ electrons.
4. [2]—If you got this question wrong, go back and read the question slower. It asks how many electrons an f **orbital** can hold. All orbitals, regardless of which sublevel they are in, can hold up to two electrons.
5. [$1s^2\ 2s^2\ 2p^6\ 3s^2\ 3p^6\ 4s^1$]—Potassium has an atomic number of 19, which means that it has 19 protons. In a neutral atom, the number of protons is equal to the number of electrons, so neutral potassium has 19 electrons.

Using the arrow diagram to determine the order for filling the sublevels, and placing 2 electrons in each s sublevel and 6 in each p sublevel, we run out of electrons when the 4s sublevel has 1 electron in it. We can check our answer by making sure that the sum of all the superscripts is equal to 19.

6. [$1s^2$ $2s^2$ $2p^6$ $3s^2$ $3p^6$ $4s^2$ $3d^{10}$ $4p^6$]—Krypton has 36 electrons and a complete octet. Notice that there are a total of 8 electrons in the fourth (valence) energy level of this atom, giving it the special stability of a noble gas.

Lesson 3–5 Review

1. [orbitals]—Remember that in some books the orbitals are represented with squares.
2. [electrons]—Slash marks are also used to represent electrons, in some books.
3. [C.]—The orbital notation shown in answer C shows 17 arrows, representing the 17 electrons in the chlorine atom.
4. [D.]—Helium, with an electron configuration of "$1s^2$," can be represented by the orbital notation shown in answer D.
5. [A.]—A neutral atom of fluorine has 9 electrons, as does the orbital notation shown in answer A.
6. [F.]—Lithium, with an electron configuration of "$1s^2$ $2s^1$," can be represented by the orbital notation shown in answer F.
7. [E.]—The orbital notation shown in answer E shows 12 arrows, representing the 12 electrons in a neutral atom of magnesium.
8. [B.]—Boron, with an electron configuration of "$1s^2$ $2s^2$," can be represented by the orbital notation shown in answer B.

Lesson 3–6 Review

1. [electrons]—More specifically, the dots represent the valence electrons.
2. [the electrons in the valence shell]—Only the outer energy level electrons as shown as dots in a Lewis dot diagram. Inner electrons, which are part of the "kernel" of the atom, are represented along with the nucleus by the elemental symbol.
3. [B.]—Helium, with the electron configuration of $1s^2$, has 2 valence electrons.
4. [D.]—Lithium, with the electron configuration of $1s^2$ $2s^1$, has only 1 valence electron.
5. [A.]—Oxygen, with the electron configuration of $1s^2$ $2s^2$ $2p^4$, has 6 valence electrons.

6. [F.]—Aluminum, with the electron configuration of $1s^2\ 2s^2\ 2p^6\ 3s^2\ 3p^1$, has 3 valence electrons.
7. [C.]—Nitrogen, with the electron configuration of $1s^2\ 2s^2\ 2p^3$, has 5 valence electrons.
8. [E.]—Fluorine, with the electron configuration of $1s^2\ 2s^2\ 2p^5$, has 7 valence electrons.

Lesson 3-7 Review

1. [atomic]—Remember: It is the atomic number that determines the atoms' identity.
2. [metalloids or semiconductors]—Semimetal tends to be better conductors than nonmetals but worse conductors than metals, hence the term *semiconductors*.
3. [groups]—Just as members of your family may share certain physical characteristics, elements in a family have similar properties.
4. [periods]—The periodic table has seven periods and room to grow.
5. [alkali metals]—Alkali metals are extremely reactive. They took a long time to be discovered because they are only found in nature as part of other compounds.
6. [metals]—You may have noticed that almost all of the metals are solids at room temperature.

Chapter 3 Examination

1. [c. electron]
2. [f. kernel]—Think of a kernel of popcorn. Valence electrons are not included in the kernel.
3. [d. mass number]—Because the electrons have essentially no mass, the protons and the neutrons make up the vast majority of the mass of an atom.
4. [l. orbital]—Orbitals are not always full, but when they are, they can hold a maximum of two electrons.
5. [i. positive]—Remember, losing negative charges (electrons) leaves an atom with "extra" positive charges.
6. [j. quantum numbers]
7. [e. atomic number]—The atomic number is also equal to the nuclear charge of an atom.
8. [g. negative]—When a neutral atom gains extra negative charges, it will have a net negative charge.
9. [a. proton]

10. [hydrogen]—H is the elemental symbol for hydrogen.
11. [1]—The atomic number is shown in the lower left-hand corner of the notation.
12. [3]—The mass number is shown in the upper left-hand corner of the notation.
13. [1]—The number of protons is equal to the atomic number.
14. [2]—The number of neutrons is equal to (the mass number – the atomic number), or $3 - 1 = 2$.
15. [0]—The number of electrons is equal to (the atomic number – the charge of atom), or $1 - (+1) = 0$.
16. [c. $1s^2$]—Although it only contains 2 electrons, helium is considered a noble gas because its valence shell is full.
17. [d. $1s^2\ 2s^1\ 2p^3$] The 2s sublevel is not full, yet there are electrons in the 2p sublevel. An electron must have "jumped up" from the 2s sublevel to the 2p sublevel.
18. [a. $1s^2\ 2s^2\ 2p^6\ 3s^1$]—With a total of 11 electrons, this neutral configuration must represent sodium.
19. [e. $1s^2\ 2s^2\ 2p^5$]—With seven valence electrons, this atom would be located in column 17. With a total of 9 electrons, it must be fluorine.
20. [b. $1s^2\ 2s^2\ 2p^2$]—The question clearly asked for a ground state configuration, so that ruled out answer d. The only other configuration that shows 4 valence electrons is b.

Bonding and Molecular Structure

Lesson 4–1: Types of Bonds

In Chapter 3 we discussed the octet rule, which states that an atom with a full valence shell (8 electrons in most cases) is considered relatively stable. Atoms tend to gain, lose, or share electrons in order to complete their valence shells and obtain the configuration of a noble gas. What an atom does in order to complete its valence shell depends upon the situation that it is in. Recall that oxygen has 6 electrons in its valence shell. In some situations, oxygen is able to gain 2 additional electrons, for a total of 8 valence electrons, to become the oxide ion (O^{2-}). This anion (negative ion) may then attach to one or more cations (positive ions), forming one or more ***ionic bonds***. Compounds formed in this way are called ***ionic compounds***, as shown in the following reactions:

$$Ca^{2+} + O^{2-} \rightarrow CaO$$
$$2Na^{+} + 0^{2-} \rightarrow Na_2O$$

In other situations, the neutral oxygen atom may share electrons with one or more other atoms, in order to "act" as though it has a complete valence shell part of the time. These shared electrons represent ***covalent bonds*** and result in the formation of ***molecular compounds***, as shown here:

$$N_2 + O_2 \rightarrow 2NO \text{ or } 2H_2 + 0_2 \rightarrow 2H_2O$$

The type of bond, and therefore the type of compound, that an atom will form in a given situation depends upon the relative ***electronegativities*** of the elements involved. The ***electronegativity*** of an element is a relative measure of its attraction for bonding electrons. The scale for electronegativity was based on the most electronegative element, fluorine, which has an electronegativity of 4.0. The electronegativity values for the elements in the main body of the periodic table are shown in Figure 4–1a.

Electronegativities of Selected Elements

H 2.2						
Li 1.0	Be 1.5	B 2.0	C 2.6	N 3.1	O 3.5	F 4.0
Na 0.9	Mg 1.2	Al 1.5	Si 1.9	P 2.2	S 2.6	Cl 3.2
K 0.8	Ca 1.0	Ga 1.6	Ge 1.9	As 2.0	Se 2.5	Br 2.9
Rb 0.8	Sr 1.0	In 1.7	Sn 1.8	Sb 2.1	Te 2.3	I 2.7
Cs 0.7	Ba 0.9	Tl 1.8	Pb 1.8	Bi 1.9	Po 2.0	At 2.2

Figure 4–1a

When atoms with very different electronegativities combine, they tend to form bonds with a high degree of ionic character, which means that the sharing of electrons is so unequal that the electrons can effectively be thought to be in the possession of one atom. This "stealing" of electrons results in one atom having extra electrons and a net negative charge, and another atom, which is missing electrons, having a net positive charge. Each pair of such ions are held together by the electrostatic force of attraction between unlike charges, which is called an ionic bond.

Bonds formed between atoms with relatively little or no difference in electronegativity show a high degree of covalent character, meaning that the sharing of electrons is closer to equal. When we say that atoms share a pair of electrons, we don't mean that they necessarily share them 50/50. It is possible, for example, that the pair of shared electrons will spend approximately 70% of the time in one atom's electron cloud and only 30% of the time in the other atom's electron cloud. How equal the sharing of the electrons is depends on the difference in the electronegativiy values of the elements involved. If the difference in electronegativity between the bonding atoms is very low, then a non-polar covalent bond is formed, which means that the atoms share the electrons essentially equally. If the electronegativity difference between the bonding atoms is moderate, then a ***polar covalent bond*** is formed. By "polar" we mean that it has a positive end and a negative end, akin to the poles on a bar magnet.

In this book, we will consider a bond that is formed between two atoms that have an electronegativity difference of 0.0–0.3 to be a non-polar covalent bond. If the electronegativity difference between the bonding elements is 0.4–1.7, we will consider the bond formed to be a polar covalent bond. When the electronegativity difference between the two elements is greater than 1.7, we will call the formed bond an ionic bond.

Major Types of Bonds Between Atoms	
Ionic bond	A bond formed between a negative ion (anion) and a positive ion (cation).
Covalent bond	A bond formed when atoms share one or more pairs of electrons.
Non-polar covalent bond	A bond formed when atoms share one or more pairs of electrons relatively equally.
Polar covalent bond	A bond formed when atoms share one or more pairs of electrons but one atom attracts the electrons more strongly, resulting in unequal sharing.

Let's try a few examples of determining the type of bond that forms between two elements.

Example 1

Determine the types of bonds that form between the following pairs of elements.

A. KCl B. H_2O C. HCl D. O_2

All we need to do is look up, and subtract, the electronegativities (E.N.) of the bonding elements. We ignore any subscripts, such as the subscript of "2" found in the H_2O.

A. From Figure 4–1a, we find:

E.N. of K = 0.8

E.N. of Cl = 3.2

Difference in E.N. = 2.4 (3.2 – 0.8 = 2.4).

Because 2.4 is greater than 1.7, this bond is **ionic**.

B. From Figure 4–1a, we find:

E.N. of H = 2.2

E.N. of O = 3.5

Difference in E.N. = 1.3 (3.5 – 2.2 = 1.3).

Because 1.3 is less than 1.7, this bond is covalent.

Because 1.3 is greater than 0.3, this bond is **polar covalent**.

C. From Figure 4–1a, we find:

E.N. of H = 2.2

E.N. of Cl = 3.2

Difference in E.N. = 1.0 (3.2 – 2.2 = 1.0).

Because 1.0 is less than 1.7 this bond is covalent.

Because 1.0 is greater than 0.3, this bond is **polar covalent.**

D. From Figure 4–1a, we find:

E.N. of O = 3.5

E.N. of O = 3.5

Difference in E.N. = 0.0 (3.5 – 3.5 = 0.0).

Because 0.0 is less than 1.7 this bond is covalent.

Because 0.0 is less than 0.3, this bond is **non-polar covalent**.

Answers:

A. KCl	ionic bond
B. H_2O	polar covalent bond
C. HCl	polar covalent bond
D. O_2	non-polar covalent bond

Before you try the practice problems on page 119, let me remind you of two things. First, because we are looking for the difference between the electronegativities, we always subtract the smaller number from the larger. Second, remember not to use the subscripts that come with formulas as multipliers for the electronegativities. For example, if you wanted to determine the type of bond that forms between hydrogen and carbon in the compound CH_4, you would **not** multiply the electronegativity of carbon by a factor of four.

Lesson 4–1 Review

Use the electronegativity values from Figure 4–1a to determine if the bonds formed in the following substances will be ionic, polar covalent, or non-polar covalent.

1. CH_4
2. NaCl
3. H_2
4. BaI
5. CO
6. NO_2

Lesson 4–2: Ionic vs. Molecular Compounds

Students who are just learning chemistry usually have a great deal of trouble distinguishing ionic compounds from molecular compounds, which is understandable, considering how similar their formulas can look. For example, KCl is an ionic compound, yet HCl is a molecular compound. It is hard to see and remember this difference, and students often become frustrated when they are corrected by their instructor after they incorrectly refer to an ionic compound as a molecule. How can you quickly tell ionic compounds from molecular ones without looking up the electronegativities, so you can avoid the embarrassment of being corrected in class? More importantly, what is the big deal about the distinction between ionic compounds and molecular compounds anyway?

Recall that ***ionic compounds*** are formed from ions, which, in turn, are formed by gaining or losing electrons. When a large number of these ions are grouped together in a solid form, we get a crystal that is large enough to see with the naked eye. Table salt (NaCl) consists of such crystals. What is both interesting and important to note is that these crystals are not made up of individual molecules. In fact, there is no special relationship between any one sodium ion and a neighboring chloride ion. Each sodium ion is equally attracted to each equidistant chloride ion. There are no special "partners," or distinct groups, that make up the ionic crystal.

You see, the formula NaCl only shows us the ratio by which the anions and cations can be found in the crystal. The fact that there are no subscripts in the formula tells us that for every 1 sodium ion in the crystal, there is 1 chloride ion. If we have a crystal of $CaCl_2$, we don't really have 2 chloride ions attached to or combined with each calcium ion in some type of special bond. All that formula really tells us is that the ratio of calcium ions to chloride ions is 1:2.

As you may have guessed from the previous examples, ionic compounds are also called ***salts***, and they tend to have the following properties.

Properties of Ionic Compounds

1. Their bonds are very strong, which means it takes a relatively high amount of energy to break them.
2. They tend to be dull, hard, brittle solids at standard temperature and pressure.
3. They usually don't conduct electricity in the solid state, but they are good conductors in molten or aqueous (solution) form.
4. They tend to have high melting and freezing points.
5. The structure of the crystals that they form depends on both the size and ratio of the ions that make them up.

Molecular compounds are formed from covalent bonds between atoms. The major difference between molecular and ionic compounds is that in molecular compounds, you have smaller discreet units making up the larger sample. In other words, in a block of ice, solid H_2O, you have groupings of 3 atoms that "belong" together. The atoms within a particular molecule share bonds that they do not share with atoms from other molecules. So, whereas the ionic formula of a salt only gives you the ratio of the elements in the compound, the molecular formula of a molecular compound actually indicates the small grouping of atoms that exist as molecules in the larger sample.

A simplistic, but helpful, analogy is to compare a crystal of an ionic compound to a large crowd of strangers, where no person in the crowd has a special relationship with any other member of the crowd. A sample of a molecular compound could be likened to a large crowd that is made up of individual couples or families. The people in the families have special relationships, which make them units within the larger group.

Although there many different molecular compounds, they tend to have certain similar properties.

Properties of Molecular Compounds

1. The strength of the bonds that make them up can vary a great deal.
2. They tend to be liquids or gases at standard temperature and pressure.

3. They are poor conductors in any state. They are good insulators of both heat and electricity.
4. They tend to have low melting and boiling points.
5. They tend to have weak forces of attraction between molecules.

You should be able to see why ionic compounds have high melting and boiling points and tend to be solids at standard temperature and pressure, whereas molecular compounds tend to have low melting and boiling points and tend to be liquids or gases at room temperature. In other words, if you put a block of salt next to a block of ice, why would the ice melt first? The reason is that the ions in an ionic compound are relatively strongly attracted to all of their neighbors of opposite charge, whereas the individual molecules of a molecular compound have relatively little attraction for their neighbors.

This is not to say that molecular compounds have no forces of attraction between particles; they do. These forces are called ***intermolecular forces***. They vary from substance to substance and are responsible for some of the varying properties between molecular compounds. For example, when a molecular compound is liquid at standard temperature and pressure, it is likely that it has stronger intermolecular forces than a molecular compound that is a gas at standard temperature and pressure.

Now, let's go back to our original question. How can you quickly tell ionic compounds from molecular ones without looking up the electronegativities? Remember that to form ionic compounds, there must be a relatively high difference between the electronegativities of the elements involved. A quick look back at the table of electronegativities from the last lesson will show you that most of the elements with very low electronegativities are the metals found in the first two columns on the left-hand side of the periodic table, in the alkali metal and alkaline earth metal groups. The elements with the very highest electronegativities are found towards the upper-right portion of the periodic table, when you leave out column 18, the noble gases. When a compound consists of one of the metals in the first two columns combined with one of the highly electronegative elements in the upper-right part of the periodic table, as in the case of KCl or NaF, the difference in their electronegativities will be relatively high, and the result is often an ionic compound. When two elements close to each other on the periodic table combine, as in CO or NO_2, the result is likely to be a molecular compound, because the electronegativity difference will not be high.

Finally, we must discuss the fact that many compounds contain **both** ionic and covalent bonds. This occurs when compounds contain special ions called polyatomic ions. ***Polyatomic ions*** are simply ions that are made up of more than one atom. How do such ions form? Think back to the example of oxygen, with six valence electrons. If an atom of oxygen "steals" two additional electrons to complete its valence shell, it becomes the oxide (O^{2-}) ion. If it shares two pairs of electrons with two hydrogen atoms, it becomes part of a neutral molecule of water (H_2O). However, if it "steals" one electron and shares one pair with an atom of hydrogen, you get a molecular compound, which is also an ion, the polyatomic ion called hydroxide (OH^-). A polyatomic ion can then attach to another ion with an opposite charge, forming an ionic compound such as sodium hydroxide (NaOH), which contains both ionic and covalent bonds.

Lesson 4–2 Review

1. ______________ compounds are formed from the electrostatic force of attraction between charged particles.
2. ______________ compounds are formed when atoms are held together by covalent bonds.
3. The relatively weak forces of attraction between the molecules of a substance are called ______________ forces.
4. ______________ ions are ions that are made up of more than one atom.
5. Which of the following compounds contain both ionic and covalent bonds?

   a) KCl b) $Al(OH)_3$ c) NO_2 d) CH_4

Lesson 4–3: Lewis Dot Diagrams for Compounds

In Lesson 3–6, we learned how to produce Lewis dot diagrams for elements. In this lesson, we will go over the process of creating Lewis dot diagrams for compounds. As you will learn in the next lesson, constructing these diagrams for molecular compounds is a good way to predict the shape of the molecules. These diagrams can also be useful when you need to determine the ratio by which elements combine to form compounds. When constructing these diagrams, you may want know which electrons come with each atom. For this reason, I will be using dots of different shades to represent electrons from different atoms in this lesson.

We will start by constructing the Lewis dot diagram for the molecular compound known as water, H_2O. Before we begin, let's refresh our memory of what the Lewis dot diagrams for oxygen and hydrogen look like.

Notice that the oxygen atom has two pairs of electrons, and two electrons that are unpaired, or lone, electrons. These lone electrons represent bonding sites. The oxygen atom can do a number of things to complete its octet. It could steal two additional electrons and become the oxide (O^{2-}) ion; it could form two single covalent bonds, sharing enough electrons to allow it to act like it has a noble gas configuration for part of the time; or it could form one double bond, by sharing two pairs of electrons with a single atom.

Lewis Dot Diagram for Oxygen and Hydrogen

O H

Oxygen has 6 valence electrons	Hydrogen has 1 valence electron

Figure 4–3a

When oxygen reacts with hydrogen to form water, each oxygen atom makes two single bonds with two different hydrogen atoms. Each hydrogen atom has only a single unpaired electron, so it can only make one single bond in order to complete its valence shell. Because hydrogen is so small, each hydrogen atom is considered complete if it obtains the valence configuration of helium, which only has two electrons in its valence shell. Sometimes a hydrogen atom steals an electron to become the hydride (H-) ion, but in the reaction to form water with oxygen, each hydrogen atom will form one single bond with an oxygen atom.

The Lewis Dot Notation for Water

Each shared pair of electrons are also known as a covalent bond.

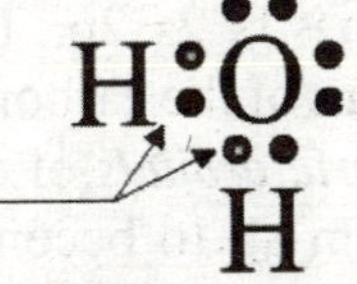

By sharing electrons, each element can gain stability by completing its valence shell. Notice, the oxygen atom is surrounded by 8 electrons, like the noble gas neon Each hydrogen atom is surrounded by 2 electrons, like the noble gas helium.

Figure 4–3b

In Figure 4–3b you will see a Lewis dot notation for water. For the sake of clarity, the electrons that came with the hydrogen atom are represented as grey dots, and the electrons that came with the oxygen atom are represented as black dots.

The lone electron for each atom acts as a

bonding site, which fit together like pieces of a puzzle. When the individual Lewis dot diagrams combine, each atom will look as if it has a complete valence shell. Oxygen will act as though it has 8 valence electrons, like neon, and hydrogen will act as if it has 2 valence electrons, like helium. Notice that the number of hydrogen atoms that could bond with the oxygen atom in this situation was dictated by the number of bonding sites available in each atom.

For our next example, let's look at how chlorine and carbon react to form the molecular compound called carbon tetrachloride. As before, we will begin by looking at the Lewis dot diagrams for the individual elements, which are based on their individual electron configurations.

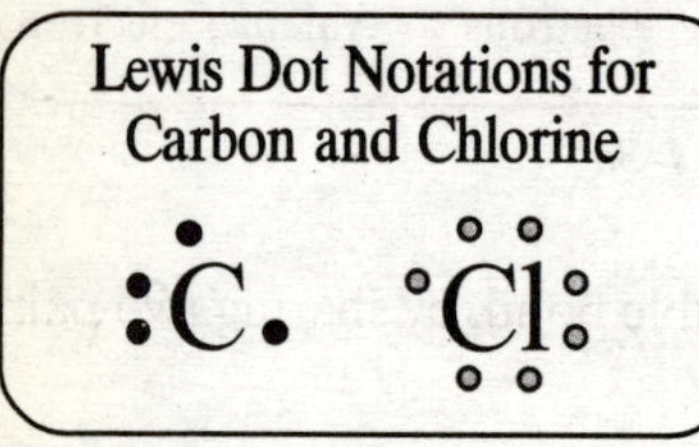

Figure 4–3c

Each chlorine atom has 7 valence electrons. In order to complete its octet, each chlorine atom needs one more electron. It can either steal an additional electron to form the chloride (Cl^-) ion, or it can form one single covalent bond and act as though it has a noble gas configuration for part of the time.

There is, however, a problem with the Lewis dot diagram for the carbon atom as it appears in Figure 4–3c. Carbon is shown with only two lone electrons, suggesting that it can only make two single covalent bonds. The problem is that sharing two pairs of electrons with chlorine would not complete its octet, and would leave it unstable, as shown in Figure 4–3d.

Predicted Lewis Dot Notation for a Compound Containing Carbon and Chlorine

Each chlorine atom appears to have a complete octet, but the carbon atom lacks the stability of a noble gas.

Figure 4–3d

Experimental evidence shows that carbon will react with chlorine in a 1:4 ratio, to form the compound called carbon tetrachloride (CCl_4). You will recall that the electron configuration for carbon is $1s^2 2s^2 2p^2$. The "2s" and "2p" electrons of the carbon atom are said to form ***hybrid orbitals*** of equal energy, allowing them all to become bonding sites. This allows each carbon atom to make four single covalent bonds, two double covalent bonds, or one triple and one single covalent bond, any of which would result in completing the carbon atom's octet.

There are times when no single Lewis dot diagram for a compound can be considered to be the only possible structure. In such a situation, a series of diagrams are used to represent all of the possible structures, which are said to exist simultaneously. The series of diagrams are collectively referred to as ***resonance*** structures.

Lewis dot diagrams are also useful for illustrating a special type of covalent bond, called a coordinate covalent bond. A ***coordinate covalent bond*** is formed when both electrons that become a shared pair of electrons between two atoms are provided by only one of the atoms. In other words, an atom with a strong attraction for electrons can combine with an atom that has already satisfied its desire for a noble gas configuration and force it to share a pair of unshared electrons. It is in this way that the neutral compound called ammonia (NH_3) can combine with a hydrogen ion (H^+) to become the polyatomic ion called the ammonium (NH_4^+) ion, as shown in Figure 4–3f.

The Correct Lewis Dot Notation for Carbon Tetrachloride

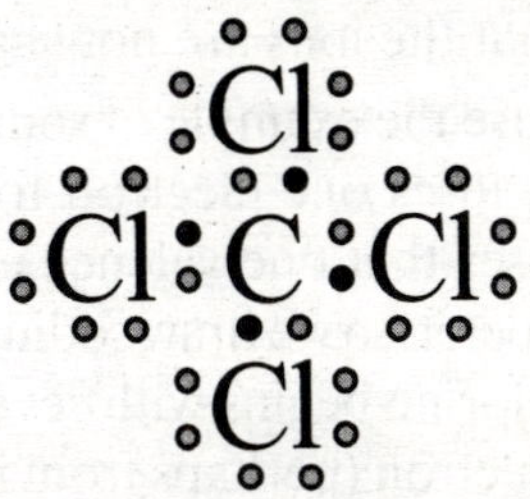

Cl C Cl Cl Cl

Each atom in the compound is surrounded by 8 electrons, completing its octet.

Figure 4–3e

Hydrogen Ion Reacting With Ammonia to Form the Ammonium Ion

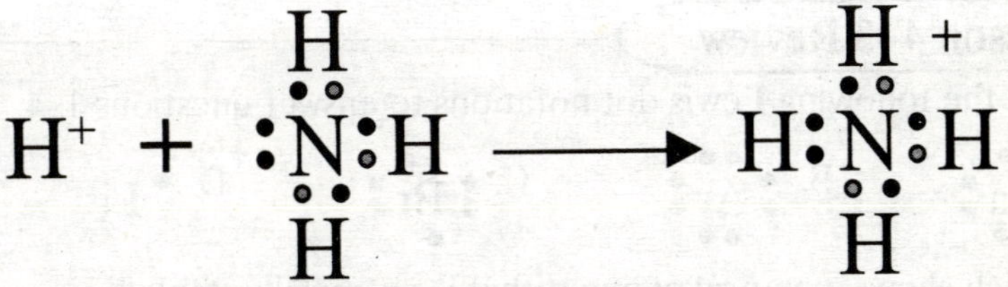

$$H^+ + :NH_3 \longrightarrow NH_4^+$$

When the ammonia molecule provides both electrons that get shared with the H^+ ion, a **coordinate covalent bond** is formed.

Figure 4–3f

Lewis dot diagrams can also be used to represent ionic compounds. When dealing with ionic compounds, it is important to remember that the individual atoms in the ionic compound complete their valence shells, by gaining or losing electrons, before you put the Lewis dot diagram together. For this reason, instead of showing elemental Lewis dot diagrams that seem to interlock to form complete octets, we draw diagrams that show complete octets that are next to each other. To make this even clearer, brackets are sometimes drawn around the individual ions, to draw attention to the fact that the ions are not sharing electrons.

Let's use the example of sodium chloride. Sodium, an alkali metal from column 1, loses one electron to form an ion with a charge of +1. When sodium loses that one valence electron, its inner energy level becomes its complete octet, so we draw sodium with eight valence electrons and a charge of +1. Chlorine begins with seven valence electrons but then steals an additional electron (perhaps from the sodium atom) to form an ion with eight valence electrons and a charge of -1. When we put these two ions together, we get the Lewis dot diagram shown in Figure 4–3g.

Lewis Dot Notation for Sodium Chloride

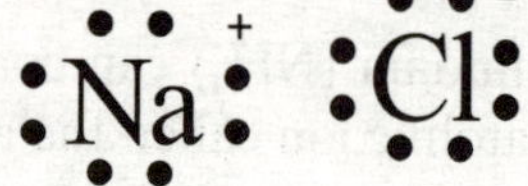

These ions don't share electrons, and their Lewis dot diagrams don't overlap. They are held together by the nature of their opposite charges.

Figure 4–3g

Lesson 4–3 Review

Use the following Lewis dot notations to answer questions 1–4.

A. Ba^{2+} (8 dots) B. Ar (8 dots) C. Br (7 dots) D. Li (1 dot)

1. Which shows a neutral element that is chemically stable?
2. Which shows an atom that has lost two electrons?
3. Which element could complete its octet by forming a single covalent bond?
4. Which neutral element is most likely to form a positive ion, with a charge of +1?

Use the following Lewis Dot Notations to answer questions 5–8.

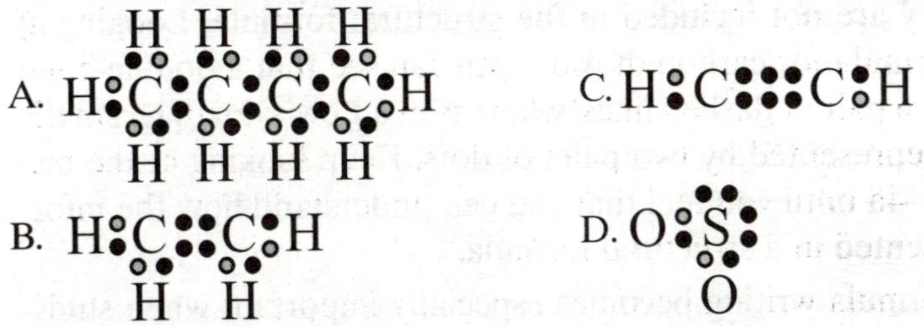

```
     H H H H
A. H:C:C:C:C:H          C. H:C:::C:H
     H H H H

                             ..
B. H:C::C:H             D. O:S:
     H  H                    ..
                             O
```

5. Which diagram shows a triple covalent bond?
6. Which diagram shows a total of 13 single covalent bonds?
7. Which diagram shows a double covalent bond?
8. Which diagram shows a compound that has the ability to form a coordinate covalent bond?

Lesson 4–4: Structural Formulas

In our last lesson, we discussed the procedures for constructing Lewis dot diagrams for compounds. Although these diagrams can be very useful, especially for smaller molecules, they can become unwieldy when used for larger molecules. It can be time-consuming to add the dots that represent valence electrons, and the detail that these diagrams show for nonbonding electrons is not always necessary. It is more common to use a structural formula to graphically represent a molecule. A ***structural formula*** is a formula that uses elemental symbols to represent atoms and dashed lines to show covalent bonds. Structural formulas show the general shape of a molecule, but the exact bond angles aren't represented.

If you have become comfortable with constructing Lewis dot diagrams for molecules, then drawing most structural formulas should seem fairly easy. For the most part, the structural formula for a molecule looks just like its Lewis dot diagram, except that pairs of bonding electrons are replaced with dashed lines and nonbonding electrons are usually left out of the diagram completely. Figure 4–4a on page 128 compares the Lewis dot diagrams and structural formulas for several common compounds.

You can learn a great deal about structural formulas by comparing the formulas in Figure 4–4a. Compare the Lewis dot diagram for water to the structural formula for the same compound. Each single bond between the oxygen atom and a hydrogen atom is represented by a pair of dots in the Lewis diagram and by a dashed line in the structural formula. Notice also that the unpaired electrons of oxygen are represented in the Lewis dot diagram, but they are not included in the structural formula. Looking at the structural formula for carbon dioxide, you can see that a double bond is represented by a pair of dashed lines, whereas in a Lewis dot diagram the double bond is represented by two pairs of dots. Keep looking at the pictures in Figure 4–4a until you feel that you can understand how the information gets presented in a structural formula.

Structural formula writing becomes especially important when studying compounds containing carbon, which are called ***organic compounds***. Carbon atoms have the ability to link together into long chains and rings,

Lewis Dot Diagrams and Structural Formulas

Compound	Lewis Dot Diagram	Structural Formula

Water (H_2O)	H:Ö:
	  H
		H–O
		   |
		   H

Methane (CH_4)	H H:C:H H	H | H–C–H | H
Carbon Dioxide (CO_2)	:Ö::C::Ö:	O=C=O
Propane (C_3H_8)	H H H H:C:C:C:H H H H	H H H | | | H–C–C–C–H | | | H H H
Propyne (C_3H_4)	H H:C:::C:C:H H	H | H–C≡C–C–H | H

Figure 4–4a

which make up the "backbones" of these organic compounds. Organic compounds in which all of the carbon atoms are bound together by single bonds are called ***saturated compounds***. If a compound contains two or more carbon atoms that are held together by double or triple bonds, the compound is said to be an ***unsaturated compound***. Figure 4–4a shows propane (C_3H_8), which is a saturated compound, and propyne (C_3H_4), which is an unsaturated compound. Why do these two compounds share the same prefix ("prop-") and have different suffixes ("-ane" and "-yne")? This has to do with the system for naming ***hydrocarbons***, molecules containing only hydrogen and carbon.

If a hydrocarbon is saturated—that is, it contains only single bonds—then it is given the suffix "-ane." If the hydrocarbon contains one or more double bonds, then it is given the suffix "-ene." A hydrocarbon is given the suffix "-yne" if it contains one or more triple bonds. The prefix associated with a hydrocarbon is determined by the number of carbon atoms that its molecule contains. For example, the prefix "meth-" is given to the hydrocarbon containing only one carbon atom. Methane is a saturated hydrocarbon (all single bonds) with only one carbon atom. "Prop-" is the prefix given to hydrocarbons containing three carbon atoms, so both the compounds C_3H_4 and C_3H_8, with three carbon atoms, are given this prefix. The chart on page 130 shows the prefixes for the first 10 hydrocarbon configurations, along with an example of each.

It may be worth the effort to study the names and formulas of the examples shown in Figure 4–4b. Do you notice any patterns between the suffixes of the compounds and the relative number of carbon and hydrogen atoms that they show? Notice all of the compounds with the ending "-ane." Can you think of a formula that you can use to determine the number of hydrogen atoms found in each of the saturated hydrocarbons?

The saturated hydrocarbons are given the group name ***alkanes***, and they have only single bonds between their carbon atoms. They are all given the suffix "-ane." The prefix tells you the number of carbon atoms that an alkane contains, and you can determine the number of hydrogen atoms that they contain by using the formula C_nH_{2n+2} where "n" is the number of carbon atoms in the compound. For example, methane contains one carbon atom, so n = 1. The number of hydrogen atoms is determined with the formula $2n+2$, so there are four hydrogen atoms in a molecule of methane $((2 \times 1) + 2 = 4)$. Propane contains three carbon atoms, so n = 3, and the number of hydrogen atoms in a molecule of propane would be eight $((2 \times 3) + 2 = 8)$.

Prefixes for Hydrocarbons		
Number of Carbon Atoms	**Prefix**	**Example of Hydrocarbon**
1	Meth-	Methane (CH_4)
2	Eth-	Ethyne (C_2H_2)
3	Prop-	Propene (C_3H_6)
4	But-	Butane (C_4H_{10})
5	Pent-	Pentene (C_5H_{10})
6	Hex-	Hexyne (C_6H_{10})
7	Hept-	Heptene (C_7H_{14})
8	Oct-	Octane (C_8H_{18})
9	Non-	Nonyne (C_9H_{16})
10	Dec-	Decane ($C_{10}H_{22}$)

Figure 4–4b

Hydrocarbons that contain a single double bond are called ***alkenes***, and they are given the suffix "-ene." The number of carbon atoms in an alkene is given by the prefix, and the number of hydrogen atoms can be determined with the formula C_nH_{2n}, where, once again, n is equal to the number of carbon atoms in the compound. Quite simply, the alkenes contain twice as many hydrogen atoms as carbon atoms. Ethene contains two carbon atoms, as indicated by the prefix "eth-" and four hydrogen atoms ($2 \times 2 = 4$). Pentene contains five carbon atoms, as indicated by the prefix "pent-" and ten hydrogen atoms ($2 \times 5 = 10$).

Hydrocarbons that contain one triple bond are called ***alkynes***, and they are given the suffix "-yne." The number of carbon atoms that an alkyne contains is indicated by its prefix, and the number of hydrogen atoms can be determined from the formula C_nH_{n-2}, where n is the number of carbon atoms. For example, butyne contains four carbon atoms, as indicated by the prefix "but-" and six hydrogen atoms ($2 \times 4 - 2 = 6$).

Let's use the information that we just reviewed to try some examples of structural formulas.

Example 1

Write the molecular formula and structural formula for a molecule of propene.

The first thing that you want to do is determine the molecular formula for the compound. The prefix "prop-" tells us (see Figure 4–4b) that we are dealing with a compound that contains three carbon atoms. The suffix "-ene" tells us that we are dealing with an alkene. Alkenes contain twice as many hydrogen atoms as carbon atoms, so propene must have six hydrogen atoms ($2 \times 3 = 6$).

A. Molecular formula: "prop-" = 3 carbon atoms, or C_3
"-ene" means the # of hydrogen atoms $= 2n = 2 \times 3 = 6$ or H_6
so, the molecular formula for propene is C_3H_6

As for the structural formula, we start by laying down the three carbon atoms in a line, as shown here.

C C C

You will recall that the electron configuration for carbon is $1s^2\,2s^2\,2p^2$, so each carbon atom, with four valence electrons, can make up to four single bonds. The suffix "-ene" tells us that there is a double bond between two of the carbon atoms. The carbon atoms involved in the double bond will use up two out of their four bonding electrons with that one double bond, and will only be able to make two additional single bonds. It doesn't matter between which two carbon atoms we show the double bond.

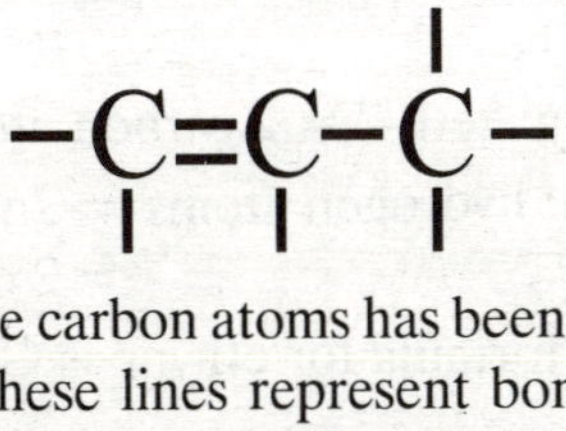

```
           |
  –C=C–C–
    |   |   |
```

Notice that each of the carbon atoms has been drawn with a total of four lines coming out of it. These lines represent bonds or bonding sites. The carbon atom on the left has made a double bond and still has two single bonding sites available for bonding. The middle carbon atom has made one double bond and one single bond and only has one single unpaired electron available for bonding. The carbon atom on the right has only made one single bond and has three unpaired electrons available for bonding.

How do we know that all of the unpaired electrons that haven't been used will be used to form single bonds? Each hydrogen atom, with only one valence electron, can only make a single covalent bond. If any of the hydrogen atoms were to bond with another one of the hydrogen atoms, they wouldn't have any bonding electrons left to attach to a carbon atom. Therefore, each hydrogen atom must use its one bonding electron to connect to a carbon atom, where there is an available bonding site. The following Example 1 summary shows the completed structural formula for propene:

A. Molecular formula: "prop-" = 3 carbon atoms, or C_3
"-ene" means the # of hydrogen atoms = 2n = 2 × 3 = 6 or H_6
so, the molecular formula for propene is C_3H_6

B. Structural formula for propene:

```
            H
            |
H–C=C–C–H
   |   |   |
   H   H   H
```

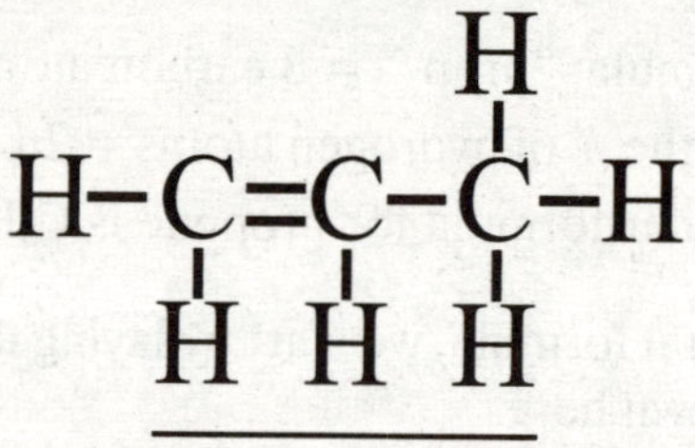

This may seem confusing at first, so try as many examples as you can. Always pay attention to the valence configurations of the elements in order to figure out how many bonds each can make. Let's try another example. I show the work for the molecular formula already completed, but I will go over the structural formula more slowly.

Example 2

Write the molecular formula and structural formula for a molecule of ethyne.

A. Molecular formula: "eth-" = 2 carbon atoms, or C_2
"-yne" means # of hydrogen atoms = 2n – 2 = (2 × 2) – 2
= 2 or H_2
so, the molecular formula for ethyne is C_2H_2

I hope that you find writing the molecular formula easy. It is as easy as looking up what the prefix means, then calculating the number of carbon atoms and then calculating the number of hydrogen atoms from the suffix.

For the structural formula, we will start by laying down the two carbon atoms and connecting them with a triple bond. How do we know that the carbon atoms are joined by a triple bond? Remember: The suffix "-yne" indicates the alkyne series, which contain a triple bond. The triple bond can't be between a carbon and hydrogen atom, because hydrogen atoms only have the ability to make a single bond. Carbon atoms have the ability to make up to four single bonds, or two double bonds, or one triple and one single bond. If we use up three bonding electrons for each carbon atom with a triple bond, then each will have enough for one more single bond, as shown here.

$$-C\equiv C-$$

You will recall that each carbon atom can make one triple and one single bond. Notice that four lines come out of each carbon atom. Each carbon atom has used up three of its four bonding electrons, leaving just one more line available for each.

Now, it should be obvious where the two hydrogen atoms go. Each of the two hydrogen atoms can make a single bond, and there are two bonding sites available. We simply add the hydrogen atoms to the spaces available, as shown in this summary:

A. Molecular formula: "eth-" = 2 carbon atoms, or C_2
"-yne" means # of hydrogen atoms $= 2n - 2 = (2 \times 2) - 2$
$= 2$ or H_2
so, the molecular formula for ethyne is C_2H_2

B. Structural formula for ethyne:

$$H-C\equiv C-H$$

Something strange seems to happen when we start constructing structural formulas for hydrocarbons with more than three carbon atoms. Suddenly, you will find that there seems to be more than one possible answer. You can create more than one structural formula for a given molecular formula. You can put all of the carbon atoms in a long chain, or you can have one or more branches coming off the main chain. Two or more compounds that have the same molecular formula but different structural formulas are called ***isomers***. Figure 4–4c on page 134 shows an example of isomerism.

For butane, as shown in Figure 4–4c, there are two possible isomers. As you add more and more carbon atoms, there become more and more possible combinations. Molecular formulas with many carbon atoms will have dozens of different possible structural formulas. For this reason, a more complex naming system is required for more complex hydrocarbons. Depending upon the course that you are taking, you might not need to know the more complex naming system, but you will almost certainly be expected to understand the concept of isomers and be able to identify examples of isomerism.

Examining Isomers

Name of Compound	Butane	Iso-butane
Molecular Formula	C_4H_{10}	C_4H_{10}

Structural Formula:

```
  H H H H              H H H
  | | | |              | | |
H-C-C-C-C-H          H-C-C-C-H
  | | | |              |   |
  H H H H              H   H
                       H-C-H
                         |
                         H
```

Figure 4–4c

Lesson 4–4 Review

Base your answers to questions 1–6 on the following structural formulas.

```
          H                   H
          |                   |
A. H-C≡C-C-H       C. H-C=C-C-H        E. H-O
          |              | | |              |
          H              H H H              H

     H
     |
B. H-C-H           D. H-C≡C-H          F. O=C=O
     |
     H
```

1. Which image represents a saturated hydrocarbon?
2. Which image shows the structural formula for the compound called "propene"?
3. What would be the name of the compound shown in image D?
4. Which image shows a member of the ***alkene*** group?
5. Which hydrocarbon follows the general formula of C_nH_{2n+2}?
6. What would be the name of the compound shown in image A?

Lesson 4–5: Geometry of Molecules

In addition to bonding, another important factor that influences the properties of a molecular substance is the shape or geometry of its particles. When atoms share electrons, they are held together by covalent bonds.

Depending on the types and numbers of atoms involved, the molecules can form many different shapes. The valence shell electron pair repulsion (VSEPR) theory suggests that the shape a molecule forms is based on the valence electrons surrounding the central atom. In this lesson, we will explore the process of predicting the shapes of molecules.

When you are dealing with a molecule that is made up of only two atoms, HCl for example, it is easy to see that the nuclei of the atoms must fall along a straight line. We call such a molecular shape ***linear***. Here are some examples.

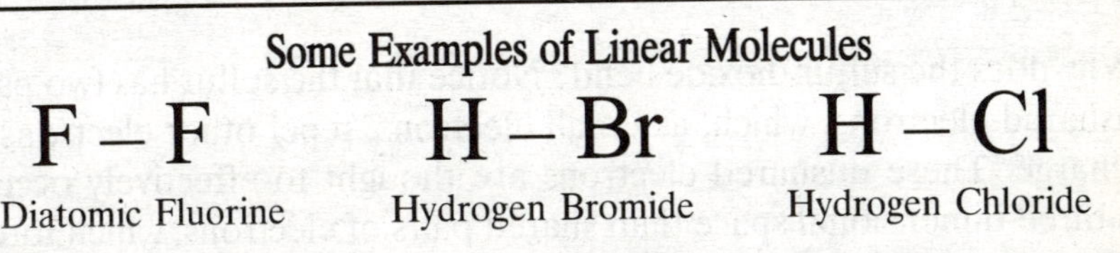

Figure 4–5a

When predicting the shape of a molecule made up of three atoms, we find that it is not as easy. CO_2, for example, is linear, with all three of its atoms lining up in a straight line. SO_2, however, with almost the same formula, is ***bent***. Why the difference? The answer lies with the electrons found in the valence shell of the central atoms in each molecule, the carbon and the sulfur. Let's start by comparing the Lewis dot diagrams of each of these atoms.

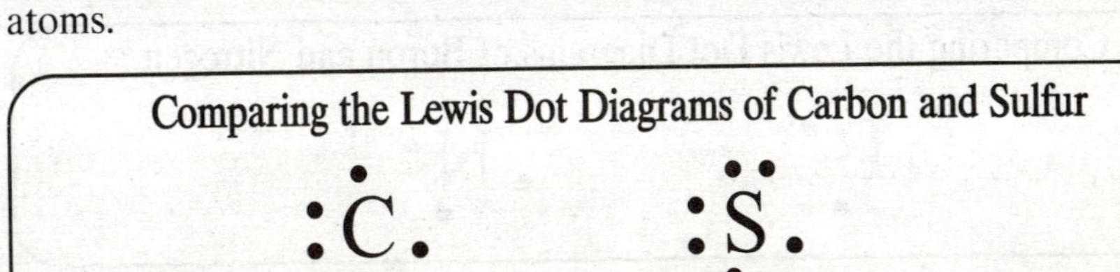

Figure 4–5b

As you can see in Figure 4–5b, the sulfur atom has only two unpaired electrons and can complete its valence shell by forming two single bonds. The carbon, on the other hand, only has four valence electrons, so making two single bonds won't satisfy the octet rule. To become stable when combining with two oxygen atoms, carbon dioxide must make two double bonds, sharing a total of eight valence electrons. Let's look at the Lewis dot diagrams for the two compounds in Figure 4–5c on page 136.

Comparing the Lewis Dot Diagrams of Carbon Dioxide and Sulfur Dioxide

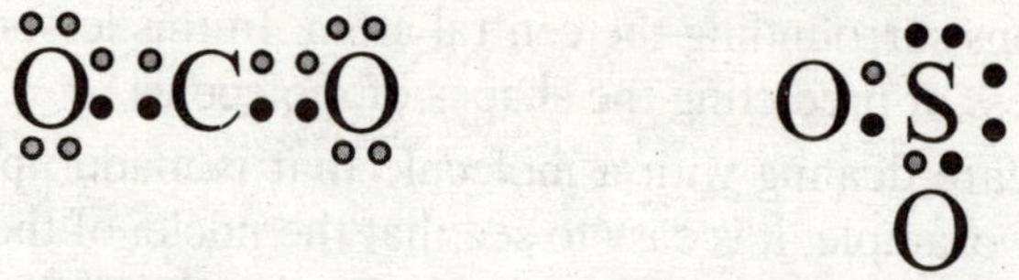

Ö::C::Ö   O:S̈:
O

Figure 4-5c

Why does the sulfur dioxide bend? Notice that the sulfur has two pairs of unshared electrons, which, as do all electrons, repel other electrons of like charge. These unshared electrons are thought to effectively occupy more three-dimensional space than shared pairs of electrons, which forces the molecule to bend as the electrons repel each other.

What if four atoms combine to form a molecule, as in the cases of BH_3 and NH_3? Again, we don't want to assume that they would have the same shape just because the formulas look so similar. We want to look at the valence shells of each of the central atoms—that is, the boron and the nitrogen—and see if there will be any unpaired electrons exerting forces on the shared pairs.

Comparing the Lewis Dot Diagrams of Boron and Nitrogen

.B˙   .N̈˙

Figure 4–5d

Notice that once the boron atom makes single bonds with each of the hydrogen atoms, it won't have any unpaired electrons distorting the shape of the molecule. When the shared pairs of electrons repel each other, the molecule will take on the shape called ***trigonal planar***. The nitrogen, on the other hand, will still have a pair of unshared electrons repelling the three single bonds that it forms with the hydrogen atoms. This will push the shared pairs of electrons away, forming a ***pyramidal*** molecule. This distinction is

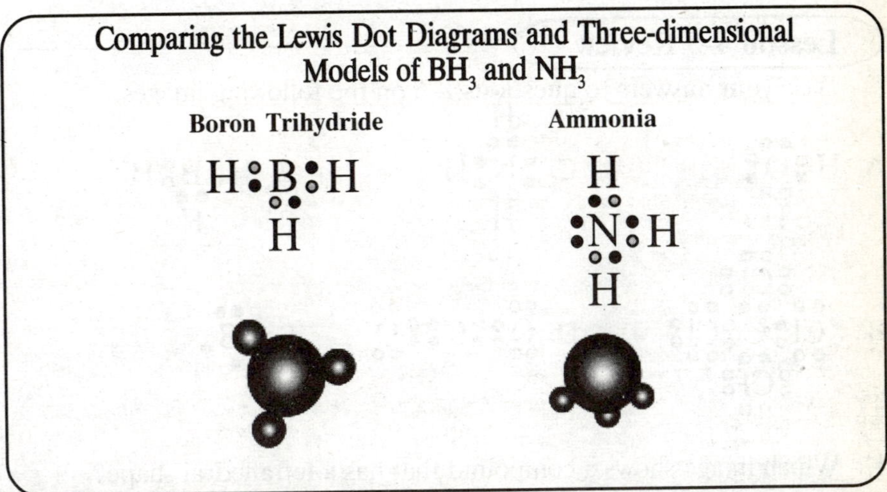

Figure 4–5e

not very clear in a Lewis dot diagram, which shows a two-dimensional representation of a molecule, but it will be more clear in a three dimensional model, such as the "ball and stick" models that you probably have in your chemistry laboratory. (See Figure 4–5e.)

There are many ways that five atoms can combine to form a molecule, and thus several different shapes are possible. When four atoms of one type surround one atom of another type, as in the example of CH_4, a ***tetrahedral*** molecule is formed. To understand the reason for this, you must think in three dimensions again. Each of the four shared pairs of electrons (covalent bonds) repel each other. In order to maximize the distance between electrons, the molecule forms a tetrahedral shape. Once again, the Lewis dot diagram fails to indicate this, but it is clear in a three-dimensional model for the molecule, as Figure 4–5f illustrates.

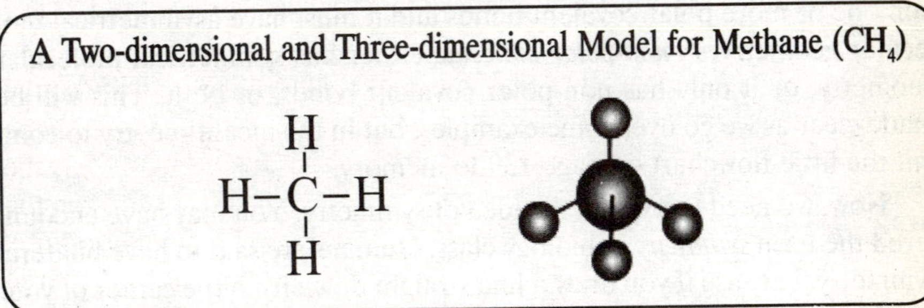

Figure 4–5f

Lesson 4-5 Review

Base your answers to questions 1–6 on the following images:

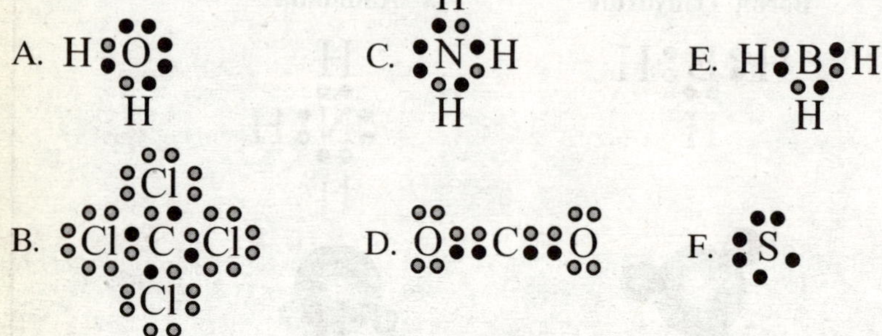

1. Which image shows a compound that has a tetrahedral shape?
2. What would cause the compound shown in image C to form a different shape than the compound shown in image E?
3. What shape does the molecule from image A show?
4. Which image shows a linear molecule?
5. What shape molecule would the atom in image F form if it acts as the central atom, bonded to two other atoms?
6. What is the shape of the molecule formed by the compound shown in figure E?

Lesson 4-6: Polarity of Molecules

When we talk about the polarity of molecules, we are talking about whether the molecule has positive and negative poles, like a bar magnet. A ***polar molecule***—that is, one with a positive and negative side—must contain one or more polar covalent bonds and it must have asymmetrical molecular geometry. A ***non-polar molecule*** either has symmetrical molecular geometry, or it only has non-polar covalent bonds, or both. This will be made clear as we go over some examples, but in the meantime, try to commit the little flowchart on page 139 to memory.

Now, we need to explain the idea of symmetry. You may have encountered the term *symmetry* in biology class. Humans are said to have bilateral symmetry, because if you draw a line straight down from the center of your head, your two "halves" look the same. If you can draw a line across a molecule and divide it into unequal looking halves, then it is asymmetrical.

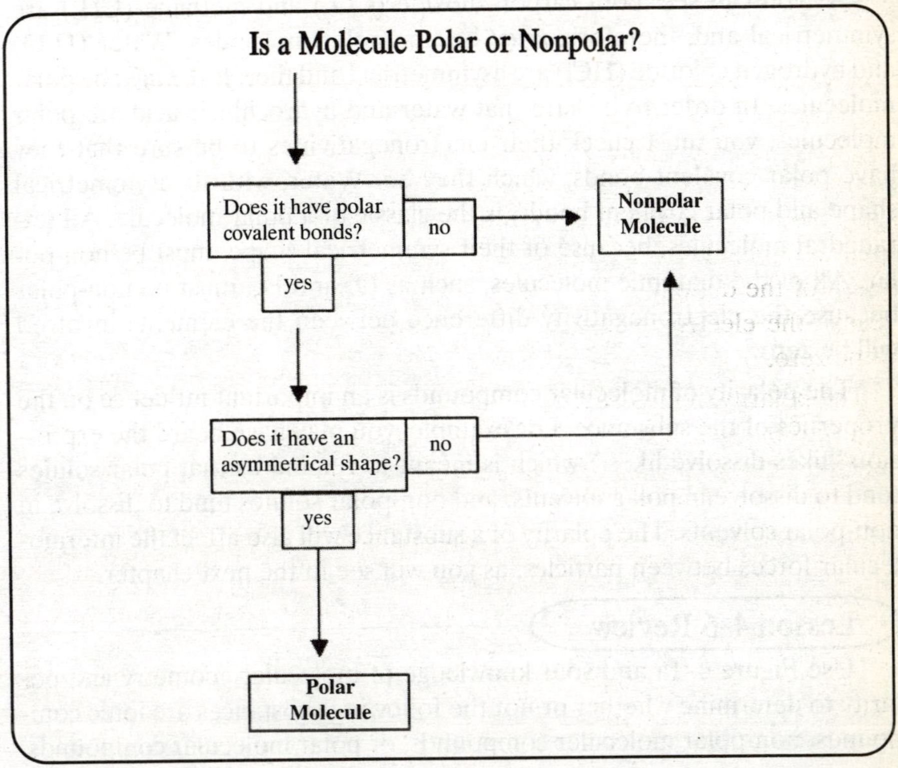

Is a Molecule Polar or Nonpolar?
Does it have polar covalent bonds?
no
Nonpolar Molecule
yes
Does it have an asymmetrical shape?
no
yes
Polar Molecule

If the halves of the molecule look the same, along both the horizontal axis and the vertical axis, then the molecule is said to be symmetrical. Figure 4–6a shows several examples of how you can use lines to determine if a molecule is symmetrical or not.

Comparing Symmetrical and Asymmetrical Molecules
Carbon Dioxide
Methane
Water
Hydrogen Chloride

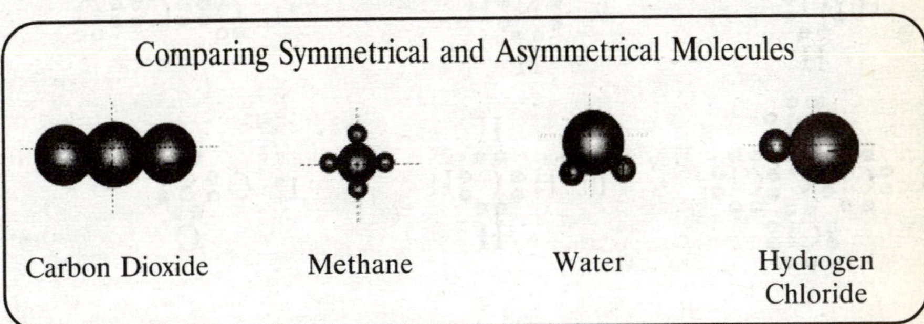

Figure 4-6a

As you can see, both carbon dioxide (CO_2) and methane (CH_4) are symmetrical and, therefore, *must* be non-polar molecules. Water (H_2O) and hydrogen chloride (HCl) are asymmetrical and therefore *might* be polar molecules. In order to be sure that water and hydrochloric acid are polar molecules, you must check their electronegativities to be sure that they have polar covalent bonds, which they do. Water, with its asymmetrical shape and polar covalent bonds, is the classic of a polar molecule. All tetrahedral molecules, because of their symmetrical shape, must be non-polar. All of the diatomic molecules, such as O_2 and H_2, must be non-polar because the electronegativity difference between the elements involved will be zero.

The polarity of molecular compounds is an important influence on the properties of the substance. For example, you may have heard the expression "likes dissolve likes," which is meant to remind us that polar solutes tend to dissolve in polar solvents, and non-polar solutes tend to dissolve in non-polar solvents. The polarity of a substance will also affect the intermolecular forces between particles, as you will see in the next chapter.

Lesson 4-6 Review

Use Figure 4–1a and your knowledge of molecular geometry and polarity to determine whether or not the following substances are ionic compounds, non-polar molecular compounds, or polar molecular compounds.

1. Li_2O
2. I_2
3. CBr_4
4. NO
5. CO
6. MgS

Look at the shapes of the molecules in the following figure and determine if they are symmetrical or asymmetrical.

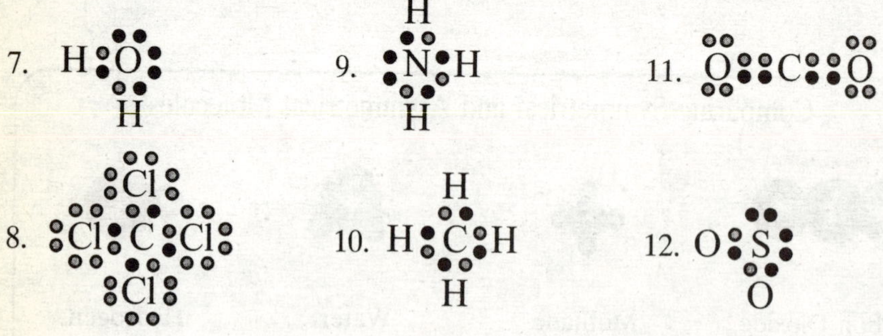

Chapter 4 Examination

Part I—Matching

Match the following terms to the definitions that follow. Not all of the terms will be used.

a. non-polar covalent bond
b. polar covalent bond
c. ionic bond
d. polyatomic ion
e. coordinate covalent bond
f. resonance
g. organic compounds
h. hydrocarbons
i. isomers

_____1. Compounds with the same molecular formula but different structural formulas.

_____2. The group of compounds containing the elements carbon and hydrogen.

_____3. A group of covalently bonded atoms, which obtain a charge.

_____4. The type of bond formed between two atoms with a difference of electronegativity of greater than 1.7.

_____5. The type of bond formed when one atom provides both of the electrons that will be shared between two atoms.

_____6. The name for the class of compounds that contain carbon.

_____7. The type of bond that forms between the atoms in a diatomic molecule, such as H_2.

Part II—Multiple Choice

Select the best answer choice for each of the following questions.

8. Which of the following represents a saturated compound?
   a) CH_4 b) C_2H_2 c) C_2H_4 d) C_3H_6

9. Which of the following represents a saturated compound?
   a) ethene b) ethyne c) propane d) propene

10. How many carbon atoms does a molecule of octane contain?
    a) 2 b) 4 c) 6 d) 8

11. How many hydrogen atoms would a molecule of hexene contain?
    a) 6 b) 8 c) 10 d) 12
12. What would be the proper name for a hydrocarbon with the formula C_2H_4?
    a) ethane b) ethene c) ethyne d) propene
13. Which of the following is the formula for an alkyne?
    a) C_2H_2 b) C_2H_4 c) C_2H_6 d) C_3H_8
14. Which of the following compounds contains only ionic bonds?
    a) H_2O b) MgF_2 c) CH_4 d) SO_2
15. Which of the following compounds contains both ionic and covalent bonds?
    a) BaI_2 b) N_2O c) $NaNO_3$ d) $CaBr_2$
16. Which of the following represents a molecular compound?
    a) $Ag_{(s)}$ b) $BaF_{2(s)}$ c) $NaCl_{(s)}$ d) $H_2O_{(s)}$
17. Which of the following molecules has a bent shape?
    a) N_2 b) H_2O c) CO_2 d) CH_4
18. Which of the following represents a polar molecular substance?
    a) NH_3 b) CCl_4 c) CO_2 d) O_2
19. Which of the following is not a polyatomic ion?
    a) OH b) SO_4^{2-} c) NH_4^+ d) Cl^-
20. Which of the following has a tetrahedral shape?
    a) BH_3 b) CBr_4 c) C_2H_6 d) SO_2

Answer Key

The actual answers will be shown in brackets, followed by the explanation. If you don't understand an explanation that is given in this section, you may want to go back and review the lesson that the question came from.

Lesson 4–1 Review

1. [polar covalent]—Carbon has an electronegativity of 2.6, and hydrogen has an electronegativity of 2.2 (2.6 – 2.2 = 0.4), giving us an E.N. difference of 0.4.
2. [ionic]—Sodium has an electronegativity of 0.9, and chlorine has an electronegativity of 3.2 (3.2 – 0.9 = 2.3), giving us an E.N. difference of 2.3.

3. [non-polar covalent]—Hydrogen has an electronegativity of 2.2, and hydrogen has an electronegativity of 2.2 (2.2 – 2.2 = 0.0), giving us an E.N. difference of 0.0.
4. [ionic]—Barium has an electronegativity of 0.9, and iodine has an electronegativity of 2.7 (2.7 – 0.9 = 1.8), giving us an E.N. difference of 1.8.
5. [polar covalent]—Carbon has an electronegativity of 2.6, and oxygen has an electronegativity of 3.5 (3.5 – 2.6 = 0.9), giving us an E.N. difference of 0.9.
6. [polar covalent]—Nitrogen has an electronegativity of 3.1, and oxygen has an electronegativity of 3.5 (3.5 – 3.1 = 0.4), giving us an E.N. difference of 0.4.

Lesson 4–2 Review

1. [ionic]
2. [molecular]
3. [intermolecular]
4. [polyatomic]—The atoms of a polyatomic ion are held together by covalent bonds, but the entire substance becomes a charged ion.
5. [B. $Al(OH)_3$]—Neutral compounds that contain polyatomic ions have both ionic and covalent bonds. Covalent bonds hold the polyatomic ion together, and then ionic bonds hold the polyatomic ion to other ions. The hydroxide ion (OH^-) is the only polyatomic ion that appears in our answer choices, so answer B must be correct.

Lesson 4–3 Review

1. [B.]—The notation for argon shows a complete octet (eight electrons), which represents a stable element. The barium ion also shows a complete octet, but the question asked for a neutral element.
2. [A.]—The charge of +2 that appears with the diagram for the barium ion indicates that it has lost two electrons.
3. [C.]—Bromine has seven valence electrons. Forming a single covalent bond will complete its octet.
4. [D.]—Lithium is an alkali metal with one valence electron. Losing its one valence electron will expose its inner energy level, giving it the configuration of helium.
5. [C.]—Answer C represents the compound ethyne, which has a molecular formula of C_2H_2. The three pairs of dots found between the carbon atoms represent a triple covalent bond.
6. [A.]—Each pair of electrons found between two atoms represents a covalent bond. Answer A shows the dot diagram for the compound butane, which has thirteen pairs of shared electrons.

7. [B.]—A double covalent bond is represented by two pairs of shared electrons, found between two atoms. Answer B shows the dot diagram for the compound called ethane.

8. [D.]—In order to form a coordinate covalent bond, one atom must provide a pair of electrons. Of the choices, only the sulfur (IV) oxide has a pair of unshared electrons that can be used for a coordinate covalent bonding site.

Lesson 4-4 Review

1. [B.]—A hydrocarbon is a compound that is composed of hydrogen and carbon, so that rules out images E and F. A saturated hydrocarbon only has single covalent bonds. Image B, representing the compound methane, shows the only saturated hydrocarbon.

2. [C.]—The prefix "pro-" indicates that the compound that we are looking for contains three atoms of carbon, as in images A and C. The suffix "-ene" indicates that the compound has a double bond like image C.

3. [ethyne]—We use the prefix "eth-" for hydrocarbons with two atoms of carbon, and the suffix "-yne" for hydrocarbons with a triple bond.

4. [C.]—The alkenes are hydrocarbons that contain a double bond, as in the compound shown in image C. Carbon dioxide has double bonds, but it is not a hydrocarbon.

5. [B.]—The general formula C_nH_{2n+2} indicates that if you take the number of carbon atoms and multiply by 2 and then add 2, you will get the number of hydrogen atoms in the compound. Image B shows the compound methane, which follows this general formula. Methane contains one carbon atom and $(2(1) + 2 = 4)$ 4 hydrogen atoms.

6. [propyne]—We use the prefix "prop" to indicate the three carbon atoms and the suffix "yne" to indicate the triple bond.

Lesson 4-5 Review

1. [B.]—Compounds with the general formula XY_4, such as CH_4 and CCl_4, tend to form the tetrahedral shape.

2. [unshared electrons]—The unshared electrons that the nitrogen atom has will repel the electron pairs in the compound further, resulting in a pyramidal shape.

3. [bent]—Water is the classic example of a bent molecule. The unshared electrons in the oxygen atom's valence shell are thought to repel the shared pairs of electrons enough to bend the molecule.

4. [D.]—The nuclei of the atoms in the molecule of carbon dioxide are arranged in a line.

5. [bent]—Because the valence shell configuration of sulfur is similar to that of oxygen, they will form similar shapes when they are the central atoms in the molecules.
6. [trigonal planar]—The atoms of this compound are arranged in a triangler shape, along the same plane.

Lesson 4–6 Review

1. [ionic compound]—The difference in electronegativity between lithium and oxygen, as shown in Figure 4–1a, is 2.5 (3.5 – 1.0 = 2.5). Because 2.5 is greater than 1.7, we would classify lithium oxide as an ionic compound.
2. [non-polar molecular compound]—All of the diatomic elements are made up of non-polar molecules, because the electronegativity of each atom involved in a molecule will always be the same, so the difference between them will be zero. As a technicality, they don't represent molecular compounds, because they are each made up of only one type of element.
3. [non-polar molecular compound]—According to Figure 4–1a, the electronegativity of carbon is 2.6 and the electronegativity of bromine is 2.9. The difference in electronegativity is 0.3, which makes the bonds between the carbon and bromine atoms non-polar covalent and the molecule non-polar. Even if the bonds were polar, the shape of this molecule is symmetrical tetrahedral, which would result in a non-polar molecule.
4. [polar molecular compound]—The electronegativities of nitrogen and oxygen are 3.1 and 3.5, respectively, making for an electronegativity difference of 0.4. This means that NO is made with polar covalent bonds. The molecule will have a linear asymmetrical shape, with a positive end and a negative end, making it polar.
5. [polar molecular compound]—The electronegativities of carbon and oxygen are 2.6 and 3.5, respectively, making for an electronegativity difference of 0.9. This means that CO is made with polar covalent bonds. The molecule will have a linear asymmetrical shape, with a positive end and a negative end, making it polar.
6. [ionic compound]—The electronegativity of potassium is 0.8. Sulfur has an electronegativity of 2.6. Subtracting 0.8 from 2.6, we get an electronegativiy difference of 1.8, which is just high enough to result in the compound being classified as ionic.
7. [asymmetrical]—Water is the classic example of an asymmetrical molecule. The fact that water is asymmetrical and polar explains many properties of water. We determine it is asymmetrical by drawing an imaginary line through the molecule to see if we can have positive poles on one side and negative poles on the other. See Figure 4–6a.

8. [symmetrical]—Tetrahedral molecules are symmetrical, because their negative charges are distributed all around the molecule. You can't draw an imaginary line that would result in the negative poles all on one side.
9. [asymmetrical]—Ammonia is asymmetrical, although it is easier to understand why when you consult a three-dimensional image of the molecule, such as the one shown in Figure 4–5e. The positive and negative poles will be on opposite sides of an imaginary line drawn through the middle of the molecule.
10. [symmetrical]—As with the compound CCl_4 from #8, the molecules of methane (CH_4) take a tetrahedral shape. Because the negative charges are distributed all around the molecule, they can't be separated into poles by an imaginary line.
11. [symmetrical]—As shown in Figure 4–6a, the two sides of the carbon dioxide molecule are mirror images of each other, making it impossible to separate poles with an imaginary line.
12. [asymmetrical]—With a shape that is bent, similar to the water molecule, an imaginary line can be used to separate the poles.

Chapter 4 Examination

1. [i. isomers]
2. [h. hydrocarbons]
3. [d. polyatomic ion]
4. [c. ionic bond]
5. [e. coordinate covalent bond]
6. [g. organic compounds]—All hydrocarbons are organic compounds, but not all organic compounds are hydrocarbons.
7. [a. non-polar covalent bond]—Any two atoms of the same element must have an electronegativity difference of 0.
8. [A. CH_4]—A saturated compound contains only single bonds. Because each carbon atom can make four single bonds, a saturated hydrocarbon must have the general formula shown by C_nH_{2n+2}. Of the choices, only methane, CH_4, fits that format.
9. [C. propane]—The suffix "ane" is used to indicate a hydrocarbon with only single bonds.
10. [D. 8]—The prefix "oct" is used to indicate 8 carbon atoms.
11. [D. 12]—Hexene is an alkene, which follow the general formula C_nH_{2n}. The prefix "hex" tells us that it has six carbon atoms. Because alkenes have twice as many hydrogen atoms as carbon atoms, it must have 12 hydrogen atoms.

12. [B. ethene]—If you try to sketch the structural formula for C_2H_4, you will see that you will need a double bond between the two carbon atoms, in order to occupy all of the bonding electrons. We use the prefix "eth" because each molecule has two carbon atoms, and the suffix "ene" because it contains a double bond.
13. [A. C_2H_2]—An alkyne must contain a triple bond. If you draw structural formulas for each of the answer choices, only choice A will have a triple bond. Alkynes also have the general formula of C_nH_{2n-2}. Only answer choice A follows that formula.
14. [B. MgF_2]—We look for an electronegativity difference of greater than 1.7, and we find that in answer B only. Magnesium has an electronegativity of 1.2 and fluorine has an electronegativity of 4.0, making the difference 2.8. Notice that these two elements are on the opposite sides of the periodic table.
15. [C. $NaNO_3$]—To answer this question correctly, we look for a compound that contains a polyatomic ion. Sodium nitrate ($NaNO_3$) contains the nitrate (NO_3^-) ion, which is held together by covalent bonds. The polyatomic ion then attaches to the sodium ion with an ionic bond.
16. [D. $H_2O_{(s)}$]—A molecular compound must contain two elements (which rules out answer choice A) with similar electronegativities. Often, you can just look for the compound that contains no metals, and be correct.
17. [B. H_2O]—Carbon dioxide and diatomic nitrogen have a linear shape. Methane has a tetrahedral shape. Water is the classic example of a bent shape.
18. [A. NH_3]—Although all four answer choices represent molecular substances, only the ammonia molecule is asymmetrical.
19. [D. Cl^-]—Of course, a polyatomic ion must contain more than one atom. The chloride ion does not.
20. [B. CBr_4]—The molecules with a tetrahedral shape follow the basic formula XY_4.

[illegible]

[illegible]

[illegible]

[illegible]

[illegible]

[illegible]

[illegible]

[illegible]

[illegible]

Chemical Formulas

Lesson 5–1: Oxidation Numbers

Two common salts, which you are probably familiar with, are sodium chloride (NaCl), which we sometimes call "table salt," and calcium chloride ($CaCl_2$), which many people use to melt the ice on driveways. Why does the chlorine have a subscript of "2" in one chemical formula but not in the other? The ratio by which elements combine in order to form compounds can often be predicted by using the oxidation numbers of the elements in the compound. The ***oxidation number*** represents the charge, or the apparent charge, of an element in the compound.

Atoms can often gain or lose electrons to form ions. When an atom loses electrons, it will have more protons than electrons and show a positive charge. If an atom of aluminum loses three electrons, it will be left with a net charge of +3, and we indicate that with the symbol Al^{3+}. When an atom gains electrons, it will have more electrons than protons and will have a net negative charge. For example, if an atom of sulfur gains two additional electrons, it will have a net charge of -2. We represent such an ion with the symbol S^{2-}.

When you are dealing with elements that only show one oxidation state, determining the ratio by which they form can be quite easy. The idea is that the elements will combine in such a ratio that there will be a net charge of zero, for a neutral compound. For example, how would we find the chemical formula for the compound of sodium and chlorine? If I told you that sodium (Na) has an oxidation number of +1 and chlorine (Cl) has an oxidation number of -1, they must combine in a 1:1 ratio because $(+1) + (-1) = 0$. The chemical formula for the compound formed by these elements, in a 1:1 ratio, would be NaCl. No subscripts are needed or used, because they are each understood to be 1.

Would calcium and chlorine also combine in a 1:1 ratio? To answer that, you would need to know the oxidation numbers of the elements involved. Chlorine, as I mentioned in the last example, has an oxidation number of -1. If I told you that the oxidation number of calcium is +2, you should see that it would take two atoms of chlorine for every atom of calcium to make a net charge of zero. You see, $(+2) + 2(-1) = 0$, so you need one atom of calcium (+2) for every two atoms of chlorine (-1) to get a net charge of 0.

How do you determine the oxidation number for an element? In most cases, you might look up the oxidation number on a periodic table or some other reference table. Often, you can predict the oxidation number of an element by its location on the periodic table. You should remember that the electron configuration, and therefore the Lewis dot notation, for many elements may be determined by locating the element on the periodic table. If you can figure out what type of ion the atom is likely to form in order to complete its valence shell, you can figure out the likely oxidation number for the element.

Sodium is in the first column of the periodic table, which, as explained in Chapter 3, gives it the electron configuration of $1s^2 2s^2 2p^6 3s^1$. If an atom of sodium loses one electron to form a positive ion (Na^+), its new configuration of $1s^2 2s^2 2p^6$ shows a complete valence shell. This charge of +1, which the sodium ion shows, is equal to the oxidation number for sodium.

Calcium, an alkali earth metal found in column 2 of the periodic table, shows the electron configuration $1s^2 2s^2 2p^6 3s^2 3p^6 4s^2$. In order to complete its octet, calcium will often give up its two valence electrons, in the fourth energy level, to expose the 8 electrons that it has in the third energy level. Through this process, the calcium atom becomes the calcium ion (Ca^{2+}) with the electron configuration $1s^2 2s^2 2p^6 3s^2 3p^6$. This +2 is not only the charge on the ion of calcium; it is also the oxidation number that we use for determining the type of compounds calcium will form.

Some elements have several different oxidation numbers, because they have more than one option for improving their stability. Carbon, for example, has four valence electrons in its neutral state. It can complete its octet by either gaining four electrons or losing four electrons. For this reason, carbon shows oxidation numbers of +4 and -4, and the number that is used in a given situation depends on whether carbon is going to form or act like a positive or a negative ion.

We can often determine the oxidation number of one element by making use of the oxidation number of another element. For example, suppose you are asked to determine the oxidation number shown by carbon in the formula CF_4. Even if you knew the possible oxidation states of carbon, as explained in the previous paragraph, can you tell if carbon is acting like it is +4 or -4 in this particular compound? The key to solving this problem is in knowing that fluorine always shows an oxidation number of -1. If each fluorine atom has an oxidation number of -1 and there are 4 fluorine atoms, as shown by the subscript 4, then you have a total apparent negative charge of -4. Remember that the net charge on each neutral compound is 0, so the apparent charge on the carbon atom must be +4, as shown in this equation:

$$+4 + 4(-1) = 0$$

Memorizing the following rules about oxidation numbers will make it easier to write chemical formulas. Even when you have access to an oxidation table, you can save a great deal of time by memorizing some of the basic facts.

Rules for Assigning Oxidation Numbers

1. In a free element (e.g., Ba or I_2) the oxidation number for each atom is 0.
2. With the exception of hydrogen, all of the elements in group 1 show an oxidation number of +1 in compounds.
3. Hydrogen has an oxidation number of +1 in all compounds except for hydrides (e.g., NaH), where it shows an oxidation number of -1.
4. The elements in group 2, the alkaline earth metals, show an oxidation number of +2 in all compounds.
5. When any of the elements in column 17, the halogens, show a negative oxidation state, it will be -1.
6. Oxygen shows an oxidation number of -2 in all compounds, except for the peroxides, where it is -1.
7. In all neutral compounds, the algebraic sum of the oxidation numbers is 0.
8. In polyatomic ions, the algebraic sum of the oxidation numbers is equal to the charge on the polyatomic ion.

Lesson 5–1 Review

Use your knowledge of the information found in the Rules for Assigning Oxidation Numbers to assign oxidation numbers to the requested elements in numbers 1–8.

1. What is the oxidation number shown by nitrogen in a molecule of N_2?
2. What is the oxidation number of chlorine in a formula unit of NaCl?
3. What is the oxidation number shown by potassium in K_2S?
4. What is the oxidation number shown by oxygen in a molecule of water?
5. What is the oxidation number of calcium in CaO?
6. What is the oxidation number of sulfur in $BaSO_4$?
7. What is the oxidation number of sulfur in Li_2SO_3?
8. What is the oxidation number of carbon in $CaCO_3$?

Lesson 5–2: Writing Chemical Formulas

We are now ready to go over the very important skill of writing chemical formulas. Recall the analogy that I made use of in Chapter 3, when I referred to the chemical formulas as the "words" that make up the chemical equations that you will need to learn how to write. These chemical formulas, in turn, are made up of the elemental symbols, which represent the "letters" in the language of chemistry. This will be one of the most important lessons that you will cover in chemistry. Learning this lesson well will allow you to develop the confidence to succeed in this subject. I urge you to take the time to master this topic by practicing as many examples that you can get your hands on.

The tools that you use for this lesson will be one or more oxidation tables. These tables look different in different books, but the information that they contain is the same. Some periodic tables will include oxidation numbers, so your instructor may ask you to work from that. It is probably more likely that your instructor will give you a separate set of tables with oxidation numbers, so I will follow that format in this text. Often, the oxidation numbers are divided into two tables, one for monatomic (single-atom) ions and one for polyatomic (multiple-atom) ions, as shown here. It doesn't matter where you get the information from, as long as you can use the oxidation numbers to write proper chemical formulas.

Symbols and Oxidation Numbers for Select Monatomic Ions					
+1		**+2**		**−1**	
Cesium	Cs^{+}	Barium	Ba^{2+}	Bromide	Br^{-}
Copper (I)	Cu^{+}	Beryllium	Be^{2+}	Chloride	Cl^{-}
Francium	Fr^{+}	Calcium	Ca^{2+}	Fluoride	F^{-}
Hydrogen	H^{+}	Cobalt (II)	Co^{2+}	Hydride	H^{-}
Lithium	Li^{+}	Copper (II)	Cu^{2+}	Iodide	I^{-}
Potassium	K^{+}	Iron (II)	Fe^{2+}		
Rubidium	Rb^{+}	Lead (II)	Pb^{2+}		
Silver	Ag^{+}	Magnesium	Mg^{2+}		
Sodium	Na^{+}	Manganese	Mn^{2+}		
		Mercury (II)	Hg^{2+}		
+3		Nickel (II)	Ni^{2+}	**−2**	
Aluminum	Al^{3+}	Strontium	Sr^{2+}	Oxide	O^{2-}
Iron (III)	Fe^{3+}	Tin (II)	Sn^{2+}	Sulfide	S^{2-}
Nickel (III)	Ni^{3+}	Zinc	Zn^{2+}		

Figure 5–2a

Formulas and Oxidation Numbers for Select Polyatomic Ions					
−1		**−2**		**−3**	
Acetate	CH_3COO^{-}	Carbonate	CO_3^{2-}	Phosphate	PO_4^{3-}
Chlorate	ClO_3^{-}	Chromate	CrO_4^{2-}		
Chlorite	ClO_2^{-}	Dichromate	$Cr_2O_7^{2-}$		
Cyanide	CN^{-}	Oxalate	$C_2O_4^{2-}$	**+1**	
Hydroxide	OH^{-}	Peroxide	O_2^{2-}	Ammonium	NH_4^{+}
Hypochlorite	ClO^{-}	Silicate	SiO_3^{2-}	Hydronium	H_3O^{+}
Nitrate	NO_3^{-}	Sulfate	SO_4^{2-}		
Nitrite	NO_2^{-}	Sulfite	SO_3^{2-}	**+2**	
Perchlorate	ClO_4^{-}	Tartrate	$C_4H_4O_6^{2-}$	Mercury (I)	Hg_2^{2+}
Permanganate	MnO_4^{-}	Tetraborate	$B_4O_7^{2-}$		
		Thiosulfate	$S_2O_3^{2-}$		

Figure 5–2b

In addition to the reference tables, the important thing to remember is that you want to construct chemical formulas with no net apparent charge. This means that you want to combine the elements in a ratio that would result in a net oxidation number of zero. It is also important to note that we will use parentheses if, and only if, we need to multiply a polyatomic ion with a subscript. The use of parentheses seems to be the thing that students have the most trouble with. Let's try several examples together.

Suppose your instructor asked you to write the proper chemical formula for aluminum chloride. The first step is to look up the elemental symbol and the oxidation number of each of the elements involved. Because this compound contains only monatomic ions, we can find both of the required oxidation numbers in the first group of oxidation numbers.

Example 1

Write the correct chemical formula for aluminum chloride.

Name of the compound—aluminum chloride
Elemental symbols and oxidation numbers—Al^{3+} and Cl^{-}

Next, we figure out in what ratio the ions must combine in order to have a net charge of zero. In this example, we need 3 chloride ions to cancel out the aluminum ion, because +3 + 3(-1) = 0. We indicate the fact that we need 3 chlorides by using a subscript of 3.

Name of the compound—aluminum chloride
Elemental symbols and oxidation numbers—Al^{3+} and Cl^{-}
Proper chemical formula—$AlCl_3$

Notice that the oxidation numbers are not part of the proper chemical formula; they are use to determine the ratio by which the elements combine, but they don't get written with the formula. Now, we will try a slightly harder example. Suppose you need to write the proper formula for aluminum sulfide. Once again, the first step will be to look up the proper elemental symbol and oxidation number for each of the elements in the compound.

Example 2

Write the correct chemical formula for aluminum sulfide.

Name of the compound—aluminum sulfide
Elemental symbols and oxidation numbers—Al^{3+} and S^{2-}

You will find that it is slightly more difficult to figure out the ratio of aluminum to sulfide that will result in a net charge of zero. In this case, we must find the lowest number that both oxidation numbers will go into equally, which turns out to be 6. We want to use subscripts which result in +6 + (-6) = 0, because that is the way to end up with a net charge of zero. Because 2(+3) + 3(-2) = 0, we get subscripts of 2 and 3 from aluminum and sulfide, respectively.

Name of the compound—aluminum sulfide
Elemental symbols and oxidation numbers—Al^{3+} and S^{2-}
Proper chemical formula—Al_2S_3

You will often encounter compounds that contain polyatomic ions, and you will simply look up the symbols and oxidation numbers from the second group of oxidation numbers. For example, let's write the proper chemical formula for the compound called calcium carbonate. Use Figures 5–2a and 5–2b to find the symbols and oxidation numbers associated with each ion.

Example 3

Write the correct chemical formula for calcium carbonate.

Name of the compound—calcium carbonate
Elemental symbols and oxidation numbers—Ca^{2+} and CO_3^{2-}

The oxidation numbers show us that we can combine each of the parts, the calcium and the carbonate, in a 1:1 ratio. We do not, however, remove the subscript of 3 that comes with the carbonate polyatomic ion. If we remove that subscript, the ion fails to be a carbonate ion! You must never alter the subscripts that are built into these polyatomic ions.

Name of the compound—calcium carbonate
Elemental symbols and oxidation numbers—Ca^{2+} and CO_3^{2-}
Proper chemical formula—$CaCO_3$

As I mentioned earlier, we use parentheses to indicate the use of multiple polyatomic ions. It is with the proper use of parentheses that I find that students have the most trouble. Let's suppose that we wanted to write the proper formula for the compound calcium nitrate. Once again, we would look up the oxidation numbers and symbols of the ions involved.

Example 4

Write the correct chemical formula for calcium nitrate.

Name of the compound—calcium nitrate
Elemental symbols and oxidation numbers—Ca^{2+} and NO_3^-

In this example, we need 2 nitrate ions to balance out the calcium ion, because +2 + 2(-1) = 0. Normally, we would just add a subscript of 2 to the negative ion. However, the nitrate already has a subscript of 3. We can't just change the existing subscript to 2, because that would simply leave us with one **nitrite** (NO_2^-) ion. Remember that I said that we cannot alter the subscripts that come with these polyatomic ions. When we want to express multiple polyatomic ions, we use parentheses, as shown below:

Name of the compound—calcium nitrate
Elemental symbols and oxidation numbers—Ca^{2+} and NO_3^-
Proper chemical formula—$Ca(NO_3)_2$

Let's try one more example that illustrates the proper use of parentheses. Find the symbols and oxidation numbers to write the proper formula for aluminum hydroxide. Remember that hydroxide isn't an element, so you will find it with the polyatomic ions.

Example 5

Write the correct chemical formula for aluminum hydroxide.

Name of the compound—aluminum hydroxide
Elemental symbols and oxidation numbers—Al^{3+} and OH^-

You can see that we will need 3 hydroxide ions to balance out the aluminum ion, because +3 + 3(-1) = 0. Many students are tempted to simply add a subscript of 3 after the hydrogen to get the formula $AlOH_3$, but this is not the correct formula! The reason is because we need 3 hydroxide ions, and adding the 3 subscript in this fashion only gives us 3 hydrogen atoms. The subscript as shown would not act as a multiplier for the oxygen. What we need to do, in order to use the 3 as a multiplier for the whole polyatomic ion, is to put the hydroxide in parentheses and put the subscript outside of them.

Name of the compound—aluminum hydroxide
Elemental symbols and oxidation numbers—Al^{3+} and OH^-
Proper chemical formula—$Al(OH)_3$

Finally, another mistake that beginners make is to use the parentheses for monatomic ions, as in the example $Ca(Cl)_2$. Please, let me be very clear on this: **Never use parentheses around monatomic ions. Only polyatomic ions will ever require the use of parentheses—and only if you need more than one of them.**

Lesson 5–2 Review

Write the proper chemical formula for each of the following compounds.

1. lithium hydroxide ______
2. barium sulfate ______
3. silver nitrate ______
4. ammonium nitrite ______
5. magnesium oxide ______
6. potassium sulfide ______
7. ammonium sulfate ______
8. barium hydroxide ______

Lesson 5–3: Naming Compounds

In many ways, the skill of naming compounds is easier than that of writing formulas. At first, you may rely heavily upon the reference tables that your instructor provides, but you will eventually become so familiar with the names of the various ions that you won't need the tables all of the time. For this lesson, we will be making use of the oxidation tables that we used in the last lesson (Figures 5–2a and 5–2b).

First, we will go over the process for naming binary compounds. Recall from Chapter 1 that ***binary compounds*** are compounds that contain two, and only two, elements. This does not mean that there are necessarily only two atoms in the compound. For example, calcium chloride ($CaCl_2$) is a binary compound, because it only contains two elements: carbon and chlorine. You may have noticed that the name, calcium chloride, consists of the names of the two elements, with the name of the second element changed from "chlorine" to "chloride." In fact, naming many binary compounds is just that simple. We put the name of the less electronegative element first, the more electronegative element second, and write the names of each element, changing the ending of the name of the second element to "ide."

Naming Simple Binary Compounds		
Chemical Formula	**Elements in Compound**	**Name of Compound**
LiBr	Lithium and bromine	Lithium bromide
CaF_2	Calcium and fluorine	Calcium fluoride
SrO	Strontium and oxygen	Strontium oxide
BaI_2	Barium and iodine	Barium iodide
MgS	Magnesium and sulfur	Magnesium sulfide

Figure 5–3a

All of the examples used here involve elements with only one possible oxidation state. When we are required to name compounds that contain elements with more than one oxidation state, we use the stock system, which uses Roman numerals to indicate the oxidation number of the less electronegative element. We can usually determine the oxidation number of the less electronegative element by considering the rules for oxidation numbers presented in Lesson 5–1. It would probably be a good idea to review those rules before you try to follow these next examples.

Suppose you were asked to name the compound with the formula FeO. Following the method for naming simple binary compounds, we would get iron oxide. The problem is that there is more than one form of iron oxide. There is another compound, for example, with the formula Fe_2O_3, which you might also call iron oxide. How do we differentiate between the names of the two compounds?

What we need to do is make use of Roman numerals, in order to indicate the oxidation number of the iron in each of the compounds. We can figure out the oxidation number of the iron by paying attention to the ratio by which it is combining with the oxygen in each compound. You will recall from the Rules for Assigning Oxidation Numbers in Lesson 5-1 that the oxidation number for oxygen (except in peroxides, and these are not peroxides) is -2. If the oxidation number of oxygen is -2, what does that make the oxidation number of the iron in each compound, if the net charge on each compound must be 0?

For the first compound, FeO, the iron and oxygen combine in a 1:1 ratio. If the oxidation number of the oxygen is -2, the oxidation number of the iron must be +2, because +2 + (-2) = 0. We indicate the fact that iron takes on an oxidation number of +2 by inserting a Roman numeral for "two" between the names of the elements. The name of the compound with the formula FeO is "iron (II) chloride," which we read as "iron two chloride."

Naming the second compound, Fe_2O_3, is only slightly more complicated. Remember that we have an oxidation number of -2 for each atom of oxygen. Because we have 3 atoms of oxygen in the formula, the total negative charge is -6 (3 × -2). Because our compound must have a net charge of 0, the iron atoms must equal a total positive charge of +6. The +6 charge is distributed between 2 atoms of iron, so each iron atom must have an oxidation number of +3. Using the stock system, we would call this compound "iron (III) oxide," which we would read as "iron three oxide."

Using the Stock System to Name Compounds

Chemical Formula	Oxidation Number of 1st Element	Name of Compound
N_2O_5	+5	Nitrogen (V) oxide
$PbCl_2$	+2	Lead (II) chloride
CuF	+1	Copper (I) fluoride
NiS	+2	Nickel (II) sulfide

Figure 5–3b

When it comes to naming ternary ionic compounds, you simply need to identify the polyatomic ion in the formula. Then write the names of the monatomic and polyatomic ions, in order. For example, suppose you were asked to name the compound with the formula $NaNO_3$. Two of the three elements involved (sodium, nitrogen, and oxygen) must make up a polyatomic ion. A quick check on Figure 5–2b will show you that there is no polyatomic ion that is made up of sodium and nitrogen, or sodium and oxygen. You will, however, be able to locate the polyatomic ion called nitrate (NO_3) with a charge of -1. To name this compound, we simply write the name of the positive ion first and the negative ion second, giving us "sodium nitrate."

Notice that we do not, in this case, change the ending of the second ion to "ide." We never change the name of a polyatomic ion. Changing "nitrate" to "nitride" would actually change NO_3^- to N^{3-}, which, as you can see, is a very different ion. Notice that the polyatomic ion called "nitrate" contains oxygen. Nitride, a monatomic ion, does not. When naming compounds, we only change the ending to "ide" for monatomic ions, although some polyatomic ions, hydroxide (OH^-) for example, already end in "ide."

When the ternary ionic compound that we are trying to name contains a cation (positive ion) with multiple oxidation states, we still need to employ the stock system. For example, if we were asked to name the compound with the formula $Cu(NO_3)_2$, we would determine that the oxidation number of the copper must be +2, because there are two nitrate ions, each with a charge of -1, and $+2 + 2(-1) = 0$. This would give us the name "copper (II) nitrate" for this compound.

Following are several more examples of compounds and their proper names. Look at each example, and see if you understand why the stock system is used for some compounds and not for others. Remember: The stock system is only used when you have a cation (or an atom with an apparent positive charge, in the case of molecular compounds) with more than one possible oxidation number.

More Examples of Formulas and Names

Formula	Name	Formula	Name
LiOH	lithium hydroxide	$FeCl_3$	iron (III) chloride
BaS	barium sulfide	CuI_2	copper (II) iodide
NH_4Cl	ammonium chloride	$Pb(NO_3)_2$	lead (II) nitrate
$CaCO_3$	calcium carbonate	NO_2	nitrogen (IV) oxide

Figure 5–3c

Lesson 5–3 Review

Name the compounds represented by the following chemical formulas.

1. Li_2CO_3 ____________
2. $CaBr_2$ ____________
3. SrS ____________
4. Cu_2O ____________
5. NO_2 ____________
6. $Ba(NO_3)_2$ ____________
7. $MgSO_3$ ____________
8. NH_4I ____________
9. NO ____________
10. NaCN ____________

Lesson 5–4: Molecular and Empirical Formulas

At this point, we should distinguish between two important types of formulas that you will encounter during your study of chemistry. Each molecular compound will have its own unique molecular formula, as well as an empirical formula, which it may share with one or more other compounds. The ***molecular formula*** of a compound represents the particles that make up the compound, as they actually exist. For example, glucose, with the molecular formula $C_6H_{12}O_6$, actually exists in 24-atom molecules, as the molecular formula suggests. The ***empirical formula*** of a compound shows the simplest whole number ratio for the elements that make up the compound. So, for example, if we reduced all of the subscript numbers in the molecular formula for glucose by dividing each by the number 6, we would get the empirical formula for glucose: CH_2O.

Notice that the empirical formula for glucose gives less information than the molecular formula, simply noting the fact that there are twice as many hydrogen atoms as either carbon or oxygen atoms. You may wonder why anyone would be interested in the empirical formula if it contains less information. The answer to that is very simple: It is easier to experimentally determine the empirical formula of a compound than the molecular formula of a compound, so that the empirical formula is often found first, as a step towards finding the molecular formula.

Comparing Molecular and Empirical Formulas

Molecular Formula	Empirical Formula	Molecular Formula	Empirical Formula
H_2O_2	HO	H_2O	H_2O
N_2O_4	NO_2	C_8H_{16}	CH_2
CH_2	CH_2	C_2H_4	CH_2

Figure 5–4a

You should notice two interesting things from Figure 5–4a. First, notice that some molecular formulas can't be reduced, as in the case of H_2O, so sometimes the molecular and empirical formulas for a compound are identical. Next, notice that some molecular compounds, such as C_2H_4 and C_8H_{16}, have exactly the same empirical formula. As I mentioned previously, each compound has its own unique molecular formula, but several compounds can share the same empirical formula.

You may be asked to determine the molecular formula of a compound, given the empirical formula and the molecular mass of the compound. This type of calculation simulates one experimental procedure for determining the molecular formula of a given compound. The key to solving this type of problem is to find the mass of the empirical formula and divide that value into the molecular mass, then multiply the result by each of the subscripts in the empirical formula.

Example 1

Determine the molecular formula for a compound with an empirical formula of CH_2 and a molecular mass of 70 u (atomic mass units).

Step 1. Use the periodic table to determine the mass of the empirical formula.

$C = 12.0$ u, $H_2 = 2.02$ u

Mass of empirical formula = (12.0 u + 2.02 u) = **14.0 u**

Step 2. Divide the molecular mass by the mass of the empirical formula.

$$\frac{70\ \not{u}}{14\ \not{u}} = 5$$

Step 3. Multiply the result by the subscripts in the empirical formula.

$$5 \times (CH_2) = C_5H_{10}$$

It is also likely that you will be asked to determine the empirical formula for a compound, if you know its percentage composition by mass. The key to solving this type of problem will be to assume that you are dealing with a 100-gram sample, which will allow you to change percentages to grams. Then divide the mass of each part by the mass values shown on the periodic table. Finally, figure out the simple whole number ratio given by the answers to these calculations.

Example 2

A particular compound has been experimentally determined to contain 71.5% calcium and 28.5% oxygen by mass. Determine the empirical formula for this compound.

Step 1. Assume that we could obtain a 100-gram sample, which allows us to work with grams.

71.5 grams of calcium and 28.5 grams of oxygen

Step 2. Next, look up the masses of each element on the periodic table, and divide the above masses by the elemental masses.

Calcium → 71.5 ~~grams~~/40.1 ~~grams~~ per part = **1.78 parts**

Oxygen → 28.5 ~~grams~~/16.0 ~~grams~~ per part = **1.78 parts**

Step 3. Use the number of parts to construct a simple, whole number ratio.

1.78:1.78 equals **1:1**

Step 4. Use the numbers in the ratio as the subscripts in the empirical formula.

Empirical formula → **CaO**

Lesson 5–4 Review

1. Which of the following pairs could represent the molecular formula and the empirical formula for the same compound?
   a) C_3H_8 and CH_2 c) C_4H_8 and CH
   b) C_3H_6 and CH_2 d) C_3H_6 and CH_3
2. Which of the following pairs could represent the molecular formula and the empirical formula for the same compound?
   a) C_4H_{12} and CH_3 c) C_3H_6 and CH
   b) C_3H_5 and CH_8 d) C_3H_8 and CH_2
3. A student determines that a compound contains 58.8% barium, 13.7% sulfur, and 27.5% oxygen by mass. Find the empirical formula for this compound.
   a) $BaSO_4$ b) $BaSO_3$ c) Ba_2SO d) BaS_2O_3
4. A laboratory experiment determines that a compound contains 57.5% sodium, 40.0% oxygen, and 2.5% hydrogen by mass. Find the empirical formula for this compound.
   a) NaOH b) $NaOH_2$ c) Na_2OH d) $Na(OH)_3$
5. A student determines that a compound contains 40.0% calcium, 12.0% carbon, and 48.0% oxygen by mass. Find the empirical formula.
   a) $CaCO_4$. b) $CaCO_3$ c) Ca_2CO d) CaC_2O_3
6. Determine the molecular formula of a compound with an empirical formula of CH_4 and a molecular mass of 16 u (atomic mass units).
   a) CH_4 b) C_2H_8 c) CH_6 d) C_5H_{11}

7. Determine the molecular formula of a compound with an empirical formula of CH_3 and a molecular mass of 30 u (atomic mass units).
   a) CH_4 b) C_2H_8 c) C_2H_6 d) C_5H_{11}
8. Determine the empirical formula and the molecular formula for a compound that contains 92.3% carbon and 7.7% hydrogen by mass and a molecular mass of 26.0 u (atomic mass units).
   a) CH_2 and C_2H_4 c) CH_3 and C_2H_6
   b) CH_4 and C_2H_8 d) CH and C_2H_2

For each of the following molecular formulas, find the proper empirical formula.

9. H_2O_2________________ 10. C_4H_{12}________________

Lesson 5–5: Molecular and Formula Mass

We have already discussed that fact that you can find the atomic masses for each of the elements on the Periodic Table of Elements. In this lesson, we will examine the process of taking these individual elemental masses and determining the mass of a compound. In reality, you have already needed to do some of these calculations, but now we will cover them formally. When dealing with molecular compounds we talk about molecular mass. The ***molecular mass*** is the mass of one molecule of a molecular compound. When we are dealing with an ionic compound, we can't use the term *molecular mass*, because ionic compounds are not made up of individual molecules. For ionic compounds, we use the term ***formula mass***, which is the mass of one formula unit. The term ***formula unit*** refers to the simplest ratio of the cations to anions that are found in the compound. NaCl, the formula unit of table salt, for example, indicates that there is one Na^+ ion for every Cl^- ion.

Although the terms *molecular mass* and *formula mass* are different, and they really do represent different concepts, the method for calculating each is identical. In fact, you might not even know if the particular compound that you are calculating the mass for is molecular or ionic, but you can still get the correct value. Don't get thrown off by the language in the questions that you are asked to answer. If the question asks for the "molecular mass," that is simply an indication that you are working with a molecular compound. If the question refers to "formula mass," it is simply noting that the compound is ionic.

Whichever type of compound you are dealing with, calculating the mass of a compound is very simple. All you need to do is add up the masses of each atom involved in the compound. Look up each of the atomic masses and multiply them by the subscripts associated with each element. Add these totals together to get the overall mass.

First, we will do an example involving the molecular mass of water.

Example 1

Find the molecular mass of water, $H_2$0.

H_2 = 1.01 u/~~atom~~ x 2 ~~atoms~~ = 2.02 u
O = 16.0 u/~~atom~~ x 1 ~~atom~~ = + 16.0 u
Total = **18.0 u**

Now, let's see an example involving the formula mass of carbon dioxide.

Example 2

Find the formula mass of carbon dioxide, CO_2.

C = 12.0 u/~~atom~~ x 1 ~~atom~~ = 12.0 u
O_2 = 16.0 u/~~atom~~ x 2 ~~atoms~~ = + 32.0 u
Total = **44.0 u**

As you can see, whether you are dealing with a molecular compound or an ionic compound, the method for solving the problem is identical. The key is to be sure that you take all of the atoms in the formula into account.

Students are sometimes confused by parentheses, so let's try a couple of examples that include them.

Example 3

Find the formula mass of carbon dioxide, $Ca(OH)_2$.

Ca = 40.1 u/~~atom~~ x 1 ~~atom~~ = 40.1 u
O_2 = 16.0 u/~~atom~~ x 2 ~~atoms~~ = 32.0 u
H_2 = 1.01 u/~~atom~~ x 2 ~~atoms~~ = + 2.02 u
Total = **74.1 u**

Notice, the subscript "2", which is found on the outside of the parentheses, acts as a multiplier for everything that is inside the parenthesis. For this reason, the formula $Ca(OH)_2$ shows 2 atoms of both oxygen and hydrogen. If you're wondering why we round our answer to 74.1 u, you may want to review the rules for rounding (Lesson 2–4). Because we are adding, our answer may only contain as many digits past the decimal point as the addend with the least number of digits past the decimal. In this case, 40.1 and 32.0 are reported to the tenth place, so we must round our answer to the tenth place.

How do we work with formulas that have subscript numbers both inside and outside of the parentheses? A subscript that is on the inside of parentheses acts as multiplier for only the number that precedes it. A subscript that is on the outside of parentheses acts as a multiplier for all of the elements represented by symbols inside.

Example 4

Find the formula mass of carbon dioxide, $Ba(NO_3)_2$.

$$\begin{aligned} Ba &= 137\ u/\cancel{atom} \times 1\ \cancel{atom} = & 137\ u \\ N_2 &= 14.0\ u/\cancel{atom} \times 2\ \cancel{atoms} = & 28.0\ u \\ O_6 &= 16.0\ u/\cancel{atom} \times 6\ \cancel{atoms} = & +\ \underline{96.0\ u} \\ & \text{Total} = & \mathbf{261\ u} \end{aligned}$$

Make sure that you can see how we determined the total number of atoms for each element in the previous example. There is no subscript associated with the barium, so it is understood that there is only one atom per formula unit. The nitrogen has no subscript directly following it, but there is a subscript of "2" on the outside of the parentheses, which acts as a multiplier for everything within the parentheses. The oxygen has a subscript of "3" following it, which means that there are 3 oxygen atoms per nitrate ion. We must multiply the 3 by the subscript of "2" that is on the outside of the parentheses, for a total of six oxygen atoms.

Lesson 5–5 Review

1. The _______________ mass is the mass of one molecule of a molecular compound.
2. The _______________ mass is the mass of one formula unit of an ionic compound.

3. The smallest complete unit of water you could have is one ______________.

4. The smallest complete unit of table salt you could have is one ______________.

Find the formula mass of the following ionic compounds:

5. CaI_2 ______________ 6. LiOH ______________

7. $Cu(NO_3)_2$ ______________ 8. $(NH_4)_2S$ ______________

Find the molecular mass of each of the following molecular compounds:

9. CO ______________ 10. NO_2 ______________

Lesson 5–6: Using Coefficients With Formulas

Up to this point in Chapter 5, we have been concerned with individual formula units or molecules. In this lesson we will discuss how we represent multiple formula units and molecules, in preparation for working with entire chemical equations. You have probably had some experience working with chemical equations before, possibly in a biology class, so this may not be entirely new to you. What may be new to you is the level of understanding that you will be expected to demonstrate, in a variety of ways, now that you are studying chemistry.

When we want to represent multiple molecules or multiple formula units, we simply write a large coefficient number before our chemical formula. Let's illustrate this by looking at the formula for photosynthesis, which you may remember from biology.

$$\mathbf{C_6H_{12}O_6 + 6O_2 \rightarrow 6CO_2 + 6H_2O}$$

glucose + oxygen → carbon dioxide + water

You will notice that the first molecular formula, which represents glucose, is not preceded by a coefficient. When this is the case, we understand there to be only one molecule or formula unit involved in the balanced chemical equation. There is a coefficient of "6" preceding each of the other chemical formulas, indicating the six molecules of each of the other substances are involved in the chemical reaction. In summary, the reaction shows us that one molecule of glucose will react with six molecules of diatomic oxygen, to form six molecules of carbon dioxide and six molecules of water.

As we look at an individual molecular compound from that equation, you should be aware that the coefficient that precedes a chemical formula acts as a multiplier for each of the elements in the formula. For example, look at the part of the equation that reads "$6CO_2$." One molecule of carbon dioxide, without a coefficient present, contains a total of three atoms, 1 of carbon and 2 of oxygen. When we place the coefficient of 6 in front of the formula for carbon dioxide, we are indicating that there are 6 molecules of carbon dioxide in the reaction. If one molecule of carbon dioxide contains 3 atoms, then 6 molecules of carbon dioxide must contain a total of 3 × 6, or 18 atoms. If one molecule of carbon dioxide contains 1 carbon atom, then 6 molecules of carbon dioxide must contain 1 × 6, or 6 carbon atoms. Finally, if one molecule of carbon dioxide contains 2 oxygen atoms, then 6 molecules of carbon dioxide must contain 2 × 6, or 12 oxygen atoms.

When you want to determine the number of atoms of a particular element found in a certain number of molecules, it is often easiest to determine the number of atoms of that element in one molecule, and then multiply by the number of molecules. The same is true when you are working with formula units and, as was the case in our previous lesson, working with formula units or molecules doesn't change the method for solving these particular problems.

Example 1

Determine the number of A. sulfur atoms and B. oxygen atoms shown in the following: $8SO_3$

A. Each molecule of sulfur (VI) oxide contains one atom of sulfur. So, 8 molecules of sulfur (VI) oxide contain 8 × 1, or 8 atoms of sulfur.

B. Each molecule of sulfur (VI) oxide contains three atoms of oxygen. So, 8 molecules of sulfur (VI) oxide contain 8 × 3, or 24 atoms of oxygen.

Now, let's look at an example using the ionic compound calcium nitrate, $Ca(NO_3)_2$. You will notice that, except for substituting the term *formula unit* for molecule, we solve this example in exactly the same way as we approached the first example.

Example 2

Determine the number of A. calcium atoms; B. nitrogen atoms; and C. oxygen atoms, shown in the following: $6Ca(NO_3)_2$

A. Each formula unit of calcium nitrate contains one atom of calcium. So, 6 formula units of calcium nitrate contain 6 × 1, or 6 atoms of calcium.

B. Each formula unit of calcium nitrate contains two atoms of nitrogen. So, 6 formula units of calcium nitrate contain 6 × 2, or 12 atoms of nitrogen.

C. Each formula unit of calcium nitrate contains six atoms of oxygen. So, 6 formula units of calcium nitrate contain 6 × 6, or 36 atoms of oxygen.

Lesson 5–6 Review

Base your answers to questions 1–5 on the following formula:

$$3Al_2(CO_3)_3$$

1. How many total aluminum atoms are represented?
2. How many total carbon atoms are represented?
3. How many total oxygen atoms are represented?
4. How many total atoms are represented?
5. How many formula units are shown?

Base your answers to questions 6–10 on the following formula:

$$3C_6H_{12}O_6$$

6. How many total atoms are represented?
7. How many molecules are represented?
8. How many carbon atoms are represented?
9. How many hydrogen atoms are represented?
10. How many oxygen atoms are represented?

Chapter 5 Examination

Part I—Matching

Match the following terms to the definitions that follow. Not all of the terms will be used.

a. oxidation number
b. binary compound
c. molecular formula
d. molecular mass
e. formula mass
f. formula unit
g. empirical formula

_____1. The mass of one formula unit of an ionic compound.

_____2. A compound made up of only two elements.

_____3. Shows the simplest whole number ratio for the elements that make up a compound.

_____4. Shows the particles that make up the compound, as they actually exist.

_____5. Represents the charge, or the apparent charge on an atom.

Part II—Matching

Match each of the following formulas to the correct name of the compound it represents.

a. $FeCl_3$
b. $FeCl_2$
c. Cu_2O
d. CuO
e. $BaSO_4$
f. $BaSO_3$
g. $LiNO_3$
h. $LiNO_2$

_____6. barium sulfite

_____7. barium sulfate

_____8. copper (I) oxide

_____9. copper (II) oxide

_____10. iron (III) chloride

_____11. iron (II) chloride

_____12. lithium nitrate

_____13. lithium nitrite

Part III—Multiple Choice

Select the best answer choice for each of the following questions.

14. Which of the following shows the correct formula for ammonium carbonate?

a) $(NH_4)_2CO_3$ b) $NH_4(CO_3)_2$ c) NH_4CO_3 d) NH_2CO_3

15. Which of the following represents the correct formula for magnesium hydroxide?
    a) MgOH b) $Mg(OH)_2$ c) $MgOH_2$ d) Mg_2OH
16. Which of the following represents the correct formula for barium chloride?
    a) $Ba(Cl)_2$ b) $BaCl_2$ c) Ba_1Cl_2 d) Ba_2Cl
17. Which of the following shows the proper empirical formula for a compound with the molecular formula, C_4H_8?
    a) C_4H_8 b) C_2H_4 c) CH_3 d) CH_2
18. How many oxygen atoms are shown in the expression $5Ca(NO_2)_2$?
    a) 2 b) 4 c) 10 d) 20
19. Which of the following compounds has a molecular mass of 44.08 u ?
    a) CO b) NO c) CCl_4 d) C_3H_8
20. The formula mass of $CaCl_2$ is closest to which of the following?
    a) 40.0 u b) 75.5 u c) 111 u d) 134 u

Part IV—Writing Formulas

Write the correct chemical formula for each of the following compounds.

21. nitrogen (V) oxide ____________________
22. strontium bromide ____________________
23. magnesium sulfide ____________________
24. copper (II) nitrate ____________________
25. barium carbonate ____________________

Answer Key

The actual answers will be shown in brackets, followed by the explanation. If you don't understand an explanation that is given in this section, you may want to go back and review the lesson that the question came from.

Lesson 5–1 Review

1. [0]—According to Rule #1, the oxidation number for each atom of a free element, such as N_2, is equal to 0.

2. [-1]—According to Rule #5, when a halogen such as chlorine appears in a compound, it shows an oxidation number of -1.
3. [+1]—According to Rule #2, the oxidation number shown by an alkali metal in a compound is +1.
4. [-2]—Rule #6 states that oxygen shows an oxidation number of -2 in all compounds except peroxides. Water is not a peroxide.
5. [+2]—Rule #4 states that the alkaline earth metals, such as calcium, show an oxidation number of +2 in all compounds.
6. [+6]—Remember that the net charge on the compound must be equal to zero. Oxygen shows an oxidation number of -2, and there are 4 atoms of oxygen for a total of 4(-2) = -8. Barium is an alkaline earth metal with an oxidation number of +2, according to Rule #4. Because there is only one barium atom, we get a total of 1(+2) = +2. After we add the +2 from the barium atom and the -8 from the oxygen atoms, we get +2 + (-8) = -6. Because the one sulfur atom is the only atom left, it must be +6 to balance the rest of the atoms.
7. [+4]—There are two atoms of lithium, for a total charge of 2(+1) = +2. The total charge of the oxygen is 3(-2) = -6. These two elements leave a net charge of +2 + (-6) = -4, so the sulfur must be +4 to balance this out.
8. [+4]—The calcium from column 2 has an oxidation number of +2. There are three atoms of oxygen for a total charge of 3(-2) = -6. These elements leave us with a net charge of +2 + (-6) = -4, so the carbon must be +4 to balance this out.

Lesson 5–2 Review

1. [LiOH]—Lithium is an alkali metal with only one valence electron, so it has an oxidation number of +1. The hydroxide ion (OH-) has an oxidation number of -1, so the two ions combine in a 1:1 ratio because +1 + -1 = 0.
2. [$BaSO_4$]—Barium is an alkaline earth metal with two valence electrons and an oxidation number of +2. The sulfate ion (SO_4^{2-}) has an oxidation number of -2. The ions combine in a 1:1 ratio because +2 + -2 = 0.
3. [$AgNO_3$]—Silver has an oxidation number of +1. The nitrate ion (NO_3^-) has an oxidation number of -1. The ions combine in a 1:1 ratio because +1 + -1 = 0.
4. [NH_4NO_2]—This compound contains two polyatomic ions. The ammonium ion (NH_4^+) has an oxidation number of +1. The nitrite ion (NO_2^-) has an oxidation number of -1. The ions combine in a 1:1 ratio because +1 + -1 = 0.

5. [MgO]—Magnesium, an alkaline earth metal with two valence electrons, has an oxidation number of +2. Oxygen, a nonmetal with six valence electrons, has an oxidation number of -2. They combine in a 1:1 ratio because +2 + -2 = 0.
6. [K_2S]—Potassium is an alkali metal with an oxidation number of +1. Sulfur, a nonmetal with six valence electrons, has an oxidation number of -2. They combine in a 2:1 ratio because 2(+1) + (-2) = 0.
7. [$(NH_4)_2SO_4$]—Ammonium (NH_4^+) is a polyatomic ion with a charge of +1. Sulfate (SO_4^{2-}) is a polyatomic ion with a charge of -2. These ions combine in a 2:1 ratio because 2(+1) + (-2) = 0. Because we need two of the ammonium ions, we must enclose the polyatomic ion in parentheses and put a subscript of 2 on the outside.
8. [$Ba(OH)_2$]—Barium is an alkaline earth metal with two valence electrons and an oxidation number of +2. The hydroxide ion (OH^-) has an oxidation number of -1, so the two ions combine in a ratio of 1:2 because (+2) + 2(-1) = 0.

Lesson 5–3 Review

1. [lithium carbonate]—Carbonate (CO_3^{2-}) is a polyatomic ion, and we never change the name of polyatomic ions. Lithium is the positive ion, and we never change the name of the positive ion. Adding the names of the ions together gives us our answer.
2. [calcium bromide]—Calcium is the positive ion, and we don't change its name. Br is bromine, but we change the ending of its name to "ide" when it forms a negative monatomic ion.
3. [strontium sulfide]—Sr is the elemental symbol for strontium, our cation. S is the elemental symbol for sulfur. We change the ending of the names of monatomic negative ions to "ide," hence our answer.
4. [copper (I) oxide]—Copper is one of the elements with multiple oxidation states, so we must indicate the oxidation number of copper in the name of the compound. Because copper combines with oxygen in a 2:1 ratio in Cu_2O, and the oxidation number of oxygen is -2, the oxidation number of copper must be +1 because 2(+1) + (-2) = 0.
5. [nitrogen (IV) oxide]—Nitrogen shows more than one possible oxidation number, so we must indicate the oxidation number of nitrogen in our formula name. Oxygen has an oxidation number of -2, and there are 2 atoms of oxygen balancing out one atom of nitrogen in the formula NO_2. Therefore, nitrogen must be showing an oxidation state of +4. Remember: (+4) + 2(-2) = 0.

6. [barium nitrate]—Barium is a positive ion, so we don't change its name. Nitrate is a polyatomic ion, so we don't change its name either. Adding the names of the two ions together gives us our answer.
7. [magnesium sulfite]—Magnesium is a cation, so we don't change its name. Sulfite is a polyatomic ion, so we don't change its name either. Adding the names of the two ions together gives us our answer.
8. [ammonium iodide]—Our negative (anion) ion is monatomic, so we change the ending of its name to "ide." We leave the name of our polyatomic positive ion as is.
9. [nitrogen (II) oxide]—Nitrogen shows more than one possible oxidation number, so we must indicate the oxidation number of nitrogen in our formula name. Oxygen has an oxidation number of -2, and the two ions combine in a 1:1 ratio. So, the oxidation number of nitrogen must be +2, because (+2) + (-2) = 0.
10. [sodium cyanide]—Cyanide is a polyatomic ion, so we don't change its name. We simply add the names of the two ions together to get our answer.

Lesson 5-4 Review

1. [B. C_3H_6 and CH_2]—If we take the molecular formula C_3H_6 and divide both of the subscripts by 3, we get the empirical formula CH_2.
2. [A. C_4H_{12} and CH_3]—If we take the molecular formula C_4H_{12} and divide both of the subscripts by 4, we get the empirical formula CH_3.
3. [A. $BaSO_4$]—The key to this type of problem is to imagine that you had a 100-gram sample of the substance, because this allows us to change % to grams. In this case, you would have 58.8 g of barium, 13.7 g of sulfur, and 27.5 g of oxygen. Next, you divide the mass of each element by its molar mass to find out how many moles you have of each. Finally, you reduce the number of moles to a simple whole number ratio, which becomes the subscripts for the formula.

A. $$\text{\# of moles of barium} = \frac{58.8 \text{ g}}{137 \text{ g/mole}} = 0.429 \text{ moles}$$

$$\text{\# of moles of sulfur} = \frac{13.7 \text{ g}}{32.1 \text{ g/mole}} = 0.427 \text{ moles}$$

$$\text{\# of moles of oxygen} = \frac{27.5 \text{ g}}{16.0 \text{ g/mole}} = 1.72 \text{ moles}$$

B. $$\text{subscript for barium} = \frac{0.429 \text{ moles}}{0.427 \text{ moles}} = 1$$

$$\text{subscript for sulfur} = \frac{0.427 \text{ moles}}{0.427 \text{ moles}} = 1$$

$$\text{subscript for oxygen} = \frac{1.72 \text{ moles}}{0.427 \text{ moles}} = 4$$

Empirical Formula = $\mathbf{BaSO_4}$

4. [A. NaOH]—Work shown here:

A. $$\text{\# of moles of sodium} = \frac{57.5\ \cancel{g}}{23.0\ \cancel{g}/\text{mole}} = 2.50\ \text{moles}$$

$$\text{\# of moles of oxygen} = \frac{40.0\ \cancel{g}}{16.0\ \cancel{g}/\text{mole}} = 2.50\ \text{moles}$$

$$\text{\# of moles of hydrogen} = \frac{2.5\ \cancel{g}}{1.01\ \cancel{g}/\text{mole}} = 2.5\ \text{moles}$$

B. $$\text{subscript for sodium} = \frac{2.50\ \cancel{\text{moles}}}{2.50\ \cancel{\text{moles}}} = 1$$

$$\text{subscript for oxygen} = \frac{2.50\ \cancel{\text{moles}}}{2.50\ \cancel{\text{moles}}} = 1$$

$$\text{subscript for hydrogen} = \frac{2.5\ \cancel{\text{moles}}}{2.50\ \cancel{\text{moles}}} = 1$$

Empirical Formula = **NaOH**

5. [B. $CaCO_3$]—Work shown here:

A. $$\text{\# of moles of calcium} = \frac{40.0\ \cancel{g}}{40.1\ \cancel{g}/\text{mole}} = 0.998\ \text{moles}$$

$$\text{\# of moles of carbon} = \frac{12.0\ \cancel{g}}{12.0\ \cancel{g}/\text{mole}} = 1.00\ \text{moles}$$

$$\text{\# of moles of oxygen} = \frac{48.0\ \cancel{g}}{16.0\ \cancel{g}/\text{mole}} = 3.00\ \text{moles}$$

B. $$\text{subscript for calcium} = \frac{0.998\ \cancel{\text{moles}}}{0.998\ \cancel{\text{moles}}} = 1$$

$$\text{subscript for carbon} = \frac{1.00\ \cancel{\text{moles}}}{0.998\ \cancel{\text{moles}}} = 1$$

$$\text{subscript for oxygen} = \frac{3.00\ \cancel{\text{moles}}}{0.998\ \cancel{\text{moles}}} = 3$$

Empirical Formula = $\mathbf{CaCO_3}$

6. [A. CH_4]—The empirical formula is CH_4, which has a total mass of 16.0 u (12.0 u + 4.0 u = 16.0 u). The molecular mass is also 16.0 u, so the empirical and molecular formula must be the same.

7. [C. C_2H_6]—The empirical formula is CH_3, which has a total mass of 15.0 u (12.0 u + 3.0 u = 15.0 u). The molecular mass is 30.0 u, which, when divided by 15.0 u, gives us 2 ($30.0\ \cancel{u} / 15.0\ \cancel{u} = 2$). We multiply each subscript in the empirical formula by 2 to get the molecular formula $2(CH_3) = C_2H_6$. To check your answer, find the mass of your molecular formula, and you will find that it is 30.0 u.

8. [D. CH and C_2H_2]—This problem can be solved by combining the work we did for a question such as #7 with the work we did for a question such as #5, as shown here:

$$\text{\# of moles of carbon} = \frac{92.3\ \cancel{g}}{12.0\ \cancel{g}/\text{mole}} = 7.69 \text{ moles}$$

$$\text{\# of moles of hydrogen} = \frac{7.7\ \cancel{g}}{1.0\ \cancel{g}/\text{mole}} = 7.7 \text{ moles}$$

$$\text{subscript for carbon} = \frac{7.69\ \cancel{\text{moles}}}{7.69\ \cancel{\text{moles}}} = 1$$

$$\text{subscript for hydrogen} = \frac{7.7\ \cancel{\text{moles}}}{7.69\ \cancel{\text{moles}}} = 1$$

Empirical Formula = **CH**

Now that we know that the empirical formula for the compound is CH, we determine the mass of the empirical formula to be 13.0 u (12.0 u + 1.01 u = 13.0 u). The problem stated that the molecular mass was 26.0 u, which is double the empirical formula mass, so we multiply each subscript by 2 to get the molecular formula of 2(CH) = C_2H_2.

9. [HO]—Divide each subscript by 2.
10. [CH_3]—Divide each subscript by 4.

Lesson 5–5 Review

1. [molecular]—This is the mass of one molecule of a substance.
2. [formula]—The fact that ionic compounds don't contain molecules makes this additional term necessary.
3. [molecule]—Each water molecule contains 3 atoms.
4. [formula unit]—The formula unit indicates the smallest whole number ratio of the ions found in the ionic compound.
5. [294 u]—Work shown here:

$$\text{Ca} = 40.1\ \text{u}/\cancel{\text{atom}} \times 1\ \cancel{\text{atom}} = 40.1\ \text{u}$$

$$\text{I} = 127\ \text{u}/\cancel{\text{atom}} \times 2\ \cancel{\text{atoms}} = +\ \underline{254\ \text{u}}$$

$$\text{Total} = \mathbf{294\ u}$$

6. [24.0 u]—Work shown here:

$$\text{Li} = 6.94\ \text{u}/\cancel{\text{atom}} \times 1\ \cancel{\text{atom}} = 6.94\ \text{u}$$

$$\text{O} = 16.0\ \text{u}/\cancel{\text{atom}} \times 1\ \cancel{\text{atom}} = 16.0\ \text{u}$$

$$\text{H} = 1.01\ \text{u}/\cancel{\text{atom}} \times 1\ \cancel{\text{atoms}} = +\ \underline{1.01\ \text{u}}$$

$$\text{Total} = \mathbf{24.0\ u}$$

7. [188 u]—Work shown here:

Cu = 63.5 u/~~atom~~ × 1 ~~atom~~ =	63.5 u
N = 14.0 u/~~atom~~ × 2 ~~atom~~ =	28.0 u
O = 16.0 u/~~atom~~ × 6 ~~atoms~~ =	+ 96.0 u
Total =	**188 u**

8. [68.2 u]—Work shown here:

N = 14.0 u/~~atom~~ × 2 ~~atom~~ =	28.0 u
H = 1.01 u/~~atom~~ × 8 ~~atom~~ =	8.08 u
S = 32.1 u/~~atom~~ × 1 ~~atoms~~ =	+ 32.1 u
Total =	**68.2 u**

9. [28.0 u]—Work shown here:

C = 12.0 u/~~atom~~ × 1 ~~atom~~ =	12.0 u
O = 16.0 u/~~atom~~ × 1 ~~atoms~~ =	+ 16.0 u
Total =	**28.0 u**

10. [46.0 u]—Work shown here:

N = 14.0 u/~~atom~~ × 1 ~~atom~~ =	14.0 u
O = 16.0 u/~~atom~~ × 2 ~~atoms~~ =	+ 32.0 u
Total =	**46.0 u**

Lesson 5–6 Review

1. [6]—Each formula unit of $Al_2(CO_3)_3$ contains 2 atoms of aluminum, so 3 formula units contains 3 × 2, or 6 atoms of aluminum.
2. [9]—Each formula unit of $Al_2(CO_3)_3$ contains 3 atoms of carbon, so 3 formula units contains 3 × 3, or 9 atoms of carbon.
3. [27]—Each formula unit of $Al_2(CO_3)_3$ contains 9 atoms of oxygen, so 3 formula units contains 3 × 9, or 27 atoms of oxygen.
4. [42]—Each formula unit of $Al_2(CO_3)_3$ contains a total of 14 atoms, so 3 formula units contains 3 × 14, or 42 atoms in total.
5. [3]—The number of formula units is indicated by the coefficient of 3 at the beginning of the expression, $3Al_2(CO_3)_3$.
6. [72]—Each molecule of $C_6H_{12}O_6$ contains a total of 24 atoms, so 3 molecules contain 72 atoms.
7. [3]—The number of molecules is indicated by the coefficient of 3 at the beginning of the expression, $3C_6H_{12}O_6$.
8. [18]—Each molecule of $C_6H_{12}O_6$ contains 6 carbon atoms, so 3 molecules contain 3 x 6, or 18 carbon atoms.

9. [36]—Each molecule of $C_6H_{12}O_6$ contains 12 hydrogen atoms, so 3 molecules contain 3 x 12, or 36 hydrogen atoms.
10. [18]—Each molecule of $C_6H_{12}O_6$ contains 6 oxygen atoms, so 3 molecules contain 3 x 6, or 18 oxygen atoms.

Chapter 5 Examination

1. [e. formula mass]—Although the calculation is identical to the calculation for molecular mass, we use this term for ionic compounds.
2. [b. binary compound]—Just as a ***bi***cycle has two wheels, a ***bi***nary compound has two elements.
3. [g. empirical formula]—We reduce the coefficients of a molecular formula to get the empirical formula.
4. [c. molecular formula]—A compound that has a molecular formula of C_2H_6 actually exits as molecules made with 2 atoms of carbon and 6 atoms of hydrogen.
5. [a. oxidation number]—Oxidation numbers can be used to determine chemical formulas.
6. [F. $BaSO_3$]—The polyatomic ion called "sulfite" contains one less oxygen atom than "sulfate."
7. [E. $BaSO_4$]—The "ates" contain more oxygen than the "ites."
8. [C. Cu_2O]—The Roman numeral gives the oxidation number of the metal. When copper has an oxidation number of +1, it takes two atoms of it to balance out the oxygen, with an oxidation number of -2.
9. [D. CuO]—When copper has an oxidation number of +2, it takes one atom of it to balance out the oxygen, with an oxidation number of -2.
10. [A. $FeCl_3$]—When iron has an oxidation number of +3, it takes three atoms of chloride (Cl^-) to create a neutral compound.
11. [B. $FeCl_2$]—When iron has an oxidation number of +2, it takes two atoms of chloride (Cl^-) to create a neutral compound.
12. [G. $LiNO_3$]—Remember, the "ates" have more oxygen than the "ites", so the nitrate is the polyatomic ion with more oxygen.
13. [H. $LiNO_2$]—The polyatomic ion called nitrite contains one less oxygen atom than nitrate.
14. [A. $(NH_4)_2CO_3$]—Ammonium, NH_4^+, with an oxidation number of +1 must combine with the carbonate, CO_3^{2-}, in a 2:1 ratio. We need to put the ammonium ion in parentheses, because it is a polyatomic ion, and we need more than one of them.

15. [B. $Mg(OH)_2$]—Magnesium has an oxidation number of +2, and the hydroxide has an oxidation number of -1, so they combine in a 1:2 ratio. We use parentheses around the hydroxide ion to indicate that the subscript acts as a multiplier for both the oxygen and the hydrogen.
16. [B. $BaCl_2$]—Barium has an oxidation number of +2 and chloride has an oxidation number of -1. They combine in a 1:2 ratio, and we don't use parentheses for monatomic ions.
17. [D. CH_2]—Divide each subscript by 4.
18. [D. 20]—Each formula unit contains (2 x 2) = 4 atoms of oxygen, so 5 formula units must contain 5 x 4, or 20 oxygen atoms.
19. [D. C_3H_8]—The total mass of the carbon atoms (3 x 12.0 u) = 36.0 u and the total of the hydrogen atoms is (8 x 1.01 u) = 8.08 u. The total mass of the molecule is 44.08 u.
20. [C. 111 u]—(40.1 u + 2(35.5 u)) = 111.1 u
21. [N_2O_5]—Nitrogen (V) has an oxidation number of +5 and oxide has an oxidation number of -2. They will both go into the number 10 evenly. We give nitrogen a subscript of 2 because 2(+5) = +10. We give the oxygen a subscript of 5 because 5(-2) = -10.
22. [$SrBr_2$]—Strontium, an alkaline earth metal, has an oxidation number of +2. Bromide has an oxidation number of -1. They combine in a 1:2 ratio, because +2 + 2(-1) = 0.
23. [MgS]—Magnesium, an alkaline earth metal, has an oxidation number of +2. Sulfide, has an oxidation number of -2. They combine in a 1:1 ratio, because +2 + (-2) = 0.
24. [$Cu(NO_3)_2$]—Copper (II) has an oxidation number of +2, as indicated by the Roman numeral. Nitrate has an oxidation number of -1. They combine in a 1:2 ratio, because +2 + 2(-1) = 0. We need parentheses because nitrate is a polyatomic ion, and we want to indicate the need for two of them.
25. [$BaCO_3$]—Barium has an oxidation number of +2 and carbonate has an oxidation number of -2, so they combine in a 1:1 ratio. No parentheses are used because we only need one carbonate.

15. (c) Mn(OH)₂ — Manganese has an oxidation number of +2 and [illegible] hydroxide ion has an oxidation number of [illegible] parentheses around the hydroxide ion [illegible] is a [illegible] in the oxygen and the hydrogen.

16. (b) HClO — [illegible] has an oxidation number of [illegible] have an oxidation number of +1. They [illegible] for [illegible].

17. (a) CH_4 — Valence of carbon is 4.

18. (b) [illegible] — Each [illegible] and [illegible] atoms of oxygen [illegible] form [illegible] must contain 3 × 4 or 12 [illegible] atoms.

19. (c) [illegible] The total mass of the carbon atoms [illegible] The total mass of [illegible] The molar mass of [illegible] molecule is [illegible].

20. (c) [illegible]

21. (c) Although N has an oxidation number of +5 and oxide has an oxidation number of [illegible] [illegible]

22. [illegible] oxidation number [illegible]

23. [illegible]

24. [illegible] oxidation number [illegible]

25. [illegible] oxidation number of +2 [illegible]

Chemical and Nuclear Reactions

Lesson 6–1: Balancing Chemical Equations

When most people see a dramatic chemical reaction, they think of it as a chaotic display, with unpredictable results. Someone with training in chemistry will see the same reaction as an orderly, predictable process. You will see that even dramatic chemical reactions, involving fire and heat, proceed in predictable ways. In Chapter 5, you studied the process of naming compounds, which we described as the "words" that make up the language of chemistry. In this chapter, we will be learning more about chemical equations, which might be thought of as the "sentences" in the language of chemistry. ***Chemical equations*** are a series of chemical formulas, numbers, and symbols that describe a chemical reaction, or the net results of several reactions. These chemical equations are essential for understanding chemistry as a **quantitative** subject.

Once again, I will draw your attention to a chemical equation that you are probably familiar with from your study of biology, the process of respiration.

$$C_6H_{12}O_6 + 6O_2 \rightarrow 6H_2O + 6CO_2$$

This chemical equation shows us that one molecule of glucose will react with 6 molecules of oxygen to form 6 molecules of water and 6 molecules of carbon dioxide. The substances on the left-hand side of the arrow are called the reactants; the substances on the right-hand side are called the products. It is okay to switch the order in which the reactants, and/or the products, appear, as long as they all stay on the proper side of the equation. In other words, if you were asked to write the proper chemical equation for respiration, it would be just as acceptable if you wrote your answer as:

$$6O_2 + C_6H_{12}O_6 \rightarrow 6CO_2 + 6H_2O$$

When we refer to a chemical equation as "balanced," we mean that the appropriate coefficients have been used to show the same number of atoms of each element on both the product and reactant sides of the arrow. If you understand conservation of matter, then you would expect the number of atoms that appear on each side of the chemical equation to be the same. In fact, we must have the same number of atoms of each of the elements, on each side of the chemical equation. To check to see if that is true, you may want to perform what I will call an "atomic tally." Tally up the number of atoms of each element on each side of the equation, and make sure that they are the same, as demonstrated in Figure 6–1a.

Atomic Tally for the Process of Respiration

	Reactants			Products		
	$C_6H_{12}O_6 + 6O_2$			→ $6H_2O + 6CO_2$		
Elements →	C	H	O	C	H	O
# of atoms →	6	12	18	6	12	18

Figure 6–1a

Can you see how we came up with each of the elemental totals shown in Figure 6–1a? Remember that the subscript numbers act as multipliers for the elements that appear before them, as described in Lesson 5–1. The coefficient numbers act as multipliers for each of the elements in the compound, so, for example, $6CO_2$ shows 6 carbon atoms and 12 oxygen atoms.

For this lesson, it will be important for you to master the skill of the atomic tally. You will need to be able to check if an equation is balanced before you can learn how to balance them correctly. Let's try a couple of examples where you can do an atomic tally, before you check the answers. The first example that we will look at represents a chemical reaction that you may conduct in chemistry, as it is a common process for generating hydrogen gas in the lab.

Example 1

Do an atomic tally for the following equation and determine if it is balanced.

$$Zn + H_2SO_4 \rightarrow ZnSO_4 + H_2$$

Try to do an atomic tally for Example 1 before checking your answer. See if our answers match.

$$Zn + H_2SO_4 \rightarrow ZnSO_4 + H_2$$

	Zn	H	S	O	Zn	H	S	O	
Elements →	**Zn**	**H**	**S**	**O**	**Zn**	**H**	**S**	**O**	
# of atoms →	1	2	1	4	1	2	1	4	**Already balanced!**

Did we get the same answers? Do you notice that this chemical equation doesn't show any coefficients? Sometimes, a chemical equation is balanced when all of the substances have a coefficient of "1." When the coefficient for a substance is "1," we don't write it, and the "1" is understood. If you are ever asked to balance a chemical reaction that is already balanced, with no coefficients written in, simply write, "Already balanced," and your instructor will know that you did your atomic tally correctly.

Let's try another example. Can you do the atomic tally for the next example?

Example 2

Do an atomic tally for the following equation and determine if it is balanced.

$$Mg + O_2 \rightarrow MgO$$

Remember: Do the tally on your own before checking your answer.

$$Mg + O_2 \rightarrow MgO$$

	Mg	O	Mg	O	
Elements →	**Mg**	**O**	**Mg**	**O**	
# of atoms →	1	2	1	1	**Not balanced!**

Can you see why I said that this equation is not balanced? On the reactant side, two atoms of oxygen appear, yet on the product side, we only see one atom of oxygen. What are we allowed to do in this situation, if we want to balance the equation? Follow the rule that follows!

Never, never, never change (add, subtract, or alter) a subscript in an attempt to balance an equation!!!!

A sure sign of a beginner is that he or she will simply add a subscript in an attempt to balance an equation. For example, if a beginner wanted to balance the equation from Example 2, he or she might write:

$$Mg + O_2 \rightarrow MgO_2$$

Can you see why this is not the correct procedure for balancing an equation? By adding a subscript of "2" to the MgO, we have changed the way that the elements combine. We have created a new compound that may not even exist! Remember that when we went over oxidation numbers in Chapter 5, we pointed out that each formula for a neutral compound must show a net charge of zero. Magnesium has an oxidation number of +2 and oxygen has an oxidation number of -2. They combine in a 1:1 ratio to form magnesium oxide, because +2 + (-2) = 0. By changing the subscript to satisfy the atomic tally, we violated a rule that we learned several chapters ago! Thus, the rule was born.

What can we do to balance a chemical equation? All we are allowed to do is add coefficients in front of the chemical formulas to achieve the proper balance. The correct way to balance the equation from Example 2 is:

$$2Mg + O_2 \rightarrow 2MgO$$

How can we be sure that this is the correct balanced equation? We can do another atomic tally and make sure that the same number of atoms of each element appear on both sides of the equation.

$$2Mg + O_2 \rightarrow 2MgO$$

	Mg	O	Mg	O	
Elements →	**Mg**	**O**	**Mg**	**O**	
# of atoms →	2	2	2	2	**Balanced!**

Let's try another example. Do an atomic tally for the following equation. If the equation is not balanced, figure out what coefficient(s) you need to add in order to achieve balance. Finally, do another atomic tally to make sure that the new equation is balanced.

Example 3

Do an atomic tally for the following equation and determine if it is balanced. If not, balance the equation using coefficients, and repeat the tally.

$$CH_4 + O_2 \rightarrow H_2O + CO_2$$

I will go through the solution slowly, and if I lose you, go back to the beginning and try to follow me again. It is very important that you learn the process of balancing equations correctly.

First, we do an atomic tally to see if our equation is already balanced.

$$CH_4 + O_2 \rightarrow H_2O + CO_2$$

	C	H	O	C	H	O	
Elements →	**C**	**H**	**O**	**C**	**H**	**O**	
# of atoms →	1	4	2	1	2	3	**Not balanced!**

We notice that our equation is not balanced. The number of carbon atoms on each side of the equation is the same, but we have twice as many hydrogen atoms on the left-hand side, and there is an additional oxygen atom on the right-hand side. Let's start by adding a coefficient of "2" to the water on the product side, in an attempt to balance the hydrogen atoms. Then, we will do another atomic tally to see where we are. To make this easier to follow, I will put the coefficients that I add in bold and italics.

$$CH_4 + O_2 \rightarrow 2H_2O + CO_2$$

Elements →	C	H	O	C	H	O	
# of atoms →	1	4	2	1	4	4	**Not balanced!**

Notice the new tally. By putting the coefficient in front of the water (H_2O) I not only changed the number of hydrogen atoms on the product side, I also changed the number of oxygen atoms. Now we have twice as many oxygen atoms on the right-hand side as on the left-hand side. It should be as simple as adding another coefficient of "2," this time in front of the oxygen on the reactant side. Let's do that, and do another atomic tally.

$$CH_4 + 2O_2 \rightarrow 2H_2O + CO_2$$

Elements →	C	H	O	C	H	O	
# of atoms →	1	4	4	1	4	4	**Balanced!**

Do you see how the skills of performing an atomic tally and balancing equations go hand in hand? To be able to do one, you need to be able to do the other. Let's try one more example together. Take a look at Example 4 and see how far you can go before you check my answer.

Example 4

Do an atomic tally for the following equation and determine if it is balanced. If not, balance the equation using coefficients and repeat the tally.

$$KClO_3 \rightarrow KCl + O_2$$

In the first step, I will do an atomic tally to get an idea of what to do next.

$$KClO_3 \rightarrow KCl + O_2$$

Elements →	K	Cl	O	K	Cl	O	
# of atoms →	1	1	3	1	1	2	**Not balanced!**

Our tally shows us that we have one more atom of oxygen on the reactant side than on the product side. Balancing this equation is not as easy as you might assume when you see that we are only off by one atom. If you look at the product side, there is no way to add only one atom to that side, because the oxygen has a subscript of "2." Putting a coefficient of "2" in front of the oxygen on the product side will leave us 4 atoms of oxygen, and we will be no closer to solving the equation. This is when some students get desperate enough to violate our rule, leading them to simply change the subscript on the reactant side to "2." Please, remember that changing the subscripts is never an option!

Our plan of attack is suggested by the subscript of "2" that comes with the O_2 on the product side. Can you see that it is impossible to get an odd number of oxygen atoms on the product side? Placing either an even or odd coefficient in front of the oxygen will still result in an even number of oxygen atoms on the right-hand side; for example, $3O_2$ = 6 atoms of oxygen. This tells us that the real problem is on the left-hand side of the equation, where we want to change the odd number of oxygen atoms to an even number, perhaps by adding a coefficient of 2 before the $KClO_3$.

$$2KClO_3 \rightarrow KCl + O_2$$

Elements →	K	Cl	O	K	Cl	O	
# of atoms →	2	2	6	1	1	2	**Not balanced!**

I know that it seems that we just moved further away from the solution, but some of these equations are akin to puzzles that need to be solved one step at a time. Now, we have an even number of oxygen atoms on the left-hand side, which we can match on the right by adding a coefficient of "3" in front of the oxygen, as I show next.

$$2KClO_3 \rightarrow KCl + 3O_2$$

Elements →	K	Cl	O	K	Cl	O	
# of atoms →	2	2	6	1	1	6	**Not balanced!**

Now, if we add a coefficient of "2" in front of the potassium chloride (KCl), we will have a balanced equation.

$$2KClO_3 \rightarrow 2KCl + 3O_2$$

	K	Cl	O	K	Cl	O	
Elements →	**K**	**Cl**	**O**	**K**	**Cl**	**O**	
# of atoms →	2	2	6	2	2	6	**Balanced!**

Before I leave you with the practice problems, let me give you one last important piece of information. If you ever find that all of the coefficients that you used to balance an equation are reducible by the same number, you must reduce your coefficients. For example, let's suppose you balanced a chemical equation and came up with this result:

$$2N_2 + 2O_2 \rightarrow 4NO$$

The equation is technically balanced, as you can show with an atomic tally, but it is still not correct. Because all of the coefficients are reducible by the number 2, you must divide each coefficient by 2, leaving you with this equation:

$$N_2 + O_2 \rightarrow 2NO$$

Lesson 6–1 Review

Do an atomic tally for each of the following equations and state whether or not they are balanced as shown.

1. $NaOH + HCl \rightarrow H_2O + NaCl$
2. $KOH + H_2SO_4 \rightarrow K_2SO_4 + H_2O$
3. $C_3H_8 + 5O_2 \rightarrow 4H_2O + 3CO_2$
4. $2Al(OH)_3 \rightarrow Al_2O_3 + H_2O$

Balance each of the following chemical equations. If an equation is already balanced, then the correct response is "already balanced."

5. $HCl + Ca(OH)_2 \rightarrow CaCl_2 + 2H_2O$
6. $C_2H_6 + O_2 \rightarrow CO_2 + H_2O$
7. $Fe + O_2 \rightarrow Fe_2O_3$
8. $Na + H_2O \rightarrow NaOH + H_2$

Lesson 6–2: Classifying Chemical Reactions

There are, quite literally, millions of types of known chemical reactions. Nobody can be expected to memorize all of these different reactions. Fortunately, many of the known chemical reactions follow patterns that are similar enough to allow classification. A scheme has been created, which classifies many of the known reactions into five major categories, namely:

- Synthesis.
- Decomposition.
- Single replacement.
- Double replacement.
- Combustion.

Some reactions don't fit neatly into one of these categories, but enough of them do that the classification scheme is extremely useful. As with all classification systems, the chemical reaction classification system allows you to memorize the characteristics of the categories, rather than the individual reactions.

A ***synthesis reaction*** is one where two or more simpler substances combine to form a more complex substance. The reactants may be elements or compounds. The product, being made up of two or more elements chemically combined, must be a compound. The general format for a synthesis reaction is given by the following:

$$A + B \rightarrow AB$$

A	+	B	→	AB
element or compound		element or compound		compound

Synthesis reactions are also sometimes called direct combination reactions, or composition reactions.

Examples of Synthesis Reactions

$2H_{2(g)} + O_{2(g)} \rightarrow 2H_2O_{(g)}$
synthesis of water

$2CO_{(g)} + O_{2(g)} \rightarrow 2CO_{2(g)}$
synthesis of carbon dioxide

$4Al_{(s)} + 3O_{2(g)} \rightarrow 2Al_2O_{3(s)}$
synthesis of aluminum oxide

$3H_{2(g)} + N_{2(g)} \rightarrow 2NH_{3(g)}$
synthesis of ammonia

A ***decomposition reaction*** occurs when a more complex substance breaks down into two or more simpler substances. In this type of reaction, the reactant must be a compound, and the products can be either elements or compounds. The general format for a decomposition reaction is:

AB	→	A	+	B
compound		element or compound		element or compound

As you can see, the format for the decomposition reaction is the opposite of the format for the synthesis reaction. Another common name for a decomposition reaction is analysis.

Examples of Decomposition Reactions

$CaCO_{3(s)} \rightarrow CaO_{(s)} + CO_{2(g)}$
decomposition of calcium carbonate

$2NH_{3(g)} \rightarrow 3H_{2(g)} + N_{2(g)}$
decomposition of ammonia

$2HgO_{(s)} \rightarrow 2Hg_{(l)} + O_{2(g)}$
decomposition of mercury (II) oxide

$2H_2O_{(l)} \rightarrow 2H_{2(g)} + O_{2(g)}$
decomposition of water

A ***single replacement reaction*** occurs when an element reacts with a compound in such a way that the element replaces an ion of a similar element from the compound. In order for the replacement to occur, the elemental reactant must be more reactive than the elemental product. The general format for a single displacement reaction, when the elemental reactant is a metal, is:

A	+	BC	→	AC	+	B
element	+	compound		compound	+	element

When the elemental reactant is a nonmetal, the general format looks more like this:

A	+	BC	→	BA	+	C
element	+	compound		compound	+	element

What is the difference, you might ask? The elemental reactant will replace the part of the compound that it is most like. You won't see a metal replacing a nonmetal to form a compound like AgCu. Remember that things with like charges will repel each other, so you won't make compounds out of things with similar oxidation numbers. Single replacement reactions are also called single displacement reactions.

Examples of Single Replacement Reactions

$K_{(s)} + NaCl_{(aq)} \rightarrow KCl + Na_{(s)}$
potassium replaces sodium

$Cl_{2(g)} + 2LiBr_{(aq)} \rightarrow 2LiCl_{(aq)} + Br_{2(l)}$
chlorine replaces bromine

$F_{2(g)} + 2NaI_{(aq)} \rightarrow 2NaF_{(aq)} + I_{2(s)}$
fluorine replaces iodine

$Zn_{(s)} + 2HCl_{(aq)} \rightarrow ZnCl_{2(aq)} + H_{2(g)}$
zinc replaces hydrogen

A ***double replacement reaction*** occurs when two ionic compounds react in an aqueous solution, and an ion from one compound replaces the similar ion from the other compound. The general format for a double reaction is:

AB	**+**	**CD**	**→**	**CB**	**+**	**AD**
compound	+	compound		compound	+	compound

Double replacement reactions can be misleading, as they are written, because the compounds that are written as "products" do not always form. You should pay attention to subscripts to see the form of the products. If the phase is solid, liquid, or gas, then the product will actually form, at least to some extent. If the subscript is "aq" (for "aqueous"), it means that the ions are dissociated in water. Double replacement reactions are also called double displacement reactions.

Examples of Double Replacement Reactions

$Ba(OH)_{2(aq)} + H_2SO_{4(aq)} \rightarrow BaSO_{4(s)} + 2H_2O_{(l)}$

$Ca(OH)_{2(aq)} + Na_2CO_{3(aq)} \rightarrow 2NaOH_{(aq)} + CaCO_{3(s)}$

$3CaCl_{2(aq)} + Al_2(SO_4)_{3(aq)} \rightarrow 3CaSO_{4(aq)} + 2AlCl_{3\,(aq)}$

$AgNO_{3(aq)} + NaCl_{(aq)} \rightarrow AgCl_{(aq)} + NaNO_{3(aq)}$

A ***combustion reaction*** occurs when a ***hydrocarbon***, which is an organic compound that contains only hydrogen and carbon, reacts with oxygen to produce water and carbon dioxide. In some ways, combustion reactions are the easiest of the types of chemical reactions to identify, because oxygen is always a reactant and the products are always carbon dioxide and water. The general form for a combustion reaction is shown here:

hydrocarbon + oxygen → carbon dioxide + water

Examples of Combustion Reactions

$CH_{4(g)} + 2O_{2(g)} \rightarrow CO_{2(g)} + 2H_2O_{(g)}$ combustion of methane	$2C_2H_{6(g)} + 7O_{2(g)} \rightarrow 4CO_{2(g)} + 6H_2O_{(g)}$ combustion of ethane
$C_2H_{4(g)} + 3O_{2(g)} \rightarrow 2CO_{2(g)} + 2H_2O_{(g)}$ combustion of ethene	$2C_2H_{2(g)} + 5O_{2(g)} \rightarrow 4CO_{2(g)} + 2H_2O_{(g)}$ combustion of ethyne

Lesson 6–2 Review

Identify the following chemical reactions as one of the five main types.

1. $2H_2O_2 \rightarrow 2H_2O + O_2$
2. $4Fe + 3O_2 \rightarrow 2Fe_2O_3$
3. $Cl_2 + 2NaBr \rightarrow 2NaCl + Br_2$
4. $Zn + H_2SO_4 \rightarrow ZnSO_4 + H_2$
5. $CH_4 + O_2 \rightarrow CO_2 + H_2O$
6. $Pb(NO_3)_2 + K_2CrO_4 \rightarrow PbCrO_4 + KNO_3$
7. $2Ag + Cl_2 \rightarrow 2AgCl$
8. $6HCl + Fe_2O_3 \rightarrow 2FeCl_3 + 3H_2O$
9. $MgCO_3 \rightarrow MgO + CO_2$
10. $2C_2H_6 + 7O_2 \rightarrow 4CO_2 + 6H_2O$

Lesson 6–3: Ionic Equations

In Lesson 6–2, I mentioned that displacement reactions are a bit misleading at times. Sometimes, a product that is shown on the product side of the equation does not really appear in the physical chemical reaction. The reason for this has to do with the solubility of ionic substances in water. If a particular product is soluble, it will stay dissolved in the aqueous solution. If a product is insoluble, it will appear as a solid ***precipitate*** in the test vessel. It is important to know which products stay dissolved in the water, so we can make proper identification of the precipitates that do form as the result of the chemical reactions. Ionic equations are more realistic representations of these reactions that take place in aqueous solution. ***Ionic equations*** show the individual ions that exist in solution. When we take an ionic displacement reaction and remove the information that is misleading, we produce a ***net ionic equation***.

Let's suppose we were to carry out a chemical reaction between aqueous solutions of lead (II) chlorate and sodium iodide. We mix the two solutions in a test tube, and we find that a solid precipitate forms and falls to the bottom of the test tube. We would want to be able to identify what the solid was. The first thing that we might want to do is make a word equation for the reaction that we think is taking place. Treating this as a double displacement reaction, we would get the word equation shown here:

lead (II) chlorate + sodium iodide → sodium chlorate + lead (II) iodide

We can assume that the precipitate that formed in our test tube must be one (or both) of the possible products from our word equation. In order to identify which of the products formed, we would need to find out which of the products is insoluble in water and would therefore precipitate out. Our most handy reference tables for this type of problem will contain the solubility rules for ionic compounds, shown here, and/or a solubility table (see Figure 6–3a).

General Solubility Rules for Ionic Compounds in Water

Compounds That Tend to be Soluble

- Compounds that contain ions of alkali metals (e.g., K^+ or Na^+).
- Compounds that contain halogens (e.g., Cl^- or Br^-).
- Compounds that contain the ammonium (NH_4^+) ion.
- Compounds that contain the sulfate (SO_4^{2-}) ion.

Compounds That Tend to be Insoluble

- Compounds that contain the carbonate (CO_3^{2-}) ion.
- Compounds that contain the hydroxide (OH^-) ion.
- Compounds that contain the phosphate (PO_4^{2-}) ion.
- Compounds that contain the sulfide (S^{2-}) ion.

If we refer to the general solubility rules, in this case, we are not really able to identify the precipitate, because there seem to be conflicting trends. We notice that compounds that contain ions of alkali metals tend to be soluble, which would seem to suggest that the sodium chlorate is probably not the precipitate, but we also read that compounds that contain halogens also tend to be soluble, which would seem to let the lead (II) iodide off the hook! How is it possible for us to have a solid precipitate and then have no suspect?

Notice that the title of the rules is "***general***" solubility rules. There are exceptions to the rules listed there. For example, although ionic compounds that contain halogens tend to be soluble, when the cation is lead (II) (as in lead (II) chloride or lead (II) iodide) the ionic compound will be insoluble in water. The rules listed are really meant to be general trends, which you may want to memorize. They will give you the ability to make fast predictions about the identity of a precipitate that is found after many ionic reactions, just not the one in our first example.

When you have more time, or access to more specific information, such as that shown in Figure 6–3a, you can make a much more accurate prediction of the identity of a precipitate. If you find the cation (positive ion) on the right-hand column and match it with the anion (negative ion) across the top of the table, you will be able to discover the solubility of the products and the identity of the precipitate.

Solubility Table

	Acetate	Bromide	Carbonate	Chlorate	Chloride	Chromate	Hydroxide	Iodide	Nitrate	Phosphate	Sulfate	Sulfide
Aluminum	S	S	-	S	S	-	I	S	S	I	S	d
Ammonium	S	S	S	S	S	S	S	S	S	S	S	S
Barium	S	S	I	S	S	I	S	S	S	I	I	d
Calcium	S	S	I	S	S	S	SS	S	S	I	SS	d
Copper (II)	S	S	I	S	S	I	I	-	S	I	S	I
Iron (II)	S	S	I	S	S	-	I	S	S	I	S	I
Iron (III)	S	S	-	S	SS	I	I	-	S	I	SS	d
Lead	S	SS	I	S	S	I	I	SS	S	I	I	I
Lead (II)	S	S	I	S	S	I	SS	I	S	I	I	SS
Magnesium	S	S	I	S	S	S	I	S	S	I	S	d
Mercury (I)	SS	I	I	S	I	SS	-	I	S	I	SS	I
Mercury (II)	S	SS	I	S	S	SS	I	I	S	I	d	I
Potassium	S	S	S	S	S	S	S	S	S	S	S	S
Silver	SS	I	I	S	I	SS	-	I	S	I	SS	I
Sodium	S	S	S	S	S	S	S	S	S	S	S	S
Strontium	S	S	SS	S	S	SS	S	S	S	I	SS	S
Zinc	S	S	I	S	S	S	I	S	S	I	S	I

Key: S = Soluble SS = Slightly Soluble I = Insoluble d = Decomposes - = Unknown

Figure 6-3a

Let's take the example that we started discussing and set it up as a problem, as shown below:

Example 1

A student mixes aqueous solutions of lead (II) chlorate and sodium iodide and finds that a white precipitate forms.

For this chemical reaction:

A. Write a word equation.

B. Write a balanced chemical reaction, with subscripts indicating state or phase.

C. Write a net ionic equation.

D. Identify the precipitate.

Most of the first two steps should be easy to you, if you studied Chapter 5. You may want to try these steps on your own, for additional practice. I will jump ahead and complete step A and most of step B. Then we will discuss how you determine the subscripts indicating the state or phase of each substance.

A. Write a word equation.

lead (II) chlorate + sodium iodide → lead (II) iodide and sodium chlorate

B. Write a balanced chemical reaction, with subscripts indicating state or phase.

$$Pb(ClO_4)_{2(aq)} + 2NaI_{(aq)} \rightarrow PbI_2 + 2NaClO_4$$

Note that I included the "state" subscript of (aq) for the reactants but not for the products. We know that the reactants are in aqueous form, because that is mentioned in the original question. To figure out if the products are aqueous or not, we need to consult either the solubility rules or Figure 6–3a. The rules tell us that chlorate salts are essentially all soluble, so we will add the (aq) subscript to the sodium chlorate. We also note that Figure 6–3a states that lead (II) iodide is insoluble, which means that it is probably the precipitate that appeared in the bottom of the test tube. We use the (s) subscript to indicate a precipitate.

Adding this information to the formula that we wrote in step B, we would get the following chemical equation:

$$Pb(ClO_4)_{2(aq)} + 2NaI_{(aq)} \rightarrow PbI_{2(s)} + 2NaClO_{4(aq)}$$

When we say that the "product" called sodium chlorate is aqueous, we mean that it doesn't really form. It exists as free ions in solution. The lead (II) iodide, on the other hand, is a "true" product of the reaction. We could pour the contents of our solution through a filter paper, and only the lead (II) iodide would be left behind. An ionic reaction is a more realistic representation of what actually goes on with the substances in the reaction vessel. Remember that the reactants were in solution, which means that they started off as free ions. After the reaction, only the precipitate came out of solution; the rest of the substances remained in ion form. Following is an ionic equation for the same double displacement reaction:

$$Pb^{2+}_{(aq)} + 2ClO^-_{4\,(aq)} + 2Na^+_{(aq)} + 2I^-_{(aq)} \rightarrow PbI_{2(s)} + 2Na^+_{(aq)} + 2ClO^-_{4\,(aq)}$$

A net ionic reaction only shows the "true" product and the ions that reacted to form the product. The ions that look the same on both sides of the equation don't change or enter into a chemical reaction. Ions that don't enter into the chemical reaction are called ***spectator ions***. To create a net ionic reaction, we remove the spectator ions, as shown here:

$$Pb^{2+}_{(aq)} + \cancel{2ClO^-_{4\,(aq)}} + \cancel{2Na^+_{(aq)}} + 2I^-_{(aq)} \rightarrow PbI_{2(s)} + \cancel{2Na^+_{(aq)}} + \cancel{2ClO^-_{4\,(aq)}}$$

This gives us the net ionic equation: $Pb^{2+}_{(aq)} + 2I^-_{(aq)} \rightarrow PbI_{2(s)}$

You may not always need to show all of these steps. That will depend upon the question that you are answering. If we look back at our original problem, it did not ask us to show the ionic equation, only the net ionic equation. We could come up with the net ionic equation by crossing out the spectator ions from the original equation. Look at what we would cross out of our chemical reaction:

$$Pb\cancel{(ClO_4)_2}_{(aq)} + 2\cancel{Na}I_{(aq)} \rightarrow PbI_{2(s)} + \cancel{2NaClO_{4(aq)}}$$

Adding the proper charges to the ions that remain, we would get the same net ionic equation that we previously derived: $Pb^{2+} + 2I^- \rightarrow PbI_{2(s)}$

Now, we are ready to fill in answers C and D. Remember: The precipitate is the solid that forms, in this case, the lead (II) iodide.

A. Write a word equation.

lead (II) chlorate + sodium iodide → lead (II) iodide and sodium chlorate

B. Write a balanced chemical reaction, with subscripts indicating state or phase.

$$Pb(ClO_4)_{2(aq)} + 2NaI_{(aq)} \rightarrow PbI_{2(s)} + 2NaClO_{4(aq)}$$

C. Write a net ionic equation.

$$Pb^{2+}_{(aq)} + 2I^-_{(aq)} \rightarrow PbI_{2(s)}$$

D. Identify the precipitate.

lead (II) iodide

Now, try an example on your own, and then check your answer with the one that I produce. In this example, we will ask for both the ionic equation and the net ionic equation. If you need help, refer to the example that I showed you. If you find that you need more help, review Chapter 5 and the other lessons in Chapter 6.

Example 2

A student mixes aqueous solutions of sodium phosphate and iron (III) chloride and finds that a yellow-white precipitate forms. For this chemical reaction:

A. Write a word equation.

B. Write a balanced chemical reaction, with subscripts indicating state or phase.

C. Write an ionic equation.

D. Write a net ionic equation.

E. Identify the precipitate.

A. The word equation should be relatively simple to write. We have both reactants already, and we simply swap the negative ions from each compound to find the products, giving us the word equation:

sodium phosphate + iron (III) chloride → sodium chloride + iron (III) phosphate

B. In order to write the balanced chemical reaction, we need to write the chemical formulas properly. The stock system tells us that the oxidation number for the iron, as in iron (III) chloride, will be +3. We look up any of the oxidation numbers that we don't know, and find this information:

iron (III) = Fe^{3+} phosphate = PO_4^{3-} chloride = Cl^- sodium = Na^+

We use this information, and our knowledge of writing chemical formulas and equations, to write the balanced chemical equation:

$$Na_3PO_{4(aq)} + FeCl_{3(aq)} \rightarrow 3NaCl + FePO_4$$

Now, we know that the reactants are in solution, because the original question told us that. You probably already know that one of the products, NaCl, is soluble, so it should also get the "aq" subscript. Notice that with this problem, you could predict that the precipitate is the iron (III) phosphate, even without the solubility rules and Figure 6–3a. To be sure, however, we consult this information when we have access to them. There we find that iron (III) phosphate is indeed insoluble.

From the information found there, we can finish the subscripts in the equation, as shown here:

$$Na_3PO_{4(aq)} + FeCl_{3(aq)} \rightarrow 3NaCl_{(aq)} + FePO_{4(s)}$$

C. We turn our balanced chemical equation into an ionic equation, by showing the aqueous compounds as free ions, as they would really exist in solution. This gives us the ionic equation:

$$3Na^{+}_{(aq)} + PO_4^{3-}{}_{(aq)} + Fe^{3+}_{(aq)} + 3Cl^{-}_{(aq)} \rightarrow 3Na^{+}_{(aq)} + 3Cl^{-}_{(aq)} + FePO_{4(s)}$$

D. We turn our ionic equation into a net ionic equation by eliminating the spectator ions, which, as you know, are the ions that don't enter the reaction. Cross out the spectator ions:

$$\cancel{3Na^{+}_{(aq)}} + PO_4^{3-}{}_{(aq)} + Fe^{3+}_{(aq)} + \cancel{3Cl^{-}_{(aq)}} \rightarrow \cancel{3Na^{+}_{(aq)}} + \cancel{3Cl^{-}_{(aq)}} + FePO_{4(s)}$$

We get the net ionic equation:

$$Fe^{3+}_{(aq)} + PO_4^{3-}{}_{(aq)} \rightarrow FePO_{4(s)}$$

E. As far as identifying the precipitate, it is the iron (III) phosphate, which is insoluble in water.

All of our results are summarized here:

A. Write a word equation.

sodium phosphate + iron (III) chloride → sodium chloride + iron (III) phosphate

B. Write a balanced chemical reaction, with subscripts indicating state or phase.

$$Na_3PO_{4(aq)} + FeCl_{3(aq)} \rightarrow 3NaCl_{(aq)} + FePO_{4(s)}$$

C. Write an ionic equation.

$$3Na^{+}_{(aq)} + PO_4^{3-}{}_{(aq)} + Fe^{3+}_{(aq)} + 3Cl^{-}_{(aq)} \rightarrow 3Na^{+}_{(aq)} + 3Cl^{-}_{(aq)} + FePO_{4(s)}$$

D. Write a net ionic equation.

$$Fe^{3+}_{(aq)} + PO_4^{3-}{}_{(aq)} \rightarrow FePO_{4(s)}$$

E. Identify the precipitate.

iron (III) phosphate

Lesson 6–3 Review

For each of the following compounds, state whether they are soluble, slightly soluble, or insoluble in water at 25 °C and 101.3 kPa of pressure.

1. aluminum acetate
2. magnesium bromide
3. zinc carbonate
4. potassium chloride
5. silver iodide
6. barium hydroxide

Lesson 6-4: Oxidation and Reduction Reactions

Historically, the term *oxidation* was used to describe the process of a substance combining with oxygen. For example, the carbon in the reaction that follows would be said to "oxidize" as it combined with oxygen to form carbon dioxide.

$$C_{(s)} + O_{2(g)} \rightarrow CO_{2(g)}$$

carbon + oxygen yields carbon dioxide

Today, this reaction would still be considered an example of oxidation, but the actual definition of oxidation has been expanded to include more reactions, including many in which oxygen plays no part. ***Oxidation*** is when a substance loses electrons or appears to lose electrons, as its oxidation number appears to increase algebraically. If we look at our reaction again, we can see that the oxidation numbers of both oxygen and carbon change. How do we assign oxidation numbers to the substances in the reaction? You should start by reviewing Lesson 5–1, where we first went over the rules for assigning oxidation numbers. For convenience, I have provided the rules here.

Rules for Assigning Oxidation Numbers

1. In a free element (e.g., Ba or I_2) the oxidation number for each atom is 0.
2. With the exception of hydrogen, all of the elements in group 1 show an oxidation number of +1 in compounds.
3. Hydrogen has an oxidation number of +1 in all compounds except for hydrides (e.g., NaH), where it shows an oxidation number of -1.
4. The elements found in group 2, the alkaline earth metals, show an oxidation number of +2 in all compounds.
5. When any of the elements in column 17, the halogens, show a negative oxidation state, it will be -1.
6. Oxygen shows an oxidation number of -2 in all compounds, except for the peroxides, where it is -1.
7. In all neutral compounds, the algebraic sum of the oxidation numbers is 0.
8. In polyatomic ions, the algebraic sum of the oxidation numbers is equal to the charge on the polyatomic ion.

Remember that, as indicated in Rule #1, an atom that is in its free elemental state has an oxidation number of 0. So, we will assign both the carbon and oxygen on the reactant side of the equation oxidation numbers of 0. As for the product side, we must recall Rule #6, which tells us that when oxygen is found in a compound it shows an oxidation number of -2, unless the compound happens to be a peroxide. Now, if each atom of oxygen in the carbon dioxide has an oxidation number of -2, and there are two atoms of oxygen, the total negative charge in the compound is -4. The carbon atom found in the carbon dioxide must have an oxidation number of +4, if the one atom is able to cancel out both atoms of oxygen, leaving the compound with a net charge of zero. Let's rewrite the equation, with the oxidation numbers included:

$$C^{0}_{(s)} + O_{2}{}^{0}_{(g)} \rightarrow C^{4+}O_{2}{}^{2-}_{(g)}$$

carbon + oxygen yields carbon dioxide

Can you see that what is happening to the carbon fits into our modern definition of oxidation? The carbon atom appears to lose four electrons, as its oxidation number goes from 0 to +4. We say that carbon has been oxidized. Because the oxidation has been caused by oxygen, in this case, we call oxygen the ***oxidizing agent***.

Did you notice that the oxidation number of oxygen has changed as well? Instead of appearing to lose electrons, each oxygen atom appears to gain two electrons, as its oxidation number goes from 0 to -2. Remember: Gaining electrons makes the charge more negative. Because the oxidation number of the oxygen goes down, we say that oxygen has been reduced. It may seem strange that the substance that appears to gain electrons is said to be "reduced," but that is because it is gaining negative charges, which reduces its oxidation number. The opposite of oxidation is reduction. ***Reduction*** is when a substance gains electrons or appears to gain electrons, as its oxidation number is reduced algebraically. Because the carbon appears to be the cause of the reduction, carbon would be considered the ***reducing agent*** in this reaction.

If an atom is going to lose electrons, they must go somewhere. We say that the charge must be conserved. There can be no oxidation without an accompanying reduction; they go hand in hand in the same reaction. For that reason, reactions of this type are called oxidation-reduction reactions, or ***redox reactions*** for short.

Let's practice a couple examples of reactions where we are required to assign oxidation numbers to the participants of the reaction and then identify the "players."

Example 1

For the following redox reaction, identify:

A. The oxidation numbers for each substance involved.

B. The substance being oxidized.

C. The oxidizing agent.

D. The substance being reduced.

E. The reducing agent.

$$2Na + Cl_2 \rightarrow 2NaCl$$

Well, that may seem like a tall order, but it is really quite easy. I will walk you through a couple examples, and then you will see how easy they can be.

A. Remember: The oxidation number of an element in its free state is 0, so both the sodium and chlorine on the reactant side will be assigned an oxidation number of 0.

$$2Na^0 + Cl_2{}^0 \rightarrow 2NaCl$$

As far as the elements in the compound NaCl, we will recall other rules for assigning oxidation numbers. Chlorine is a halogen from column 17 on the periodic table and it is acting as a negative ion in the compound sodium chloride. Rule #5 tells us that the oxidation number of a halogen, when it is acting as a negative ion, is -1. Rule #2 tells us that the oxidation number of an element from group 1, such as sodium, will be +1 in compounds. These two rules seem to agree with each other, because sodium and chlorine combine in a 1:1 ratio in this compound. We can now finish the oxidation numbers for our equation.

$$2Na^0 + Cl_2{}^0 \rightarrow 2Na^{1+}Cl^{1-}$$

B. The substance that is being oxidized is the substance that appears to lose electrons. Na goes from an oxidation number of 0 to +1, so it appears to lose one electron. **Sodium, Na, is the substance being oxidized.**

C. The oxidizing agent is the substance that causes the oxidation or, in other words, takes the electrons. **Chlorine, Cl_2, is the oxidizing agent.**

D. The substance that is reduced is the substance that gains electrons or appears to gain electrons, as its oxidation number is algebraically decreased. That definition describes what happens to chlorine in this reaction. **Chlorine (Cl_2) is the substance being reduced.**

E. The reducing agent is the substance that causes the reduction to take place, by giving up the electrons. **Sodium (Na) is the reducing agent.**

Example 1 results summary:

$$2Na + Cl_2 \rightarrow 2NaCl$$

A. The oxidation numbers for each substance involved:

$$2Na^0 + Cl_2^0 \rightarrow 2Na^{1+}Cl^{1-}$$

B. The substance being oxidized:
sodium (Na)

C. The oxidizing agent:
chlorine (Cl_2)

D. The substance being reduced:
chlorine (Cl_2)

E. The reducing agent:
sodium (Na)

Notice that the oxidizing agent is also the substance that is reduced, and the reducing agent is also the substance that is being oxidized. Also, notice that the element oxygen didn't even appear in this redox reaction, so make sure that you understand the modern definition of oxidation.

Let's try a somewhat more complicated example next.

Example 2

For the following redox reaction, identify:

A. The oxidation numbers for each substance involved.

B. The substance being oxidized.

C. The oxidizing agent.

D. The substance being reduced.

E. The reducing agent.

$$SnCl_2 + 2FeCl_3 \rightarrow 2FeCl_2 + SnCl_4$$

A. We can start assigning oxidation numbers based on Rule #5, which tells us that chlorine, being a halogen that is acting as a negative ion in a compound, must show an oxidation number of -1 in all four compounds, as shown here:

$$SnCl_2^{1-} + 2FeCl_3^{1-} \rightarrow 2FeCl_2^{1-} + SnCl_4^{1-}$$

Remember: Each atom of chloride has a -1 charge. To get the total negative charge on each compound, we must multiply by the subscripts for the chloride ion. The first compound ($SnCl_2^{1-}$), for example, will have a charge of (-1 × 2) = -2 on the negative side. The tin (Sn) on the reactant side must have a charge of +2 in order to ensure that the compound has a net charge of 0. Following this pattern, we are able to assign oxidation numbers to all of the elements in the equation, as shown here:

$$Sn^{2+}Cl_2^{1-} + 2Fe^{3+}Cl_3^{1-} \rightarrow 2Fe^{2+}Cl_2^{1-} + Sn^{4+}Cl_4^{1-}$$

B. The substance that is being oxidized is the one that appears to lose electrons, as its oxidation number is algebraically increased. The oxidation number of tin (Sn) changes from +2 to +4. **Tin (Sn^{2+}) is the substance being oxidized.**

C. The oxidizing agent is the substance that is taking the extra electrons, allowing the tin to be oxidized. **Iron (Fe^{3+}) is our oxidizing agent.**

D. The oxidizing agent is the substance that is reduced. The oxidation number of iron changes from +3 to +2. **Fe^{3+} is the substance that is being reduced.**

E. We also mentioned that the substance that is being oxidized is also the reducing agent. Tin provides the electrons for the iron to become reduced. **Tin (Sn^{2+}) is the reducing agent.**

Example 2 summary:

$$SnCl_2 + 2FeCl_3 \rightarrow 2FeCl_2 + SnCl_4$$

A. The oxidation numbers for each substance involved

$$Sn^{2+}Cl_2^{1-} + 2Fe^{3+}Cl_3^{1-} \rightarrow 2Fe^{2+}Cl_2^{1-} + Sn^{4+}Cl_4^{1-}$$

B. The substance being oxidized:

tin (Sn^{2+})

C. The oxidizing agent:

iron (Fe^{3+})

D. The substance being reduced:

iron (Fe^{3+})

E. The reducing agent:

tin (Sn^{2+})

Now, try the following review problems. Be sure to check your answers at the end of the chapter, and we will continue our discussion of redox reactions in our next lesson.

Lesson 6–4 Review

1. ____________ is when a substance gains electrons, or appears to gain electrons, as its oxidation number is reduced algebraically.
2. The ____________ agent in a redox reaction is the substance that causes another substance to become oxidized.
3. ____________ is when a substance loses electrons, or appears to lose electrons, as its oxidation number appears to increase algebraically.
4. The ____________ agent in a redox reaction is the substance that causes another substance to become reduced.

Use the rules for assigning oxidation numbers to assign the correct oxidation numbers to each element in each of the following redox reactions. Then, identify the substance that is being oxidized, the substance that is being reduced, the reducing agent, and the oxidizing agent for each reaction.

5. $3H_2 + N_2 \rightarrow 2NH_3$
6. $Fe_2O_3 + 3CO \rightarrow 2Fe + 3CO_2$
7. $Zn + H_2SO_4 \rightarrow ZnSO_4 + H_2$
8. $2Mg + O_2 \rightarrow 2MgO$

Lesson 6–5: Nuclear Reactions

Take a look at the following reaction.

$$^{40}_{19}K \rightarrow \, ^{40}_{20}Ca + \, ^{0}_{-1}e$$

At first glance, you might mistake this for a chemical equation. The format of the equation is similar to the format of a chemical equation, including elemental symbols and a "yields" arrow. The elemental notation that is shown for both potassium and calcium is the same as the elemental notation that we covered in Lesson 3–2. However, there is an important difference between this equation and the chemical equations that we have been discussing up until now. Notice that we have different elements on either side of the equation! This is evidence that we are looking at a nuclear equation, rather than a chemical one.

The equation shows an atom of potassium transforming into an atom of calcium and an additional particle that we haven't learned yet. When we discussed creating an atomic tally to balance a chemical equation in Lesson 6–1, we said that you must have the same number of atoms of each element on each side of a chemical equation, or else it violates the Law of Conservation of Mass. The equation shown here does not represent a chemical reaction, because it does violate the Law of Conservation of Mass!

Way back in Chapter 1 I mentioned that Einstein combined the laws of conservation of mass and energy into one law, called the Law of Conservation of Mass-Energy to account for the fact that in a nuclear reaction, a small amount of mass is converted into energy. Perhaps the most interesting aspect of nuclear reactions is the fact that elements actually transform into other elements. Sometimes, when we hear stories of the ancient alchemists trying to turn lead into gold, we scoff at their "primitive" notations, which seem to blur magic and science. What is interesting is that humans now have the ability to do what the alchemists dreamed of doing: We can change one element into another. In fact, it happens in nature all of the time!

So, when you see an equation such as ${}^{40}_{19}K \rightarrow {}^{40}_{20}Ca + {}^{0}_{-1}e$, where you have different elements on each side of the equation, you should know right away that you are looking at a nuclear equation. Notice that both the mass number ${}^{40}K \rightarrow {}^{40}Ca + {}^{0}e$ and the nuclear charge ${}_{19}K \rightarrow {}_{20}Ca + {}_{-1}e$ of the particles that appear in the equation are conserved, because this will be the key to balancing this type of equation. In reality, you should realize that nuclear reactions do involve mass-energy conversions, so you can't expect the original Law of Conservation of Mass to apply to nuclear reactions. Rather, it is the sum of the mass and the energy that is conserved in a nuclear reaction. Such a small amount of matter is converted into energy that you won't notice the missing mass in the actual nuclear equation.

Before we learn more about the types of nuclear reactions, you should become familiar with the types of notation that are used for the particles that are common to nuclear reactions shown in Figure 6–5a on page 205.

In addition to the symbols shown in Figure 6–5a, you should also become familiar with the symbol γ, which represents another type of radiation called gamma rays or gamma radiation. The symbol for gamma radiation wasn't included in Figure 6–5a, simply because gamma rays are a form of electromagnetic radiation and are **usually** not thought of as particles. (You may get to wait until you study physics until you open this particular can of worms.)

Nuclear processes are often categorized as ***natural radioactivity*** or ***artificial radioactivity***. As you might imagine, the term *natural radioactivity* is applied to those nuclear processes that occur in nature. Many of the elements have unstable isotopes that undergo spontaneous nuclear decay.

Symbols for Particles Common to Nuclear Equations	
$^{4}_{2}\mathrm{He}$	**Alpha Particle**—The alpha particle is identical in composition (with 2 protons and 2 neutrons) to the nucleus of a helium atom. It is also important to note that, like a bare helium nucleus, the alpha particle is positively charged and will be attracted to a negatively charged object.
$^{0}_{-1}\mathrm{e}$	**Beta Particle**—The beta particle is identical in composition (with a mass of zero and a charge of –1) to an electron. Being negatively charged, it is repelled by other negatively charged objects.
$^{1}_{0}\mathrm{n}$	**Neutron**—Sometimes neutrons are absorbed or released as elements transform into other elements. Even more interesting is the fact that protons appear to turn into neutrons (and vice versa) in some nuclear reactions.
$^{0}_{+1}\mathrm{e}$	**Positron**—The positron (sometimes called a "beta plus particle") has the same mass as the electron, but has the opposite charge.
$^{1}_{1}\mathrm{p}$	**Proton**—The proton is identical in composition (with a mass of 1 and a nuclear charge of +1) to the nucleus of a hydrogen (protium) atom. The nuclear charge of a proton is understood to be positive, so the + sign is not always shown in the notation, as it is with the positron.

Figure 6-5a

If you studied the concept of carbon dating in biology, then you know that carbon-14 is an example of such an unstable isotope, which naturally decays into nitrogen-14 according to the following nuclear equation:

$$^{14}_{6}\mathrm{C} \rightarrow {}^{14}_{7}\mathrm{N} + {}^{0}_{-1}\mathrm{e}$$

This reaction is an example of ***beta decay***, because the ***parent nucleus*** $^{14}_{6}\mathrm{C}$ decays into the ***daughter nucleus*** $^{14}_{7}\mathrm{N}$ as it releases a beta particle.

Artificial radioactivity is observed when scientists take stable nuclei and try to combine them in various ways in an attempt to produce stable isotopes of "new," or at least previously undiscovered, elements. All of the elements on the periodic table with atomic numbers of 93 or greater were produced with the aid of instruments called particle accelerators, which are used to fuse nuclei of known elements together in order to produce still heavier nuclei. All of these ***transuranium elements*** have proven to be radioactive, and they decay into more stable elements after varying periods of time.

The following equation shows an example of an artificial transmutation, as curium-244 is bombarded with carbon-12 to form nobelium-254:

$$^{12}_{6}C + ^{244}_{96}Cm \rightarrow ^{254}_{102}No + 2\,^{1}_{0}n$$

Perhaps you are familiar with the terms *nuclear fission* and *nuclear fusion*. ***Nuclear fission*** is the process by which a relatively massive nucleus is divided into smaller nuclei and one or more neutrons. Nuclear fission is the process that generates so much power in nuclear power plants and in certain types of nuclear weapons, such as the bombs that were dropped on Japan in 1945. The following equation shows an example of nuclear fission (the nuclear fission of uranium-235):

$$^{235}_{92}U + ^{1}_{0}n \rightarrow ^{90}_{38}Sr + ^{143}_{54}Xe + 3\,^{1}_{0}n$$

Notice there are neutrons on both sides of the equation. The neutron on the left side of the equation can be thought of as the projectile that causes the larger atom to "split." The three neutrons that are released on the right side of the equation can then, under the proper conditions, go on to "split" more atoms, releasing still more neutrons in the process. This sets up the process that is called a ***chain reaction***.

Nuclear fusion is the process of less massive nuclei combining into more massive ones. Nuclear fusion is the process that powers stars, including our sun, with hydrogen nuclei combining to produce helium nuclei, according to a series of reactions represented by the equations that follow:

$$^{1}_{1}H + ^{1}_{1}H \rightarrow ^{2}_{1}H + ^{1}_{+1}e$$

$$^{1}_{1}H + ^{2}_{1}H \rightarrow ^{3}_{2}He$$

$$^{3}_{2}He + ^{3}_{2}He \rightarrow ^{4}_{2}He + 2\,^{1}_{1}H$$

$$^{3}_{2}He + ^{1}_{1}H \rightarrow ^{4}_{2}He + ^{0}_{+1}e$$

You will probably be called upon to balance nuclear reactions. One common type of question is shown in the following.

Example 1

Identify the particle represented by X and complete each of the following nuclear equations.

A. $^{238}_{92}U \rightarrow ^{234}_{90}Th + X$

B. $^{26}_{12}Mg + ^{1}_{1}p \rightarrow ^{4}_{2}He + X$

C. $^{239}_{94}Pu + X \rightarrow ^{240}_{94}Pu$

The trick to solving these problems is to algebraically add both the nuclear charges and the mass numbers on each side of the equation and see what values remain in order to equal out both sides. Let's go through the solutions for Example 1.

A. $^{238}_{92}U \rightarrow ^{234}_{90}Th + X$—Looking at the mass numbers, which are the upper numbers in the elemental notations, we can see that the substance represented by X must have a mass number of 4, because 238 = 234 + 4. This number doesn't give us the identity of X, because several different substances could have a mass number of 4. Looking at the atomic numbers on the bottom of each notation, we can determine that the nuclear charge of X must be 2, because 92 = 90 + 2. The nuclear charge or atomic number of 2 identifies X as the element helium ($^{4}_{2}He$), as a quick glance at the periodic table will show. Filling this notation into our original problem, we get the answer.

Answer: $^{238}_{92}U \rightarrow ^{234}_{90}Th + ^{4}_{2}He$

This type of reaction is called ***alpha decay***, because the ***parent nucleus*** $^{238}_{92}U$ loses an alpha particle.

B. $^{26}_{12}Mg + ^{1}_{1}p \rightarrow ^{4}_{2}He + X$—When we add up the mass numbers on the left-hand side of the equation we get 27 (26 + 1 = 27). We must also have a total of 27 for the mass numbers on the right side of the equation. The alpha particle has a mass of 4, so X must have a mass number of 23 (27 – 4 = 23). As far as the atomic number goes, we have 13 (12 + 1 = 13) on the left-hand side of the equation, and 2 + X on the right-hand side. X must have an atomic number of 11 (13 – 2 = 11). Looking up the atomic number of 11 on the periodic table, we find that X must be the element sodium, giving us the notation $^{23}_{11}Na$. Substituting this notation into the original equation we get the answer.

Answer: $^{26}_{12}Mg + ^{1}_{1}p \rightarrow ^{4}_{2}He + ^{23}_{11}Na$

C. $^{239}_{94}Pu + X \rightarrow ^{240}_{94}Pu$—This equation shows X on the left side of the arrow. On the right side of the arrow, we have a mass number of 240. The total of the mass numbers on the left side must also add up to 240, making the mass number of X 1 (240 – 239 = 1). The atomic number of X must be 0, because we already have 94 on both sides of the equation. Can you figure out what $^{1}_{0}X$ represents? You will not find an element with an atomic number of zero on the periodic table, so this is one of the times where it is better to think of the atomic number as the nuclear charge.

The neutron is a particle with a mass number of 1 and a nuclear charge of 0, giving us the answer shown here:

$$^{239}_{94}Pu + ^{1}_{0}n \rightarrow ^{240}_{94}Pu$$

We can now summarize our answers:

A. $^{238}_{92}U \rightarrow ^{234}_{90}Th + X \quad = \quad ^{238}_{92}U \rightarrow ^{234}_{90}Th + ^{4}_{2}He$

B. $^{26}_{12}Mg + ^{1}_{1}p \rightarrow ^{4}_{2}He + X \quad = \quad ^{26}_{12}Mg + ^{1}_{1}p \rightarrow ^{4}_{2}He + ^{23}_{11}Na$

C. $^{239}_{94}Pu + X \rightarrow ^{240}_{94}Pu \quad = \quad ^{239}_{94}Pu + ^{1}_{0}n \rightarrow ^{240}_{94}Pu$

Now, try the following problems and check your answers before moving on to the chapter exam.

Lesson 6–5 Review

Identify the particle represented by X and complete each of the following nuclear equations.

1. $^{8}_{4}Be + X \rightarrow ^{12}_{6}C$ 2. $^{12}_{6}C + ^{4}_{2}He \rightarrow X$ 3. $^{37}_{18}Ar + X \rightarrow ^{37}_{17}Cl$

Chapter 6 Examination

Part I—Matching

Match the following terms to the definitions that follow. Not all of the terms will be used.

a. spectator ion	f. hydrocarbon	k. double displacement
b. precipitate	g. synthesis	l. word formula
c. ionic equation	h. combustion	m. oxidation
d. net ionic equation	i. decomposition	n. reduction
e. chemical equation	j. single displacement	o. reducing agent

_____1. This type of reaction always produces water and carbon dioxide.

_____2. A compound that contains only hydrogen and carbon.

_____3. In this type of reaction, two or more simpler substances combine into a more complex one.

_____4. A solid formed as the result of an ionic reaction in an aqueous solution.

_____5. An ion that doesn't take place in a chemical reaction.

Part II—Fill in the Blank

Determine the missing substance (element or compound) in each of the following balanced equations.

6. _____ + $2O_2 \rightarrow CO_2 + 2H_2O$
7. 2_____ $\rightarrow N_2 + 3H_2$
8. Br_2 + 2_____ $\rightarrow$ 2KBr + I_2
9. 2Mg + $O_2 \rightarrow$ 2_____
10. _____ + $Na_2SO_4 \rightarrow BaSO_4 + 2NaNO_3$
11. Zn + 2_____ $\rightarrow ZnCl_2 + H_2$

Part III—Classification

Classify each of the following as one of the five main types of chemical reactions.

12. $H_2O \rightarrow 2H_2 + O_2$ ____________________
13. K + NaCl $\rightarrow$ KCl + Na ____________________
14. $2C_2H_6 + 7O_2 \rightarrow 4CO_2 + 6H_2O$ ____________________
15. $2CO + O_2 \rightarrow 2CO_2$ ____________________
16. element + compound $\rightarrow$ compound ____________________

Part IV—Balancing Equations

Balance each of the following chemical equations. If an equation is already balanced, then the correct response is "already balanced."

17. $AlPO_4 + Ca(OH)_2 \rightarrow Al(OH)_3 + Ca_3(PO_4)_2$
18. $C_2H_4 + O_2 \rightarrow CO_2 + H_2O$
19. $Fe_3O_4 + H_2 \rightarrow Fe + H_2O$
20. $Al(OH)_3 \rightarrow Al_2O_3 + H_2O$

Part V—Redox Reactions

Assign oxidation numbers to each of the elements in the following reactions.

21. $H_2 + Cl_2 \rightarrow 2HCl$

22. $AgNO_3 + Cu \rightarrow Cu(NO_3)_2 + Ag$

Part VI—Ionic Equations

For each of the following:

A. Write a word equation.

B. Write a balanced chemical equation.

C. Write an ionic equation.

D. Write a net ionic equation.

E. Identify the likely precipitate

23. Aqueous solutions of calcium nitrate and ammonium carbonate reacting.
24. Aqueous solutions of potassium bromide and silver nitrate reacting.
25. Aqueous solutions of aluminum iodide and potassium phosphate reacting.

Answer Key

The actual answers will be shown in brackets, followed by the explanation. If you don't understand an explanation that is given in this section, you may want to go back and review the lesson that the question came from.

Lesson 6–1 Review

1. [equation is balanced]—$NaOH + HCl \rightarrow H_2O + NaCl$

Elements →	Na	O	H	Cl	Na	O	H	Cl	
# of atoms →	1	1	2	1	1	1	2	1	**Balanced**

2. [not balanced]—$KOH + H_2SO_4 \rightarrow K_2SO_4 + H_2O$

Elements →	K	O	H	S	K	O	H	S	
# of atoms →	1	5	3	1	2	5	2	1	**Not balanced**

3. [equation is balanced]—$C_3H_8 + 5O_2 \rightarrow 4H_2O + 3CO_2$

Elements →	C	H	O	C	H	O	
# of atoms →	3	8	10	3	8	10	**Balanced**

4. [not balanced]—$2Al(OH)_3 \rightarrow Al_2O_3 + H_2O$

Elements →	Al	O	H	Al	O	H	
# of atoms →	2	6	6	2	4	2	**Not balanced**

5. [$2HCl + Ca(OH)_2 \rightarrow CaCl_2 + 2H_2O$]—The atomic tally showed that we only had 3 hydrogen on the reactant side, and 4 on the product side. By adding a coefficient of 2 to the hydrogen chloride, the entire equation balanced out.
6. [$2C_2H_6 + 7O_2 \rightarrow 4CO_2 + 6H_2O$]—We can only have an even number of oxygen on the reactant side, so we needed to put an even-numbered coefficient in front of the water, or we would have an odd number of oxygen on the product side.
7. [$4Fe + 3O_2 \rightarrow 2Fe_2O_3$]—Again, we needed to put an even-numbered coefficient on the product side, just to get an even number of oxygen there.
8. [$2Na + 2H_2O \rightarrow 2NaOH + H_2$]—This time, I knew that I needed to add an even-numbered coefficient to the sodium hydroxide (NaOH) in order to have an even number of hydrogen on the product side. After that, the rest falls into place.

Lesson 6–2 Review

1. [decomposition]—This reaction shows a more complex substance breaking down into two simpler substances.
2. [synthesis]—Here we see two simpler substances combining into a more complex substance.
3. [single replacement]—The chlorine replaces the bromine; think of it as if someone "cutting in" on a dancing couple.
4. [single replacement]—Here we see zinc replacing hydrogen.
5. [combustion]—When a hydrocarbon reacts with oxygen to form water and carbon dioxide, we call it combustion.
6. [double replacement]—The ions seem to switch partners, as the lead and potassium replace each other.
7. [synthesis]—The two elements, silver and chlorine, combine to form a compound.
8. [double replacement]—The hydrogen and iron switch places. When the hydrogen combines with oxygen, it forms water.
9. [decomposition]—The magnesium carbonate is breaking down into simpler substances.
10. [combustion]—Notice that the products are water and carbon dioxide.

Lesson 6–3 Review

1. [soluble]—Simply cross-reference the ions in the solubility rules.
2. [soluble]—Compounds containing halogens tend to be soluble.
3. [insoluble]—If a reaction forms zinc carbonate, it will precipitate out.

4. [soluble]—Compounds containing halogens tend to be soluble.
5. [insoluble]—Here, you see an exception to the solubility of halogens.
6. [soluble]—Hydroxides are bases.

Lesson 6–4 Review

1. [reduction]—Although the substance is gaining electrons, its charge is being reduced.
2. [oxidizing]—It is also the substance that gets reduced.
3. [oxidation]—The process of oxidation is no longer limited to processes involving oxygen.
4. [reducing]—The reducing agent gets oxidized by losing electrons.
5. [$3H_2^0 + N_2^0 \rightarrow 2N^{3+}H_3^{1-}$—Nitrogen is being oxidized, and it is also the reducing agent. Hydrogen is being reduced, and it is also the oxidizing agent.]
6. [$Fe_2^{3+}O_3^{2-} + 3C^{2+}O^{2-} \rightarrow 2Fe^0 + 3C^{4+}O_2^{2-}$—Iron is being reduced, and it is also the oxidizing agent. Carbon is being oxidized, so it is also the reducing agent.]
7. [$Zn^0 + H_2^{1+}S^{6+}O_4^{2-} \rightarrow Zn^{2+}S^{6+}O_4^{2-} + H_2^0$—Hydrogen is being reduced, so it is also the oxidizing agent. Zinc is being oxidized, so it is also the reducing agent.]
8. [$2Mg^0 + O_2^0 \rightarrow 2Mg^{2+}O^{2-}$—Oxygen is being reduced, so it is also the oxidizing agent. Magnesium is being oxidized, and it is also the reducing agent.]

Lesson 6–5 Review

1. [$^{8}_{4}Be + ^{4}_{2}He \rightarrow ^{12}_{6}C$]—The mass number on the right side of the equation was 4 greater than the mass shown on the left side, so we know the mass number of the unknown is 4 (12 – 8 = 4). We know that the atomic number of X has to be 2 (6 – 4 = 2), in order for the atomic number to be balanced on both sides of the equation. The element with an atomic number of 2 is helium.
2. [$^{12}_{6}C + ^{4}_{2}He \rightarrow ^{16}_{8}O$]—The left hand side of the equation was showing a total mass of 16 (12 + 4 = 16). The total of the atomic numbers on the left side of the equation was 8 (6 + 2 = 8). This gave us an unknown with an atomic number of 8, which could only be oxygen.
3. [$^{37}_{18}Ar + ^{0}_{-1}e \rightarrow ^{37}_{17}Cl$]—This was an interesting example. The mass number was already balanced, so X must have a mass of 0 (37 – 37 = 0). The atomic number on the left was already greater than the atomic number on the right, so we had to add a negative number to the atomic number in order to balance the nuclear charge, because 18 + (-1) = 17. X, with a mass of 0 and a charge of -1, could only be an electron.

Chapter 6 Examination

1. [h. combustion]—Perhaps the easiest way to identify a combustion reaction is to look for these products.
2. [f. hydrocarbon]
3. [g. synthesis]
4. [b. precipitate]
5. [a. spectator ion]
6. [CH_4]—Begin by looking at which elements are missing, then concentrate on how many atoms there are.
7. [NH_3]—After awhile, many of these reactions will start to look familiar to you.
8. [KI]—A single replacement reaction.
9. [MgO]—A synthesis reaction.
10. [$Ba(NO_3)_2$]—A double replacement reaction.
11. [HCl]—Acids react with certain metals to generate hydrogen gas.
12. [decomposition]—The electrolysis of water.
13. [single replacement]—Notice how potassium replaces the other metal.
14. [combustion]—The hydrocarbon in this reaction is ethane.
15. [synthesis]—The synthesis of carbon dioxide.
16. [synthesis]—Of the five main types of reactions, the synthesis reaction is the only one with a general formula that shows only one substance as a product.
17. [$2AlPO_4 + 3Ca(OH)_2 \rightarrow 2Al(OH)_3 + Ca_3(PO_4)_2$]—Start by trying to balance the calcium, then look at the phosphate and hydroxide ions.
18. [$C_2H_4 + 3O_2 \rightarrow 2CO_2 + 2H_2O$]—Start by balancing either the hydrogen or the carbon. Balance the oxygen last.
19. [$Fe_3O_4 + 4H_2 \rightarrow 3Fe + 4H_2O$]—Start by adding the coefficient to the water, in order to balance the oxygen. The rest should be easy.
20. [$2Al(OH)_3 \rightarrow Al_2O_3 + 3H_2O$]—Start by adding the 2 coefficients in front of the aluminum hydroxide, then focus on balancing the oxygen.
21. [$H_2{}^0 + Cl_2{}^0 \rightarrow 2H^{1-}Cl^{1-}$]—The reactants are both free elements, so they each get an oxidation number of 0. On the product side, the chloride is a halogen, so it must be -1. Hydrogen must be +1 to balance the chloride in a 1:1 ratio.
22. [$Ag^{1+}N^{5+}O_3{}^{2-} + Cu^0 \rightarrow Cu^{2+}(N^{5+}O_3{}^{2-})_2 + Ag^0$]—It helps to remember that the nitrate ion (NO_3^-) has a net charge of -1. Therefore, if the oxygen atoms have a combined charge of -6, that leaves a charge of +5 for the nitrogen.

23. [A. calcium nitrate + ammonium carbonate → calcium carbonate + ammonium nitrate
    B. $Ca(NO_3)_2 + (NH_4)_2CO_3 \rightarrow CaCO_3 + 2NH_4NO_3$
    C. $Ca^{2+}_{(aq)} + 2NO_3^-{}_{(aq)} + 2NH_4^+{}_{(aq)} + CO_3^{2-}{}_{(aq)} \rightarrow$ $CaCO_{3(s)} + 2NO_3^-{}_{(aq)} + 2NH_4^+{}_{(aq)}$
    D. $Ca^{2+}_{(aq)} + CO_3^{2-}{}_{(aq)} \rightarrow CaCO_{3(s)}$
    E. The precipitate is $\mathbf{CaCO_{3(s)}}$]

24. [A. potassium bromide + silver nitrate → potassium nitrate + silver bromide
    B. $KBr + AgNO_3 \rightarrow KNO_3 + AgBr$
    C. $K^+_{(aq)} + Br^-_{(aq)} + Ag^+_{(aq)} + NO_3^-{}_{(aq)} \rightarrow$ $K^+_{(aq)} + AgBr_{(s)} + NO_3^-{}_{(aq)}$
    D. $Br^-_{(aq)} + Ag^+_{(aq)} \rightarrow AgBr_{(s)}$
    E. The precipitate is $\mathbf{AgBr_{(s)}}$]

25. [A. aluminum iodide + potassium phosphate → aluminum phosphate + potassium iodide
    B. $AlI_3 + K_3PO_4 \rightarrow AlPO_4 + 3KI$
    C. $Al^{3+}_{(aq)} + 3I^-_{(aq)} + 3K^+_{(aq)} + PO_4^{3-}{}_{(aq)} \rightarrow AlPO_{4(s)} + 3I^-_{(aq)} + 3K^+_{(aq)}$
    D. $Al^{3+}_{(aq)} + PO_4^{3-}{}_{(aq)} \rightarrow AlPO_{4(s)}$
    E. The precipitate is $\mathbf{AlPO_{4(s)}}$]

Stoichiometry

Lesson 7–1: The Mole

The key to successfully learning the important skill of stoichiometry lies in developing an understanding of the concept of the ***mole***. Unfortunately, many students never fully understand the mole, and therefore, they must rely on memorization tricks to try to compensate for this lack of understanding. The truth is that the mole is really not a very hard concept to grasp, as long as it is explained carefully.

When you encounter the word *mole*, you should think of the word *dozen*, because if you can understand what a dozen is, you can understand what a mole is. A mole simply represents a set or grouping, much in the same way that a dozen represents a set of twelve and a score represents a set of twenty. It doesn't matter what the items are. Just as you can have a dozen cans of soda or a dozen doughnuts, you can have a mole of stars or a mole of water molecules. The only difference is that, because the mass of individual atoms is so small, we need to group them in a large set in order to get a significant mass. Therefore, the mole represents a much larger number than twelve.

How big of a number does the mole represent? Promise not to panic—one mole represents a set of 602,000,000,000,000,000,000,000 items! So, just as a dozen eggs represents twelve eggs, a mole of carbon atoms represents 602,000,000,000,000,000,000,000 carbon atoms. This is an incredibly large number, but I assure you that such a large number is entirely necessary when dealing with atoms. For example, the last time that you took a sip of water, you likely held more than 602,000,000,000,000,000,000,000 atoms in your mouth!

One mole of water molecules occupies about 18 ml, so 1/3 of a mole of water molecules occupies just 6 ml. As you should know by now, each molecule of water (H_2O) is made up of three atoms. Therefore, 1/3 of a mole of water molecules will contain a full mole of atoms. So 6 ml of water, which is really just a small sip, will contain about 602,000,000,000,000,000,000,000 atoms!

Perhaps you still can't picture the size of a 6 ml sample of water. Think, instead, of a 1-liter bottle of water. Let's calculate the number of moles of water molecules in 1 liter of water. One liter is equal to 1000 ml. We already said that one mole of water occupies about 18 ml. We need to divide 1000 ml by 18 ml to determine how many moles of water molecules there are in one liter of water. Let's solve the problem using the factor-label method, which you'll remember from Chapter 2.

$$1\cancel{L} \times \frac{1000\ \cancel{ml}}{1\cancel{L}} \times \frac{1\text{ mole of }H_2O}{18.0\ \cancel{ml}} = 55.6\text{ moles }H_2O$$

So, a 1-liter bottle of water contains 55.6 moles of water molecules. This represents about 33,500,000,000,000,000,000,000,000 molecules (55.6 ~~moles~~ × 602,000,000,000,000,000,000,000 molecules/~~mole~~)! Remember: Because each water molecule contains 3 atoms, one liter of water represents approximately 100,000,000,000,000,000,000,000,000 atoms (33,500,000,000,000,000,000,000,000 ~~molecules~~ × 3 atoms/~~molecule~~)! If a liter of water contains this many atoms, could we ever hope to measure the mass of a dozen, a score, or even a million atoms on our laboratory balances? The answer is no, and that is why we group them in such large sets.

This discussion should have not only convinced you of the need for having a very large number represent the mole, but also of the need to use scientific notation when dealing with such large numbers. Rather than writing 602,000,000,000,000,000,000,000 every time we have a mole of something, we will use 6.02×10^{23}. I didn't mention this earlier, because some students are more intimidated when they see the number in scientific notation. Don't let the notation scare you off. We are just writing the large number in a more convenient format, so that we don't have to write out all of the zeros. We can also use scientific notation for the calculations in the next lesson, so you should review this topic.

To add to this convenience, the number 602,000,000,000,000,000,000,000 is also given a name. It is called "Avogadro's number," in honor of Amedeo Avogadro, a famous Italian chemist. So, if you are speaking about the total number of atoms in 1 mole of water molecules, you can simply say "three times Avogadro's number," or $3 \times (6.02 \times 10^{23})$, which is 1.81×10^{4} atoms.

So now you are convinced that we need a really large number to represent a group of atoms, and you are convinced that we are better off using scientific notation when dealing with these large numbers, but are you still wondering why we use that specific large number? The masses of the elements on the periodic table were calculated using carbon-12 as the base or standard. The mass of one atom of that particular isotope of carbon was set at 12 atomic mass units (u). There are 6.02×10^{23} atomic mass units (u) in one gram. This conversion allows us to use the masses listed on the periodic table for both atomic mass and molar mass. Atomic mass, which is the mass of one atom, is measured in atomic mass units (u). Molar mass, which is the mass of one mole of atoms or molecules, is measured in grams. We can use the periodic table to find the number that goes with the units for each of these quantities.

Atomic and Molar Masses

Element and Symbol	Atomic Mass—Mass of 1 Atom	Molar Mass—Mass of 6.02×10^{23} Atoms
Carbon (C)	12.0 u	12.0 g
Helium (He)	4.00 u	4.00 g
Copper (Cu)	63.5 u	63.5 g
Potassium (K)	39.1 u	39.1 g

This is the point where some students might say, "Hey, I thought that the mole is always 6.02×10^{23}. How can the molar masses of carbon and helium be different?" Remember: A mole always represents 6.02×10^{23} items, but the mass of these items can certainly be different. You wouldn't expect a dozen eggs to have the same mass as a dozen cars. It is the number of items in a mole that is always the same, not the mass, weight, or volume of those items.

The periodic table can also be used to calculate the molar mass of molecules and formula units as well. If you can add up the mass of all of the atoms in a molecule to find the molecular mass in atomic mass units, the molar mass of the same molecular compound will have the same value with the unit, grams (g). Following are some examples:

Molecular and Molar Masses		
Compound Name and Molecular Formula	**Molecular Mass (mass of one molecule)**	**Molar Mass (mass of 6.02×10^{23} molecules)**
Water (H_2O)	18.0 u	18.0 g
Carbon Dioxide (CO_2)	44.0 u	44.0 g
Glucose ($C_6H_{12}O_6$)	180 u	180 g

Now you know what the mole is and why it is so important to chemistry. If you find yourself getting confused in the future, just think about the "dozen." If you don't understand what to do when someone asks you to "find the mass of 2.0 moles of carbon dioxide molecules," just substitute the word *dozen* for *mole*. If someone asked you to find the mass of 2.0 dozen carbon dioxide molecules, you would simply determine the mass of one molecule of carbon dioxide and multiply by 24 (2 dozen = 2 × 12 = 24). If you need to find the mass of 2.0 moles of carbon dioxide molecules, you could simply determine the mass of one molecule of carbon dioxide and multiply by 2 × Avogadro's number ($2 \times (6.02 \times 10^{23})$). Look at these two solutions shown with the factor-label method here:

Example 1

Determine the mass of 2 dozen carbon dioxide molecules.

$$\frac{44.0\ \text{u}}{\cancel{\text{molecule}}} \times \frac{12\ \cancel{\text{molecule}}}{\cancel{\text{doz}}} \times \frac{2\ \cancel{\text{doz}}}{} = 1056\ \text{u} = 1060\ \text{u}$$

That should seem easy enough. Now, apply the same technique to finding the mass of 2 moles of carbon atoms.

Example 2

Determine the mass of 2 moles of carbon dioxide molecules.

$$\frac{44.0\ \text{u}}{\cancel{\text{molecule}}} \times \frac{6.02 \times 10^{23}\ \cancel{\text{molecule}}}{\cancel{\text{mole}}} \times \frac{2\ \cancel{\text{mole}}}{} = 5.30 \times 10^{25}\ \text{u}$$

This number is so large that you can see why it is preferable to change atomic mass units to grams when dealing with molar mass. Example 3 will show the same calculation with an added conversion to change atomic mass units to grams.

Example 3

Determine the mass (in grams) of 2.00 moles of carbon dioxide molecules.

$$\frac{44.0\ \cancel{u}}{\cancel{\text{molecule}}} \times \frac{6.02 \times 10^{23}\ \cancel{\text{molecule}}}{\cancel{\text{mole}}} \times \frac{2\ \cancel{\text{mole}}}{} \times \frac{1\ \text{g}}{6.02 \times 10^{23}\ \cancel{u}} = 88.0\ \text{g}$$

Did you notice that we multiplied and divided by Avogadro's number? These two steps cancel each other out! Perhaps you now see the beauty of Avogadro's number. Using the number of atomic mass units in one gram eliminates our need to make this conversion every time we do one of these calculations. Example 4 shows the same calculation, except it starts with the molar mass of carbon dioxide, which has the same value as the molecular mass, except with different units.

Example 4

Determine the mass (in grams) of 2.00 moles of carbon dioxide molecules.

$$\frac{44.0\ \text{g}}{\cancel{\text{mole}}} \times \frac{2\ \cancel{\text{mole}}}{} = 88.0\ \text{g}$$

And there you have it. When you need to find the mass of a certain number of moles of molecules, simply multiply by the molar mass. Practice with the review questions here and then move on to the next lesson for more mole calculations.

Lesson 7–1 Review

Refer to the periodic table to help answer the following questions.

1. Find the molar mass of LiOH
2. Find the molar mass of $Cu(NO_2)_2$
3. Find the molar mass of $(NH_4)_2S$

Pay careful attention to the wording of the following questions.

4. How many **molecules** does one mole of water (H_2O) represent?
5. How many **atoms** does one mole of water (H_2O) molecules represent?
6. How many **molecules** do four moles of CH_4 represent?
7. How many **atoms** do four moles of CH_4 molecules represent?

Lesson 7–2: Molar Conversions

Now that you understand the mole, this lesson will show you how you can use this knowledge to analyze real-life samples in the laboratory. For example, you can figure out how many atoms you have in a sample of aluminum that you mass or in a volume of hydrogen gas you collect in the lab. Some students resort to "tricks" or mechanical aids, such as the "mole map" that is common in many classrooms. Although these aids, and the mole map in particular, can be very helpful, they stress mechanical problem solving over logical problem solving. I would caution you not to try to replace understanding with these aids. Rather, try to truly understand each question and solve it logically. "Tricks" tend to have very limited applications, but good problem-solving techniques will serve you well in unlimited ways.

Changing the Mass to Number of Moles

Suppose your instructor gave you a piece of pure aluminum foil and asked you to determine how many atoms were in the sample. Would you know how to go about solving the problem? What information do you have to start with? What other information could you easily determine in your laboratory setting?

To start with, you could find the mass of your aluminum by placing it on your laboratory balance. Let's say, for the sake of argument, that the mass of your sample was 60.0 g. You could then find the molar mass of aluminum by simply looking at the periodic table. The table tells us that the atomic mass of aluminum is 27.0 u/atom. Therefore the molar mass of aluminum is 27.0 g/mole (as explained in Lesson 7–1).

Let's summarize what we know:

The mass of the sample of aluminum = 60.0 g

The molar mass of aluminum is 27.0 g/mole

Do we have more than one mole of aluminum or less than one mole of aluminum? Remember: One mole of aluminum has a mass of 27.0 grams, and you have 60.0 grams. So clearly you have more than one mole of aluminum. Exactly how many moles of aluminum do you have?

If you don't see how to solve this right away, try thinking in terms of money. If you went to a store with \$60.00 and you wanted to buy shirts that cost \$27.00, how many shirts could you buy? You probably figured that (assuming there isn't too much tax) you could buy two shirts. Analyze how you solved the problem. If you had to write an equation for figuring out how many shirts you could buy, it might look like the following:

$$\#\text{ of shirts you can buy} = \frac{\text{total money you have}}{\text{cost of 1 shirt}} = \frac{\cancel{\$}60.00}{\cancel{\$}27.00/\text{shirt}} = 2.22 \text{ shirts}$$

Of course, we didn't include tax, which varies from place to place, and the store probably won't sell you a fraction of a shirt. Moles, however, do come in fractions. Using this model, let's see how we can solve our original mole problem.

Example 1

How many moles does 60.0 g of aluminum represent?

$$\#\text{ of moles of a substance} = \frac{\text{total mass of the sample}}{\text{molar mass of substance}} = \frac{60.0 \text{ g}}{27.0 \text{ g/mole}} = 2.22 \text{ moles}$$

As you can see, making this type of conversion is quite easy when dealing with elements. Dealing with compounds is only slightly harder, as illustrated in our next example.

Suppose you need to find the number of moles represented by 80.0 grams of water. In this problem, the only additional step would be adding up the molar masses (you can review this topic in Chapter 5) of the elements that make up the water, as shown here:

The molar mass of water (H_2O) is:

Hydrogen (H_2) = 1.01 g/mole × 2 mole = 2.02 g

Oxygen (O) = 16.0 g/mole × 1 mole = + 16.0 g

Molar mass of water = **18.0 g**

Now, we can go about solving the problem in the same way we solved Example 1.

Example 2

How many moles does 80.0 g of water represent?

$$\#\text{ of moles of a substance} = \frac{\text{total mass of the sample}}{\text{molar mass of substance}} = \frac{80.0 \cancel{\text{g}}}{18.0 \cancel{\text{g}}/\text{mole}} = 4.44 \text{ moles}$$

Changing the Number of Moles to Mass

Now, let's look at a slightly different scenario. Let's suppose your instructor told you that you would need 0.750 moles of $CuCl_2$ to begin a reaction in the lab. How would you figure out how many grams of $CuCl_2$ to start with? First, you would determine the molar mass of $CuCl_2$, as shown here:

Copper (Cu) = 63.5 g/mole × 1 mole = 63.5 g

Chlorine (Cl_2) = 35.5 g/mole × 2 mole = + 71.0 g

Molar mass of $CuCl_2$ = **134.5 g**

Next, take the formula that we used for Example 1 and algebraically isolate the unknown for this problem, which is the number of grams.

Starting with:

$$\text{\# of moles of a substance} = \frac{\text{total mass of the sample}}{\text{molar mass of substance}}$$

Multiply both sides by "molar mass of substance" to get:

Total mass of the sample = # of moles of a substance × molar mass of substance

You now have everything you need to solve the problem.

Example 3

Convert 0.750 moles of $CuCl_2$ to mass in grams.

Total mass of the sample = # of moles of a substance × molar mass of substance

Total mass of $CuCl_2$ = 0.750 ~~mole~~ × 134.5 g/~~mole~~ = <u>101 g of $CuCl_2$</u>

Note that 0.750 mole is less than 1 mole, so it makes sense that the mass that we need is less than the molar mass of $CuCl_2$.

Changing the Number of Moles to the Number of Particles

Let's say your instructor wants you to calculate the number of molecules found in 3.0 moles of carbon dioxide. If you get confused, remember to go back to thinking about the "dozen." If you were asked to figure out how many eggs there are in 3.0 dozen, you would instantly know enough to multiply 3.0 × 12, because 1 dozen equals 12 items. The mole, as is the dozen, is simply a set or a group of items. All you would need to do is multiply the number of moles times Avogadro's number. Look how similar these two calculations actually are.

Example 4

How many eggs does 3.0 dozen eggs represent?

$$\text{\# of eggs} = 3.0\ \cancel{\text{doz}} \times \frac{12}{1\ \cancel{\text{doz}}} = 36 \text{ eggs}$$

Example 5

How many molecules does 3.00 moles of carbon dioxide represent?

$$\text{\# of molecules} = 3.00\ \cancel{\text{moles}} \times \frac{6.02 \times 10^{23}\ \text{molecule}}{\cancel{\text{mole}}} = 1.81 \times 10^{23}\ \text{molecules of } CO_2$$

Other than the fact that the answer to Example 5 is large, and in scientific notation, it is just as simple as Example 4!

Let's suppose the question from Example 5 was worded somewhat differently. How would you find the number of **atoms** in 3.0 moles of carbon dioxide? To answer this question, you must understand what it is asking. This question is asking for "parts of the whole," the "whole" being molecules and the "parts" being atoms. Each molecule of carbon dioxide (CO_2) is made up of three atoms, 1 carbon and 2 oxygen. You would find out the number of molecules, and then multiply by 3 atoms/molecule to get your final answer. This question can be likened to asking, "How many tires (not counting spares) are there in 3 dozen cars?" Let's answer both of these questions.

Example 6

How many tires (not counting spares) are there in 3 dozen cars?

$$\text{\# of tires} = 3.0\ \cancel{\text{doz}}\ \cancel{\text{car}} \times \frac{12}{1\ \cancel{\text{doz}}} \times \frac{4\ \text{tires}}{\cancel{\text{car}}} = 144\ \text{tires}$$

Example 7

How many atoms does 3.0 moles of carbon dioxide represent?

$$\text{\# of atoms} = 3.0\ \cancel{\text{moles}} \times \frac{6.02 \times 10^{23}\ \cancel{\text{molecules}}}{\cancel{\text{mole}}} \times \frac{3\ \text{atoms}}{\cancel{\text{molecule}}} = 5.4 \times 10^{24}\ \text{atoms}$$

Changing the Number of Particles to the Number of Moles

Suppose a particular chemical reaction results in the production of 1.5×10^{23} molecules of hydrogen (H_2) and you want to determine the number of moles this represents. The solution to this problem may seem obvious to you, but if not, think again about a dozen. If you had six doughnuts, how many dozen doughnuts would you have? Dividing 6 by 12 you would find that you have 0.5 (or 1/2) a dozen doughnuts. We would solve the mole problem the same way!

Example 8

How many moles does 1.5 x 10^{23} molecules of hydrogen represent?

$$\# \text{ of moles} = \frac{\# \text{ of particles in sample}}{\# \text{ of particles in a mole}} = \frac{1.5 \times 10^{23} \text{ \sout{molecules}}}{6.02 \times 10^{23} \text{ \sout{molecules}/mole}} = 0.25 \text{ moles}$$

Changing the Volume of a Gas at STP to Number of Moles

Avogadro's principle tells us that "equal volumes of gases, at equal temperatures and pressures, contain an equal number of particles." It is from this principle that chemists have developed the concept of the "molar volume" of gases. It has been determined that one mole of any gas at standard temperature and pressure (a temperature of 273 K and 101.3 kPa of pressure) will occupy 22.4 dm^3 of volume. This allows us to determine the number of moles in a gas, provided we know the volume, temperature, and pressure of the sample.

It is very important to remember that this type of calculation will only work with gaseous samples. Don't try to use the molar volume of a gas for determining the number of moles in a liquid or solid sample.

Let's start with an easy example of how the molar volume of a gas can be used in a calculation. Imagine that a student generates a sample of hydrogen gas in the laboratory. She could use the Combined Gas Law, which we will cover in Chapter 8, to adjust the volume of her gas sample to standard atmospheric temperature and pressure (STP). If she concludes that the sample of hydrogen would occupy 15.7 dm^3 at STP, how many moles of hydrogen gas did she generate? Note that the specific type of gas has no effect on solving the problem. The molar volume of any gas at STP is the same, so we would get the same answer for any type of gas.

Example 9

How many moles does 15.7 dm^3 of gas at STP represent?

$$\# \text{ of moles} = \frac{\text{volume of gas sample at STP}}{\text{molar volume of gas at STP}} = \frac{15.7 \text{ \sout{dm}}^{\sout{3}}}{22.4 \text{ \sout{dm}}^{\sout{3}}\text{/mole}} = 0.701 \text{ moles}$$

That should have been easy enough, so I am going to risk a harder problem, which will combine molar conversions with some math that you will learn in Chapter 8. Although we learn how to do each of these problems in isolation, in real-life situations, you are often called upon to combine different formulas to solve a problem. If you haven't studied enough chemistry to do these calculations yet, don't get frustrated. Simply skip them for now, and come back to them after you have studied the gas laws.

Here we have a student who collects a 2.30 dm^3 sample of carbon dioxide with a pressure of 92.7 kPa and a temperature of 17.0 °C. If he wants to determine the number of molecules of carbon dioxide that his sample contains, what would he need to do?

First, he would have to convert °C to Kelvin, because the gas law calculations aren't based on the Celsius scale.

Step 1—Converting Celsius to Kelvin.

$$K = {}^{\circ}C + 273 = 17.0\ {}^{\circ}C + 273 = 290\ K$$

Next, he would use the Combined Gas Law (review Lesson 8–5) to adjust the volume of his gas to standard temperature and pressure.

Step 2—Adjusting the volume of the gas at room temperature to STP.

$$V_2 = \frac{V_1P_1T_2}{P_2T_1} = \frac{(2.30\ dm^3)(92.7\ \cancel{kPa})(273\ \cancel{K})}{(101.3\ \cancel{kPa})(290\ \cancel{K})} = 1.98\ dm^3$$

Now, he would use the molar volume of a gas to determine the number of moles of gas he collected.

Step 3—Determining the number of moles represented by 1.98 dm^3 of a gas at STP.

$$\#\text{ of moles} = \frac{\text{volume of gas sample at STP}}{\text{molar volume of gas at STP}} = \frac{1.98\ dm^3}{22.4\ dm^3/\text{mole}} = 0.088\text{ moles}$$

If you still remember the original question, he wanted to know how many molecules of carbon dioxide he generated. Therefore, his last move would be to convert moles to number of particles, by multiplying by Avogadro's number.

Step 4—Determining the number of molecules in 0.088 moles.

$$\#\text{ of molecules} = 0.088\ \cancel{\text{moles}} \times \frac{6.02 \times 10^{23}\text{ molecules}}{\cancel{\text{mole}}} = 5.3 \times 10^{22}\text{ molecules}$$

And we have our answer. Imagine that! Chemistry gives us the ability to collect an invisible gas and count the microscopic particles that are too small for us to even see!

Perhaps you thought of another way to solve the same problem, using one less step. One of the most enjoyable things about problem solving in the sciences is that there are often a number of paths and formulas that you can follow and still get to the correct answer. If you are the type of person who enjoys problem solving, then getting there is half the fun. Let's take the same original problem, but solve it using the Ideal Gas Law.

This path is actually shorter and easier, but for a beginner, it may be a little harder to understand when you see it at first. If it seems confusing, just slow down and read it a few times. We will be using the Ideal Gas Law from Lesson 8–7, so you may want to review it.

The original problem was this: A student collected a 2.30 dm^3 sample of carbon dioxide with a pressure of 92.7 kPa and a temperature of 17.0 °C. If he wants to determine the number of molecules of carbon dioxide that his sample contains, what would he need to do?

The first step is going to be the same, because he needs to convert °C to Kelvin.

Step 1—Converting Celsius to Kelvin.

$$K = {}^{\circ}C + 273 = 17.0\ {}^{\circ}C + 273 = 290\ K$$

Now, instead of using the Combined Gas Law, he can use the Ideal Gas Law to find out how many moles of carbon dioxide he has.

Step 2—Determining the number of moles of carbon dioxide he has.

Given: Pressure (P) = 92.7 kPa Volume (V) = 2.30 dm^3
Temperature (T) = 290 K Constant (R) = 8.31 $dm^3 \times kPa/moles \times K$

Find: Number of moles (n)

Formula: $PV = nRT$

Isolate: $\frac{PV}{RT} = \frac{n\cancel{RT}}{\cancel{RT}}$ so, $n = \frac{PV}{RT}$

Solve: $n = \frac{PV}{RT} = \frac{(92.7\ \cancel{kPa})(2.30\ dm^3)}{(8.31\ dm^3 \times \cancel{kPa}/moles \times \cancel{K})(290\ \cancel{K})} = 0.088\ moles$

Notice that our answer for the number of moles is the same as the answer that we got after Step 3, using the previous method. We still have one more step, and that is to convert the number of moles to the number of atoms of carbon dioxide. Again, this step will be identical to Step 4 in our previous method.

Step 3—Determining the number of molecules in 0.088 moles.

$$\#\ of\ molecules = 0.088\ \cancel{moles} \times \frac{6.02 \times 10^{23}\ molecules}{\cancel{mole}} = 5.3 \times 10^{22}\ molecules$$

And we arrive at the same answer as last time, only this time we did it with one less step.

Changing the Number of Moles of a Gas to Volume at STP

There are times when you might want to know the size of a potential sample of gas before you carry out an experiment. For example, if you were going to react a metal with an acid, you can collect the hydrogen gas that will be generated. In order to figure out how big a vessel must be to hold the gas, you can calculate the size of the gas, at room pressure, ahead of time. Let's do an example where you need to calculate the volume of a 0.0023 mole sample of a gas. Again, the specific type of gas does not factor into the problem, because the molar volume of any gas at STP is 22.4 dm^3/mole.

Example 10

Convert 0.0023 moles of a gas to volume at STP.

$$\text{Volume of gas at STP} = 0.0023 \text{ moles} \times \frac{22.4 \text{ dm}^3}{\text{mole}} = 0.052 \text{ dm}^3$$

Notice that our answer, although correct, is exactly the type of answer that makes an inexperienced chemistry student nervous. They might be more comfortable with an incorrect answer that was greater than 1. Don't doubt your answer, just because it is less than 1. Obviously, it is possible to collect less than 1 dm^3 of a gas, and the fact that you started out with much less than 1 mole of a gas should suggest to you that the volume should also be low.

Combination Problems

In addition to the basic mole conversion problems that we have gone over, there are all sorts of problems that combine these conversions in multiple steps. As I mentioned earlier, we learn to do each of these problems in isolation, for the sake of easy learning. However, real life and/or your instructor will make it necessary to mix and match the formulas that you learn to solve more complex situations. Don't be afraid of these challenges! Rather, treat each problem like a puzzle, and derive satisfaction from selecting the correct set of formulas to go with each situation.

The key to being able to handle any problem that comes your way is to develop good problem-solving strategies. Avoid falling into the habit of mechanically solving problems. Rather, approach each problem logically and make sure that you use common sense. Stop and think about what you are doing in each step. For example, would it make sense to try and divide the mass of a sample by the molar volume of a gas, if you are trying to solve for number of moles? If I told you that John was 36 years old and Mike was 6 feet tall, does that make John six times as old as Mike? We can't divide quantities with different units, yet this is just the type of mistake that mechanical problem solvers will make.

Now, practice the following problems until you feel comfortable with mole conversions, and then move on to the next lesson.

Lesson 7–2 Review

1. Find the volume that 4.50 moles of nitrogen gas will occupy at STP.
2. Calculate the mass of 2.50 moles of NaBr.
3. How many moles of carbon would 50.0 g of carbon represent?
4. How many molecules does 1.88 moles of water represent?
5. How many moles of helium gas would occupy 75.0 dm^3 at STP?
6. How many moles of molecules would 1.81×10^{24} molecules of glucose represent?
7. How many molecules would 45.0 g of water represent?
8. How much space would 85.0 g of neon gas occupy at STP?
9. How many molecules of carbon dioxide gas would occupy 67.5 dm^3 at STP?
10. What would be the mass of 9.65×10^{25} molecules of water?

Lesson 7–3: Mole-Mole Problems

Mole-mole problems are sort of like "introductory," or "skill-building," problems that will help you practice using the molar ratios given by balanced chemical reactions. The harder stoichiometry problems, which we will begin in the next lesson, all make use of mole-mole problems as a step in the problem-solving process. This lesson will give you an opportunity to become comfortable with the molar ratio without worrying about more complex problems at the same time.

Look at the following problem.

Example 1

How many moles of oxygen (O_2) are required to react completely with 3 moles of glucose ($C_6H_{12}O_6$) according to the balanced chemical equation shown here?

$$C_6H_{12}O_6 + 6O_2 \rightarrow 6H_2O + 6CO_2$$

glucose + oxygen → water + carbon dioxide

The trick to solve this problem is to look at the coefficients in front of each of the substances in the reaction. These coefficients give you the molar ratios that you need. For example, the fact that both water and carbon dioxide have a "6" in front means that the same number of moles of each will be generated as the products of this reaction. The ratio 6:6 can be reduced to 1:1, so if you produced 1 mole of water, you would produce 1 mole of carbon dioxide. If you produced 1.56 moles of water, you would produce 1.56 moles of carbon dioxide.

This question isn't about the products of the reaction, so I will cross them out. It asks about the reactants, the glucose and oxygen. The coefficients in front of these substances tell us that the ratio of glucose to oxygen is 1:6. This means that if we wanted to react 1 mole of glucose, we need 6 moles of oxygen. If we wanted to react 2 moles of glucose, we would need 12 moles of oxygen. If we want to react 3 moles of glucose, we would need 18 moles of oxygen. Look how we can use a simple ratio to solve these problems.

$$C_6H_{12}O_6 + 6O_2 \rightarrow 6H_2O + 6CO_2$$

glucose + oxygen → water + carbon dioxide

$$\frac{\text{coefficients} \rightarrow}{\text{\# of moles} \rightarrow} \quad \overset{\text{glucose}}{\frac{1}{3}} = \overset{\text{oxygen}}{\frac{6}{x}}$$

Now, cross-multiplying, we get $x = 3 \times 6 = 18$. So, it takes 18 moles of oxygen to react completely with 3 moles of glucose.

Let's use the same chemical reaction and figure out how many moles of oxygen it would take to react completely with 0.25 moles of glucose. You may not be ready to do this example in your head, but it is easy enough when you use the ratio.

Example 2

How many moles of oxygen (O_2) are required to react completely with 0.25 moles of glucose ($C_6H_{12}O_6$) according to the balanced chemical equation shown here?

$$C_6H_{12}O_6 + 6O_2 \rightarrow 6H_2O + 6CO_2$$

glucose + oxygen → water + carbon dioxide

$$\frac{\text{coefficients} \rightarrow}{\text{\# of moles} \rightarrow} \quad \overset{\text{glucose}}{\frac{1}{0.25}} = \overset{\text{oxygen}}{\frac{6}{x}}$$

Now, cross-multiplying, we get x = 0.25 × 6 = 1.5. So, it takes 1.5 moles of oxygen to react completely with 0.25 moles of glucose.

For our last example, let's try another chemical reaction.

Example 3

How many moles of water (H_2O) can be produced from the combustion of 5 moles of methane (CH_4) according to the balanced chemical reaction shown here?

$$CH_4 + O_2 \rightarrow 2H_2O + CO_2$$

methane + oxygen → water + carbon dioxide

Here, the molar ratio given by the coefficient of methane and water is 1:2. This means that for every 1 mole of methane you burn, you produce 2 moles of water. You probably can see right away that if you burn 5 moles of methane you will get 10 moles of water, but I will go through the formality of showing the ratio, just in case.

$$CH_4 + O_2 \rightarrow 2H_2O + CO_2$$

methane + oxygen → water + carbon dioxide

$$\begin{array}{ccc} & \text{methane} & \text{water} \\ \dfrac{\text{coefficients} \rightarrow}{\text{\# of moles} \rightarrow} & \dfrac{1}{5} & = \dfrac{2}{x} \end{array}$$

Now, cross-multiplying, we get x = 5 × 2 = 10. So, the combustion of 5 moles of methane will produce 10 moles of water.

Practice the following problems before moving on to Lesson 7–4.

Lesson 7–3 Review

Base your answers to questions 1–5 on the following balanced chemical equation.

$$C_3H_8 + 5O_2 \rightarrow 4H_2O + 3CO_2$$

propane + oxygen → water + carbon dioxide

1. How many moles of propane would be required to react completely with 10 moles of oxygen?
2. How many moles of oxygen would be required to react completely with 3 moles of propane?
3. How many moles of water would be produced if 3 moles of propane react with an excess of oxygen?

4. How many moles of propane must react completely to generate 30 moles of carbon dioxide?
5. How many moles of oxygen must react with an excess of propane to generate 8 moles of water?

Base your answers to questions 6–10 on the following balanced chemical equation:

$$3H_2 + N_2 \rightarrow 2NH_3$$

hydrogen + nitrogen → ammonia

6. How many moles of nitrogen would be required to react completely with 9 moles of hydrogen?
7. How many moles of hydrogen would be required to react completely with 5 moles of nitrogen?
8. How many moles of hydrogen would be required to produce 1 mole of ammonia?
9. How many moles of nitrogen would be required to produce 10 moles of ammonia?
10. How many moles of ammonia could be generated if you started with 9 moles of hydrogen and 2 moles of nitrogen?

Lesson 7–4: Mass-Mass Problems

Now you are ready to try your first complete stoichiometry problems, where you quantitatively analyze chemical reactions. The mass-mass problem is where you either know the mass of the product that you want to produce and calculate the mass of the reactant(s) you start with, or you know the mass of the reactant(s) you start with and calculate the mass of the product you will end up with. As with mole conversion problems, there are a variety of these types of problems, with a varying range of difficulty. We will start off with some of the easier types and work our way up to harder problems.

Magnesium is a metal that emits a blinding white light when it burns in the presence of oxygen. The sparklers that you see on the Fourth of July and the road flares that you see at the scene of an accident contain magnesium. In the chemistry lab, you may have seen magnesium in the form of a metallic strip or ribbon. Let's begin with a problem where you want to calculate the amount of magnesium oxide you will produce when you burn 2.00 g of pure magnesium ribbon in an excess of oxygen. When we say an "excess" of oxygen, it means that there is plenty of it. The reaction won't stop when we run out of oxygen (because we won't!); it stops when we run out of pure magnesium. In such a case, the reactant that will get completely used up is called the ***limiting reactant***, because it limits the amount of product that will be produced.

Some of the examples that we will go over in this lesson will require you to know how to balance and identify the types of chemical reactions discussed in Chapter 6. Go back and review these reactions if you find that you need to.

Example 1

How many grams of magnesium oxide (MgO) are formed when 2.00 g of magnesium (Mg) react with an excess of oxygen, according to the balanced chemical reaction shown here?

$$2MgO + O_2 \rightarrow 2MgO$$

Where do you begin? As in all examples of problem solving, start by reading the question carefully and then taking the important information out of the question. I recommend writing the appropriate information below the equation, so that you don't have to keep referring back to the written problem. The following shows the original problem with the important information italicized. Notice how we should label the problem.

How many grams of magnesium oxide (MgO) are formed when *2.00 g of magnesium (Mg)* react with an *excess of oxygen*, according to the balanced chemical reaction shown here?

$$2MgO + \cancel{O_2} \rightarrow 2MgO$$

2.00 g   excess   x g

Because the problem says, "How many grams of magnesium oxide," I know that the magnesium oxide is the unknown, which I must solve for. Therefore, I put "x g" under the magnesium oxide, indicating that I need to find this mass. The oxygen, which is the excess reactant, plays no part in the calculation. Therefore, I crossed it out, so that I will remember to ignore it. I started with 2.00 grams of magnesium, my limiting reactant, so I put 2.00 g under it to remind me of what I have.

Now, I know that the balanced chemical reaction is based on a molar ratio, not a mass ratio, so I will need to change 2.00 grams of magnesium into moles of magnesium using the calculation from Lesson 7–2. I divide the mass of my sample by the molar mass of magnesium (24.3 g/mole), which I can find on the periodic table.

$$2MgO + \Theta_2 \rightarrow 2MgO$$
$$2.00\text{ g} \quad \text{excess} \quad x\text{ g}$$

$$\#\text{ of moles of magnesium} = \frac{\text{mass of sample}}{\text{molar mass}} = \frac{2.00\text{ g}}{24.3\text{ g/mole}} = 0.0823\text{ moles of Mg}$$

So, we start with 0.0823 moles of magnesium and an excess of oxygen, and we want to find out the mass of the magnesium oxide produced. The next step is to compare the molar ratio shown by the coefficients in the balanced chemical reaction. Do you notice that both the magnesium and the magnesium oxide are preceded by coefficients of 2? This means that for every 2 moles of magnesium that you burn, you will produce 2 moles of magnesium oxide, provided you have an excess of oxygen. In other words, the molar ratio of Mg to MgO is 2:2 or 1:1. This tells us that if we put 0.0823 moles of Mg into the reaction, we will get 0.0823 moles of MgO out of the reaction.

Now that we know how many moles of MgO we will produce, we must convert that to grams by multiplying the number of moles of MgO (0.0823 moles) by the molar mass of MgO (24.3 g/mole + 16.0 g/mole = 40.3 g/mole), which we get by adding the molar mass of magnesium and oxygen.

$$2MgO + \Theta_2 \rightarrow 2MgO$$
$$2.00\text{ g} \quad \text{excess} \quad x\text{g}$$

$$\#\text{ of moles of magnesium} = \frac{\text{mass of sample}}{\text{molar mass}} = \frac{2.00\text{ g}}{24.3\text{ g/mole}} = 0.0823\text{ moles of Mg}$$

$$\text{Mass of MgO} = \#\text{ of moles} \times \text{molar mass} = 0.0823\text{ moles} \times 40.3\text{ g/mole} = 3.32\text{ g}$$

So, 3.32 g of MgO are produced when 2.00 g of Mg react with an excess of oxygen.

In summary, the steps for solving this type of problem are:

1. Begin with a balanced chemical reaction.
2. Carefully label the equation with the information from the word problem.
3. Convert the given quantity to number of moles.
4. Use the molar ratio shown by the coefficients to determine the number of moles of the unknown.
5. Convert this number of moles to the quantity (grams, in the case of this example) the question asks for.

Not all of these problems will deal with one product and one reactant. Sometimes, everything that you need to work with is on the same side of the reaction arrow. Let's try another problem of this type.

Example 2

How many grams of water would be required to react completely with 3.25 grams of sodium in a reaction that produces only hydrogen gas and sodium hydroxide?

Well, you weren't given a balanced chemical reaction this time, so that is the place to start. You do have enough information to produce the following word equation (see Lesson 6–3).

sodium + water → hydrogen and sodium hydroxide

You can then use your knowledge of elemental symbols (Lesson 3–2), writing chemical formulas (Lesson 5–2), and balancing chemical reactions (Lesson 6–1) to turn the word equation into a balanced chemical reaction.

water + sodium → hydrogen + sodium hydroxide

becomes

$$2H_2O + 2Na \rightarrow H_2 + 2NaOH$$

Now, let's go back to the original problem and use the given information to label the equation.

How many grams of water would be required to react completely with *3.25 grams of sodium* in a reaction that produces only hydrogen gas and sodium hydroxide?

$$\underset{x\text{ g}}{2H_2O} + \underset{3.25\text{ g}}{2Na} \rightarrow \cancel{H_2} + 2\cancel{NaOH}$$

Notice that you were neither given, nor asked for, a quantity of hydrogen or sodium hydroxide. These were mentioned only for the sake of producing a balanced chemical reaction. Therefore, you can treat them in the same way that we treated oxygen, the excess reactant, in the last problem.

Now, you need to convert 3.25 grams of sodium into moles by dividing by the molar mass of sodium (23.0 g/mole), which you get from the periodic table.

$$\underset{x\text{ g}}{2H_2O} + \underset{3.25\text{ g}}{2Na} \rightarrow \cancel{H_2} + 2\cancel{NaOH}$$

$$\#\text{ of moles of Na} = \frac{\text{mass of sample}}{\text{molar mass}} = \frac{3.25\ \cancel{\text{g}}}{23.0\ \cancel{\text{g}}/\text{mole}} = 0.141\text{ moles}$$

Once you know that you are starting with 0.141 moles of sodium, you use the molar ratio given by the coefficients to determine the number of moles of water you will need to react completely with the sodium. Again, the ratio between the two substances in question is 2:2, which simplifies to 1:1. This means that it takes 1 mole of water to react completely with 1 mole of sodium. It would take 0.141 moles of water to react completely with 0.141 moles of sodium.

Then, you only have to multiply the number of moles of water with the molar mass of water (18.0 g/mole) to find the mass of water required.

$$2H_2O + 2Na \rightarrow \cancel{H_2} + \cancel{2NaOH}$$

x g   3.25 g

$$\text{\# of moles of Na} = \frac{\text{mass of sample}}{\text{molar mass}} = \frac{3.25\ \cancel{g}}{23.0\ \cancel{g}/\text{mole}} = 0.141 \text{ moles}$$

$$\text{Mass of water} = \text{\# of moles} \times \text{molar mass} = 0.141\ \cancel{\text{moles}} \times 18.0\ \text{g}/\cancel{\text{mole}} = 2.54\ \text{g}$$

So, it would take 2.54 g of water to react completely with 3.25 g of sodium.

I don't want to give you the impression that the molar ratio given by the coefficients will always be 1:1, so let's try another example, where the ratio is not 1:1.

Example 3

How many grams of water are produced as 3.00 g of carbon dioxide are generated from the combustion of methane (CH_4)?

This question didn't give you a balanced chemical reaction to start with. You must use your knowledge of combustion reactions to fill in the information. As you surely know by now, having studied Lesson 6–2, the typical combustion reaction takes on the following form:

hydrocarbon + oxygen → water and carbon dioxide

That makes the word equation for this specific combustion reaction:

methane + oxygen → water and carbon dioxide

Use your knowledge of chemical formulas to change the word equation into the following balanced equation:

$$CH_4 + 2O_2 \rightarrow 2H_2O + CO_2$$

Now let's insert this equation into the problem and label it based on the information given.

How many grams of water are produced as *3.00 g of carbon dioxide* are generated from the combustion of methane (CH_4)?

$$\cancel{CH_4} + \cancel{2O_2} \rightarrow 2H_2O + CO_2$$
$$\text{x g} \quad 3.00\text{ g}$$

This problem only gives, or asks for, specific information about the two products of the reaction. The reactants are important for the purpose of balancing the reaction, because that is how you determine the molar ratio between the products, but they serve no purpose after that point and have been crossed out.

Now, you need to convert the mass of carbon dioxide to the number of moles. For that, use the periodic table to find the molar mass (12.0 g + 32.0 g = 44.0 g) of carbon dioxide.

$$\cancel{CH_4} + \cancel{2O_2} \rightarrow 2H_2O + CO_2$$
$$\text{x g} \quad 3.00\text{ g}$$

$$\#\text{ of moles of } CO_2 = \frac{\text{mass of sample}}{\text{molar mass}} = \frac{3.00\ \cancel{g}}{44.0\ \cancel{g}/\text{mole}} = 0.0682\text{ moles}$$

You need to use the molar ratio to determine the number of moles of water that are produced when 0.0682 moles of carbon dioxide are generated. The answer should be fairly obvious, when you look at the coefficients associated with the water and the carbon dioxide. From these coefficients, you can see that the molar ratio of water to carbon dioxide is 2:1. This means that 2 moles of water are produced for every mole of carbon dioxide generated. Clearly, this means that if you generate 0.0682 moles of carbon dioxide, you must produce 0.136 moles of water ($0.0682 \times 2 = 0.136$).

If this doesn't seem all that obvious to you, then I suggest that you set up a ratio, as shown here:

$$\begin{array}{lcc} & H_2O & CO_2 \\ \text{coefficients} \rightarrow & 2 & 1 \\ \hline \#\text{ of moles} \rightarrow & x & 0.0682\text{ moles} \end{array}$$

$$\frac{2}{x} = \frac{1}{0.0682\text{ moles}}$$

Cross-multiplying and solving for x (the number of moles of water), you get $x = 2 \times 0.0682 = 0.136$ moles of water. This method may not have been necessary for this particular example, but it will come in handy for less obvious problems.

All that remains for you to do is turn the number of moles of water to mass, by multiplying by the molar mass of water, which is 18.0 g.

$$\cancel{CH_4} + \cancel{2O_2} \rightarrow 2H_2O + CO_2$$
$$\quad\quad\quad\quad x\text{ g} \quad 3.00\text{ g}$$

$$\#\text{ of moles of } CO_2 = \frac{\text{mass of sample}}{\text{molar mass}} = \frac{3.00\ \cancel{g}}{44.0\ \cancel{g}/\text{mole}} = 0.0682 \text{ moles}$$

$$\text{Mass of water} = \#\text{ of moles} \times \text{molar mass} = 0.136\ \cancel{\text{moles}} \times 18.0\text{ g}/\cancel{\text{mole}} = 2.45\text{ g}$$

So, 2.45 g of H_2O are produced as 3.00 g of CO_2 are generated from the combustion of methane.

Now, complete the practice problems, and then move on to Lesson 7–5.

Lesson 7–4 Review

Use the following balanced chemical equation to answer questions 1–3:

$$2Mg_{(s)} + O_{2(g)} \rightarrow 2MgO_{(s)}$$

magnesium + oxygen → magnesium oxide

1. How many grams of magnesium would be required to produce 2.50 g of magnesium oxide?
2. How many grams of oxygen would be required to react compietely with 1.18 g of magnesium?
3. How many grams of magnesium oxide can be formed if 0.875 g of magnesium react with an excess of oxygen?

Use the following balance chemical equation to answer questions 4–6:

$$2KClO_3 \rightarrow KCl + 3O_2$$

potassium chlorate → potassium chloride + oxygen

4. How many grams of potassium chloride can be produced from the decomposition of 5.00 g of potassium chlorate?
5. How many grams of potassium chlorate are required to produce 3.00 g of oxygen?
6. How many grams of oxygen are produced from enough potassium chlorate to produce 4.00 g of potassium chloride?

Complete all of the required steps to answer the following questions:

7. How many grams of water can be produced from the complete combustion of 255 g of methane gas?
8. How many grams of water can be produced from the complete combustion of 255 g of propane gas?

Lesson 7–5: Mixed Mass-Volume Problems

In problems involving a gas, it is easier to measure the volume of the gas than the mass. Mixed mass-volume problems include both mass-volume problems, where you are given a mass and asked for a volume, and volume-mass problems, where you are given a volume and asked for a mass. You will need to do the molar conversions involving the molar volume of a gas, so review Lesson 7–2 if necessary. The basic steps for solving this type of problem remain the same:

1. Begin with a balanced chemical reaction.
2. Carefully label the equation with the information from the word problem.
3. Convert the given quantity to number of moles.
4. Use the molar ratio shown by the coefficients to determine the number of moles of the unknown.
5. Convert this number of moles to the quantity the question asks for.

Let's try an example.

Example 1

How many grams of zinc must be completely reacted with an excess of sulfuric acid according to the balanced reaction shown here, in order to generate 500 cm^3 of hydrogen gas, at STP?

$$Zn + H_2SO_4 \rightarrow ZnSO_4 + H_2$$

You have a balanced chemical reaction, which takes care of Step 1, and the fact that all of the coefficients are 1 means that the step with the molar ratio will be easy. There is one potential tricky part to this problem, however. Can you see what it is? The volume of the gas that they want to generate is reported in cm^3. The molar volume of a gas is 22.4 dm^3/mole. You must convert 500 cm^3 to dm^3 before you can convert it to the number of moles.

You should remember that there are 1000 cm^3 in one dm^3, so you can make the conversion with the factor-label method, as shown here:

$$500\ \cancel{cm^3} \times \frac{1\ dm^3}{1000\ \cancel{cm^3}} = 0.5\ dm^3$$

Now, label the problem, which is the second step in this type of problem. Notice how I cross out the original volume, and replace it with its equivalent in dm^3.

How many grams of zinc must be completely reacted with an excess of sulfuric acid according to the balanced reaction shown here, in order to generate ~~500 cm³~~ $0.5\ dm^3$ *of hydrogen gas*, at STP?

$$\begin{array}{lcr} Zn + \cancel{H_2SO_4} \rightarrow \cancel{ZnSO_4} + H_2 \\ x\ g \qquad\qquad\qquad\qquad 0.5\ dm^3 \end{array}$$

We crossed out the sulfuric acid (H_2SO_4) because it was the excess reactant and doesn't enter into our calculation. We crossed out the zinc sulfate ($ZnSO_4$) because it was never mentioned and doesn't enter into the calculation either.

Now, move on to Step 3, which is to convert the given quantity into number of moles. This conversion represents a volume-mole conversion, the type we discussed in Lesson 7–2. We convert the volume of a gas at STP to number of moles, by dividing by the molar volume ($22.4\ dm^3$/mole) of a gas. As always, the specific type of the gas is irrelevant.

$$\begin{array}{lcr} Zn + \cancel{H_2SO_4} \rightarrow \cancel{ZnSO_4} + H_2 \\ x\ g \qquad\qquad\qquad\qquad 0.5\ dm^3 \end{array}$$

$$\#\text{ of moles of } H_2 = \frac{\text{volume of gas at STP}}{\text{molar volume of gas at STP}} = \frac{0.5\ \cancel{dm^3}}{22.4\ \cancel{dm^3}/\text{mole}} = 0.02\text{ moles}$$

As we mentioned before, we can see that the molar ratio between zinc and hydrogen is 1:1. This means that 1 mole of zinc would react with an excess of sulfuric acid to produce 1 mole of hydrogen gas. Thus, 0.02 moles of zinc are required to produce 0.02 moles of hydrogen gas. Now that we know how many moles of zinc we need (0.02 moles), we can easily convert this to grams by multiplying by the molar mass of zinc 65.4 g/mole), which we find on the periodic table.

$$\begin{array}{lcr} Zn + \cancel{H_2SO_4} \rightarrow \cancel{ZnSO_4} + H_2 \\ x\ g \qquad\qquad\qquad\qquad 0.5\ dm^3 \end{array}$$

$$\#\text{ of moles of } H_2 = \frac{\text{volume of gas at STP}}{\text{molar volume of gas at STP}} = \frac{0.5\ \cancel{dm^3}}{22.4\ \cancel{dm^3}/\text{mole}} = 0.02\text{ moles}$$

$$\text{Mass of Zn} = \#\text{ of moles} \times \text{molar mass} = 0.02\ \cancel{\text{moles}} \times 65.4\ \text{g}/\cancel{\text{mole}} = 1.308\text{ g} = 1\text{ g}$$

So, approximately 1 g of Zn must react completely to generate $500\ cm^3$ of hydrogen gas at STP.

Next is an example where you start with a known mass and must calculate the volume of a gas that will be generated at STP.

Example 2

What volume of oxygen gas at STP will be generated from the decomposition of 35.0 g of water?

Again, you have to come up with the balanced chemical reaction. Based on what you should know about decomposition reactions (see Lesson 6–2) and balancing equations (see Lesson 6–1), this shouldn't be too much trouble. Complete Step 1, starting with a balanced equation, and Step 2, labeling the equation.

What volume of oxygen gas at STP will be generated from the decomposition of *35.0 g of water*?

$$\begin{array}{ccc} 2H_2O \rightarrow & 2H_2 + & O_2 \\ 35.0\text{ g} & & x\text{ dm}^3 \end{array}$$

Notice that this time the unknown is labeled with the units dm^3. We don't use the unit "g" this time, because the problem makes it quite clear that a volume is being sought. Of the various units of volume, dm^3 is the natural choice, because this is the unit that we have been using with the molar volume of gases.

For Step 3, you must convert the given quantity into the number of moles. Because the given quantity is in grams, you'll use the molar mass of water to convert to moles. Once again, find the molar mass of water (18.0 g/mole) on the periodic table.

$$\begin{array}{ccc} 2H_2O \rightarrow & \cancel{2H_2} + & O_2 \\ 35.0\text{ g} & & \text{dm}^3 \end{array}$$

$$\#\text{ of moles of } H_2O = \frac{\text{mass of sample}}{\text{molar mass}} = \frac{35.0\ \cancel{\text{g}}}{18.0\ \cancel{\text{g}}/\text{mole}} = 1.94\text{ moles}$$

Next, make use of the molar ratio of water to oxygen to determine the number of moles of oxygen we would produce. The coefficients show that the ratio is 2:1, which means that it takes 2 moles of water to produce 1 mole of oxygen. It follows that 1.94 moles of water will produce 0.970 moles of oxygen. If you don't see that, look at the ratio here:

$$\begin{array}{rcc} & 2H_2O & O_2 \\ \frac{\text{coefficients} \rightarrow}{\#\text{ of moles} \rightarrow} & \frac{2}{1.94} & = \frac{1}{x} \end{array}$$

We cross-multiply to find that 2x=1.94, which means that x (the number of moles of O_2) = 0.970 moles.

Finally, multiply the number of moles of oxygen by the volume of each mole of oxygen at STP, in dm^3.

What volume of oxygen gas at STP will be generated fiom the decomposition of *35.0 g of water*?

$$2H_2O \rightarrow \cancel{2H_2} + O_2$$
$$35.0\text{ g} \qquad\qquad x\text{ dm}^3$$

$$\#\text{ of moles of }H_2O = \frac{\text{mass of sample}}{\text{molar mass}} = \frac{35.0\ \cancel{\text{g}}}{18.0\ \cancel{\text{g}}\text{/mole}} = 1.94\text{ moles}$$

$$\text{Volume of }O_2 = \#\text{ of moles} \times \text{molar volume of gas at STP}$$
$$= 0.970\ \cancel{\text{moles}} \times 22.4\text{ dm}^3/\cancel{\text{mole}} = 21.7\text{ dm}^3$$

You find that 35.0 g of water will decompose to form 21.7 dm^3 of oxygen, at STP. Before you go on to the practice problems, let's add something to the previous problem. How many dm^3 of hydrogen at STP would be produced from the decomposition of 35.0 g of water? Can you quickly see why the answer must be 43.4 dm^3? You see, the coefficient of "2" in front of the hydrogen tells you that you are going to produce twice as many moles of hydrogen as oxygen. Or, another way to look at it, you are going to get as many moles of hydrogen out as you put of water in. So, you will generate 1.94 moles of hydrogen gas, which represents 43.4 ~~dm~~^{3} (1.94 moles $\times$ 22.4 ~~dm~~^{3}/mole). The fact that doubling the volume of oxygen ($2 \times 21.7 dm^3 = 43.4 dm^3$) gives us this same answer is a clue to a shortcut that I will teach you in the next lesson.) Answer these practice problems before moving on to the next lesson.

Lesson 7–5 Review

Use the following balanced chemical equation to answer questions 1–3:

$$3H_{2(g)} + N_{2(g)} \rightarrow 2NH_{3(g)}$$
$$\text{hydrogen} + \text{nitrogen} \rightarrow \text{ammonia}$$

1. How many dm^3 of hydrogen gas at STP would be required to produce 50.0 g of ammonia?
2. How many grams of ammonia can be produced when 23.5 dm^3 of nitrogen gas at STP react with an excess of hydrogen?
3. How many dm^3 of hydrogen gas at STP will react with exactly 56.0 grams of nitrogen gas?

Use the following balanced chemical equation to answer questions 4–6:

$$2H_2O_{(l)} \rightarrow 2H_{2(g)} + O_{2(g)}$$
$$\text{water} \rightarrow \text{hydrogen} + \text{oxygen}$$

4. How many grams of water would be required to produce 30.0 dm^3 of oxygen gas at STP?

5. How many dm^3 of hydrogen gas at STP will be produced from the decomposition of 15.0 g of water?
6. How many grams of oxygen will be produced from the decomposition of enough water to generate 22.4 dm^3 of hydrogen gas at STP?

Complete all of the required steps to answer the following questions.

7. How many dm^3 of oxygen at STP can react completely with 3.50 g of magnesium in a synthesis reaction, which forms magnesium oxide?
8. How many grams of water will form when 1.00 dm^3 of ethane at STP reacts with an excess of oxygen in a combustion reaction?

Lesson 7–6: Volume-Volume Problems

At the end of Lesson 7–5, I mentioned a shortcut that I would teach you for doing volume-volume problems. The shortcut is so logical and useful that it would not make any sense to solve volume-volume problems without it. However, there is a very real danger associated with learning this shortcut. Some students learn the shortcut and then try to apply it to mass-mass and mass-volume problems. Be warned: You cannot do this! The shortcut only works for volume-volume problems, and trying to use it on the other types of problems will result in the wrong answer!

To introduce the shortcut, I will solve the first volume-volume problem the long way. If you look at my work carefully, you should not only be able to figure out what the shortcut is, but you should be able to understand why it works for this—and only this—type of problem.

Volume-volume problems are used when you are only interested in the gaseous components of a chemical reaction. The problems only give you the volume of a substance and only ask for the volume of another substance. The following is an example of this type of problem.

Example 1

How many dm^3 of hydrogen gas are needed to react with exactly 30.0 dm^3 of oxygen, in order to synthesize water, assuming both gases are at STP?

Well, we aren't given very much to start with, but it is easy enough to complete Step 1, starting with a balanced equation based on our knowledge of synthesis reactions from Lesson 6–2. Let's complete the first two steps.

$$2H_2 + O_2 \rightarrow \cancel{2H_2O}$$
$$\text{hydrogen} + \text{oxygen} \rightarrow \text{water}$$
$$x\ dm^3 \quad 30.0\ dm^3$$

Now we need to convert the volume of the oxygen to number of moles, by dividing by the molar volume of a gas at STP, 22.4 dm^3/mole.

$$2H_2 + O_2 \rightarrow \cancel{2H_2O}$$
$$\text{hydrogen} + \text{oxygen} \rightarrow \text{water}$$
$$x\ dm^3 \quad 30.0\ dm^3$$

$$\text{\# of moles of oxygen} = \frac{\text{volume of gas at STP}}{\text{molar volume of gas}} = \frac{30.0\ \cancel{dm^3}}{22.4\ \cancel{dm^3}/\text{mole}} = 1.34 \text{ moles}$$

Next, we need to use the molar ratio, given by the coefficients, to determine the number of moles of hydrogen that will react with 1.34 moles of oxygen. The molar ratio is 2:1, which means that it would take 2 moles of hydrogen to react completely with 1 mole of oxygen. It follows that it will take 2.68 moles of hydrogen to react with 1.34 moles of oxygen, as shown by this ratio:

$$\begin{array}{rcl} & \text{hydrogen} & \text{oxygen} \\ \dfrac{\text{coefficients} \rightarrow}{\text{\# of moles} \rightarrow} & \dfrac{2}{x} = & \dfrac{1}{1.34 \text{ moles}} \end{array}$$

Cross-multiplying, we get x = 2.68 moles.

All that is left for us to do is Step 5, to convert the number of moles to the desired quantity, which, in this case, is dm^3. To do this, we have to multiply the number of moles of hydrogen (2.68 moles) by the molar volume of a gas at STP.

$$2H_2 + O_2 \rightarrow \cancel{2H_2O}$$
$$\text{hydrogen} + \text{oxygen} \rightarrow \text{water}$$
$$x\ dm^3 \quad 30.0\ dm^3$$

$$\text{\# of moles of oxygen} = \frac{\text{volume of gas at STP}}{\text{molar volume of gas}} = \frac{30.0\ \cancel{dm^3}}{22.4\ \cancel{dm^3}/\text{mole}} = 1.34 \text{ moles}$$

$$\text{Volume of } H_2 = \text{\# of moles of } H_2 \text{ at STP} \times \text{molar volume of gas}$$
$$= 2.68\ \cancel{\text{moles}} \times 22.4\ \text{g}/\cancel{\text{mole}} = 60.0\ dm^3$$

So it takes 60.0 dm^3 of hydrogen to react completely with 30.0 dm^3 of oxygen, when both gases are at STP. Now, did you figure out what the shortcut is? In Step 3, we converted the volume of oxygen to moles by dividing by 22.4 dm^3/mole. In Step 5, we converted the number of moles of hydrogen to volume by multiplying by 22.4 dm^3/mole. These steps are opposite operations, so they actually cancel each other out! This means that we can skip Steps 3 and 5 when we do volume-volume problems! That means that the procedure for this type of problem becomes:

1. Begin with a balanced chemical reaction.
2. Carefully label the equation with the information from the word problem.
3. Use the molar ratio shown by the coefficients to determine the volume of the unknown.

Now let's try the same problem with the shortcut.

$$2H_2 + O_2 \rightarrow 2H_2O$$

hydrogen + oxygen → water

x dm³ 30.0 dm³

$$\frac{\text{coefficients} \rightarrow}{\text{\# of moles} \rightarrow} \quad \frac{2}{x}\ (\text{hydrogen}) = \frac{1}{30.0\ dm^3}\ (\text{oxygen})$$

Cross-multiplying, we get a volume of 60.0 dm³ for the hydrogen. The same answer for less than half the work!

Let's try a couple more problems, using the shortcut.

Example 2

How many dm³ of oxygen are required to react completely with *5.0 dm³ of propane* when both gases are at STP, according to the balanced chemical reaction shown here?

$$5O_2 + C_3H_8 \rightarrow 4H_2O + 3CO_2$$

oxygen + propane → water + carbon dioxide

x dm³ 5.0 dm³

$$\frac{\text{coefficients} \rightarrow}{\text{\# of moles} \rightarrow} \quad \frac{5}{x}\ (\text{oxygen}) = \frac{1}{5.0\ dm^3}\ (\text{propane})$$

Cross-multiplying, we find that the volume of oxygen is 25 dm³

Example 3

How many dm³ of methane would react completely with *40 dm³ of oxygen* gas when both substances are at STP, according to the balanced chemical equation shown here?

$$CH_4 + 2O_2 \rightarrow 2H_2O + CO_2$$

methane + oxygen → water + carbon dioxide

x dm^3 40 dm^3

	methane		oxygen
coefficients →	1	=	2
# of moles →	x		40 dm^3

Cross-multiplying, we get $2x = 40\ dm^3$, so the volume of methane is 20 dm^3.

Before you move on to the practice problems, let me remind you one more time that this shortcut may only be used for the volume-volume problems. The reason for this should be clear. All gases have a molar volume of 22.4 dm^3/mole at STP; however, substances don't all have the same mass at any temperature or pressure.

Now, on to the practice problems.

Lesson 7–6 Review

Answer questions 1–3 based on the following balanced chemical equation. Assume all volumes are at STP.

$$C_3H_8 + 5O_2 \rightarrow 4H_2O + 3CO_2$$

propane + oxygen → water + carbon dioxide

1. How many dm^3 of oxygen are required to completely react with 5.0 dm^3 of propane?
2. How many dm^3 of carbon dioxide will be produced if 15 dm^3 of oxygen react with an excess of propane?
3. How many dm^3 of water vapor will be produced with 90.0 dm^3 of carbon dioxide?

Complete all of the required steps to answer the following questions. Assume STP for all volumes of gases.

4. How many dm^3 of hydrogen will react with 4.0 dm^3 of nitrogen in the synthesis reaction that produces ammonia?
5. How many dm^3 of oxygen are required to react with an excess of glucose in order to produce 5.0 dm^3 of carbon dioxide through the process of respiration?

Chapter 7 Examination

Part I—Multiple Choice

Select the best answer choice for each of the following questions.

1. How many molecules does 2 moles of water (H_2O) represent?
   a) 6 b) 6.02×10^{23} c) 1.20×10^{24} d) 1.81×10^{24}
2. How many atoms can be found in 1 mole of water (H_2O) molecules?
   a) 3 b) 6.02×10^{23} c) 1.20×10^{24} d) 1.81×10^{24}
3. How many atoms are there in one molecule of propane, C_3H_8?
   a) 3 b) 11 c) 1.20×10^{24} d) 6.62×10^{24}
4. How many atoms of hydrogen are in one molecule of propane, C_3H_8?
   a) 8 b) 11 c) 1.81×10^{24} d) 4.82×10^{24}
5. How many atoms of hydrogen are in one mole of propane, C_3H_8?
   a) 8 b) 11 c) 1.81×10^{24} d) 4.82×10^{24}
6. How many atoms of carbon are there in one mole of propane, C_3H_8?
   a) 3 b) 11 c) 1.81×10^{24} d) 4.82×10^{24}
7. How many atoms of carbon are there in 5 moles of propane, C_3H_8?
   a) 9.03×10^{24} b) 3.01×10^{24} c) 1.81×10^{24} d) 4.82×10^{24}
8. How much space would 1.50 moles of neon gas occupy at STP?
   a) 22.4 dm^3 b) 30.3 dm^3 c) 33.6 dm^3 d) 44.8 dm^3
9. What is the mass of 2.50 moles of H_2O?
   a) 7.50 g b) 56.0 dm^3 c) 1.51×10^{24} g d) 45.0 g
10. Which of the following represents 0.500 moles of carbon monoxide (CO) molecules?
    a) 28.0 g c) 33.6 dm^3 at STP
    b) 3.01×10^{23} molecules d) 6.02×10^{23} molecules

Part II—Molar Conversions

Perform the following calculations.

11. How many molecules would 3.45 moles of methane (CH_4) represent?
12. How much space would 0.0334 moles of argon gas occupy at STP?
13. What would be the mass of 5.43 moles of CO_2, with a molar mass of 44.0 g/mole?

14. How many moles of molecules would 3.50×10^{24} molecules represent?
15. How many moles would 57.5 dm^3 of hydrogen gas at STP represent?
16. How many atoms of carbon could be found in 5.00 g of this element?
17. What would be the mass of 4.56×10^{23} water molecules?
18. How much space would 39.9 g of nitrogen gas occupy at STP?
19. What would be the mass of a 33.5 dm^3 sample of O_2 at STP?
20. How many molecules of carbon dioxide gas would occupy 10.0 dm^3 at STP?

Part III—Stoichiometry

Base your answers to questions 21–23 on the following balanced chemical equation. Assume all gas volumes are at STP.

$$C_3H_8 + 5O_2 \rightarrow 3CO_2 + 4H_2O$$

propane + oxygen → carbon dioxide + water

21. How many dm^3 of oxygen at STP would be required to react completely with 38.8 g of propane?
22. How many dm^3 of oxygen at STP must react with an excess of propane, in order to generate 20.0 dm^3 of carbon dioxide at STP?
23. How many grams of water can be generated from the combustion of 40.0 g of propane in an excess of oxygen?

Base your answers to questions 24–25 on the following balanced chemical equation. Assume all gas volumes are at STP.

$$2Mg + O_2 \rightarrow 2MgO$$

magnesium + oxygen → magnesium oxide

24. How many grams of magnesium oxide can be produced if 30.0 g of magnesium are allowed to react with 12.0 dm^3 of oxygen gas at STP?
25. How many grams of magnesium oxide can be produced if 25.0 g of magnesium are allowed to react with 28.0 g of oxygen?

Answer Key

The actual answers will be shown in brackets, followed by the explanation. If you don't understand an explanation that is given in this section, you may want to go back and review the lesson that the question came from.

Lesson 7–1 Review

1. [24.0 g]—6.94 g + 16.0 g + 1.01 g = **24.0 g**
2. [155.5 g]—63.5 g + 2(14.0 g) + 4(16.0 g) = **155.5 g**
3. [68.2 g]—2(14.0 g) + 8(1.01 g) + 32.1 g = **68.2 g**
4. [6.02×10^{23} molecules]—Just as one dozen of anything represents 12 items, one mole of anything represents 6.02×10^{23} items.
5. [1.81×10^{24} atoms]—This is a "parts" question. Just as one dozen cars has more than one dozen tires, one mole of molecules has more than one mole of atoms. Each water molecule contains three atoms, so one dozen water molecules would contain 3(12) = 36 atoms. One mole of water molecules contains $3(6.02 \times 10^{23}) = 1.81 \times 10^{24}$ atoms. It may help to see the work with all of the units, as shown here:

$$1\ \cancel{\text{mole}} \times \left(\frac{6.02 \times 10^{23}\ \cancel{\text{molecules}}}{\cancel{\text{moles}}}\right) \times \left(\frac{3\text{ atoms}}{\cancel{\text{molecule}}}\right) = 3(6.02 \times 10^{23})\text{ atoms} = 1.81 \times 10^{24}\text{ atoms}$$

6. [2.41×10^{24} molecules]—Four dozen doughnuts would be 4(12) = 48 doughnuts. Four moles of CH_4 represents $4(6.02 \times 10^{23}) = 2.41 \times 10^{24}$ molecules.
7. [1.20×10^{25} atoms]—Each molecule of CH_4 represents five atoms, so four moles of molecules would contain $5 \times 4(6.02 \times 10^{23})$ atoms, as shown here.

$$4\ \cancel{\text{moles}} \times \frac{6.02 \times 10^{23}\ \cancel{\text{molecules}}}{\cancel{\text{mole}}} \times \frac{5\text{ atoms}}{\cancel{\text{molecule}}} = 1.20 \times 10^{25}\text{ atoms}$$

Lesson 7–2 Review

1. [101 dm^3]—Volume of nitrogen $= 4.50\ \cancel{\text{moles}} \times \frac{22.4\text{ dm}^3}{\cancel{\text{mole}}} = 101\text{ dm}^3$
2. [257 g]—Molar mass of NaBr is 102.9 g/mole (23.0 g + 79.9 g).
   Mass of NaBr $= 2.50\ \cancel{\text{moles}} \times \frac{102.9\text{ g}}{\cancel{\text{mole}}} = 257\text{ g}$
3. [4.17 moles]—Molar mass of carbon = 12.0 g/mole
   $\#\text{ of moles of carbon} = \frac{\text{mass of sample}}{\text{molar mass}} = \frac{50.0\ \cancel{\text{g}}}{12.0\ \cancel{\text{g}}/\text{mole}} = 4.17\text{ moles}$
4. [1.13×10^{24} molecules]—# of molecules = # of moles × Avogadro's number
   $= 1.88\ \cancel{\text{moles}} \times \frac{6.02 \times 10^{23}\text{ molecules}}{\cancel{\text{mole}}} = 1.13 \times 10^{24}\text{ molecules}$
5. [3.35 moles]— $\#\text{ of moles} = \frac{\text{volume of gas}}{\text{molar volume}} = \frac{75.0\text{ dm}^3}{22.4\text{ dm}^3/\text{mole}} = 3.35\text{ moles}$

6. [3.01 moles]—
$$\text{\# of moles} = \frac{\text{\# of molecules}}{\text{Avogadro's number}} = \frac{1.81 \times 10^{24}\ \cancel{\text{molecules}}}{6.02 \times 10^{23}\ \cancel{\text{molecules}}/\text{mole}} = 3.01 \text{ moles}$$

7. [1.51×10^{24} molecules]

a. First, find the molar mass of water: 2.02 g + 16.0 g = 18.0 g/mole

b. $\text{\# of moles of water} = \frac{\text{mass}}{\text{molar mass}} = \frac{45.0\ \cancel{\text{g}}}{18.0\ \cancel{\text{g}}/\text{mole}} = 2.50 \text{ moles}$

c. # of molecules = # of moles × Avogadro's number

$= 2.50\ \cancel{\text{moles}} \times \frac{6.02 \times 10^{23} \text{ molecules}}{\cancel{\text{mole}}} = 1.51 \times 10^{24} \text{ molecules}$

8. [94.3 dm^3]—The molar mass of neon, from the periodic table, is 20.2 g/mole.

a. $\text{\# of moles} = \frac{\text{mass of sample}}{\text{molar mass}} = \frac{85.0\ \cancel{\text{g}}}{20.2\ \cancel{\text{g}}/\text{mole}} = 4.21 \text{ moles}$

b. $\text{Volume} = 4.21\ \cancel{\text{moles}} \times \frac{22.4 \text{ dm}^3}{\cancel{\text{mole}}} = 94.3 \text{ dm}^3$

9. [1.81×10^{24} molecules]—

a. $\text{\# of moles} = \frac{\text{volume}}{\text{molar volume}} = \frac{67.5 \text{ dm}^3}{22.4 \text{ dm}^3/\text{mole}} = 3.01 \text{ moles}$

b. $\text{\# of molecules} = 3.01\ \cancel{\text{moles}} \times \frac{6.02 \times 10^{23} \text{ molecules}}{\cancel{\text{mole}}} = 1.81 \times 10^{24} \text{ molecules}$

10. [$288\bar{0}$ g]—The molar mass of water is 18.0 g/mole.

a. $\text{\# of moles} = \frac{9.65 \times 10^{25}\ \cancel{\text{molecules}}}{6.02 \times 10^{23}\ \cancel{\text{molecules}}/\text{mole}} = 16\bar{0} \text{ moles}$

b. $\text{Mass} = \text{\# of moles} \times \text{molar mass} = 16\bar{0}\ \cancel{\text{moles}} \times \frac{18.0 \text{ g}}{\cancel{\text{mole}}} = 2880 \text{ g}$

Lesson 7–3 Review

1. [2 moles]—The ratio of propane to oxygen is 1:5. It would take 1 mole of propane to react completely with 5 moles of oxygen, and 2 moles of propane to react with 10 moles of oxygen. The adjacent ratio may help.

	propane	oxygen
coefficients →	1	5
moles →	x	10

$\frac{1}{x} = \frac{5}{10}$

5x = 10, x = 2

2. [15 moles]—

	propane	oxygen
coefficients →	1	5
moles →	3	x

$\frac{1}{3} = \frac{5}{x}$

x = 15

3. [12 moles]—

	propane	water
coefficients →	1	4
moles →	3	x

$\frac{1}{3} = \frac{4}{x}$

x = 12

4. [10 moles]—

	propane	carbon dioxide
coefficients →	1	3
moles →	x	30

$\frac{1}{x} = \frac{3}{30}$

3x = 30, x = 10

5. [10 moles]—

oxygen water

$$\frac{\text{coefficients} \rightarrow}{\text{moles} \rightarrow} \quad \frac{5}{x} = \frac{4}{8}$$

$4x = 40, x = 10$

6. [3 moles]—

hydrogen nitrogen

$$\frac{\text{coefficients} \rightarrow}{\text{moles} \rightarrow} \quad \frac{3}{9} = \frac{1}{x}$$

$3x = 9, x = 3$

7. [15 moles]—

hydrogen nitrogen

$$\frac{\text{coefficients} \rightarrow}{\text{moles} \rightarrow} \quad \frac{3}{x} = \frac{1}{5}$$

$x = 15$

8. [1.5 moles]—

hydrogen ammonia

$$\frac{\text{coefficients} \rightarrow}{\text{moles} \rightarrow} \quad \frac{3}{x} = \frac{2}{1}$$

$2x = 3, x = 1.5$

9. [5 moles]—

nitrogen ammonia

$$\frac{\text{coefficients} \rightarrow}{\text{moles} \rightarrow} \quad \frac{1}{x} = \frac{2}{10}$$

$2x = 10, x = 5$

10. [4 moles]—You must determine the "limiting reactant," which means that you must figure out how much ammonia you could generate with each of the elements, if you had plenty of the other, and figure out which reactant allows you to produce the least.

hydrogen ammonia

$$\frac{\text{coefficients} \rightarrow}{\text{moles} \rightarrow} \quad \frac{3}{9} = \frac{2}{x}$$

or $x = 6$

nitrogen ammonia

$$\frac{\text{coefficients} \rightarrow}{\text{moles} \rightarrow} \quad \frac{1}{2} = \frac{2}{x}$$

or $x = 4$

The hydrogen would allow us to make 6 moles of ammonia, but we only have enough nitrogen to make 4 moles of ammonia. The nitrogen would be called our limiting reactant, because it is limiting the amount of ammonia that we can make to four moles. The hydrogen would be called our excess reactant, because we have more than we can actually use.

Lesson 7–4 Review

1. [1.51 g of Mg]— $2Mg_{(s)} + O_{2(g)} \rightarrow 2MgO_{(s)}$

x g 2.50 g

A. # of moles of MgO = $\frac{\text{mass}}{\text{molar mass}} = \frac{2.50 \text{ g}}{40.3 \text{ g/mole}}$ = 0.0620 moles

B. # of moles of Mg can be found with the molar ratio:

Mg MgO

$$\frac{\text{coefficients} \rightarrow}{\text{moles} \rightarrow} \quad \frac{3}{9} = \frac{2}{x}$$

\# of moles of Mg (x) = 0.0620 moles

C. Mass of Mg = # of moles × molar mass = 0.0620 moles × 24.3 g/mole = 1.51 g of Mg

2. [0.778 g of O_2]— $2Mg_{(s)} + O_{2(g)} \rightarrow 2MgO_{(s)}$

1.18 g x g

A. # of moles of MgO = $\frac{\text{mass}}{\text{molar mass}} = \frac{1.18 \text{ g}}{24.3 \text{ g/mole}}$ = 0.0486 moles

B. # of moles of O_2 can be found with the molar ratio:

	Mg		O_2
coefficients →	2	=	1
# of moles →	0.0486		x

# of moles of O_2 (x) = 0.0243 moles

C. Mass of O_2 = # of moles × molar mass = 0.0243 moles × 32.0 g/mole = 0.778 g

3. [1.45 g of MgO]— $2Mg_{(s)} + O_{2(g)} \rightarrow 2MgO_{(s)}$

0.875 g x g

A. # of moles of Mg = $\frac{\text{mass}}{\text{molar mass}} = \frac{0.875 \text{ g}}{24.3 \text{ g/mole}}$ = 0.0360 moles

B. # of moles of MgO can be found with the molar ratio:

	Mg		MgO
coefficients →	2	=	2
# of moles →	0.0360		x

# of moles of MgO (x) = 0.0360 moles

C. Mass of MgO = # of moles × molar mass = 0.0360 moles × 40.3 g/mole =1.45 g

4. [1.52 g of KCl]— $2KClO_3 \rightarrow KCl + 3O_2$

5.00 g x g

A. # of moles of $KClO_3$ = $\frac{\text{mass}}{\text{molar mass}} = \frac{5.00 \text{ g}}{123 \text{ g/mole}}$ = 0.0407 moles

B. # of moles of KCl can be found with the molar ratio:

	$KClO_3$		KCl
coefficients →	2	=	1
# of moles →	0.0407		x

# of moles of KCl (x) = 0.0204 moles

C. Mass of KCl = # of moles × molar mass = 0.0204 moles × 74.6 g/mole = 1.52 g

5. [7.69 g of $KClO_3$]— $2KClO_3 \rightarrow KCl + 3O_2$

x g 3.00 g

A. # of moles of O_2 = $\frac{\text{mass}}{\text{molar mass}} = \frac{3.00 \text{ g}}{32.0 \text{ g/mole}}$ = 0.0938 moles

B. # of moles of $KClO_3$ can be found with the molar ratio:

	$KClO_3$		O_2
coefficients →	2	=	3
# of moles →	x		0.0938

# of moles of $KClO_3$ (x) = 0.0625 moles

C. Mass of $KClO_3$ = # of moles × molar mass = 0.0625 moles × 123 g/mole = 7.69 g

6. [63.0 g of O_2]— $2KClO_3 \rightarrow KCl + 3O_2$

49.0 g x g

A. # of moles of KCl = $\frac{\text{mass}}{\text{molar mass}} = \frac{49.0 \text{ g}}{74.6 \text{ g/mole}}$ = 0.657 moles

B. # of moles of O_2 can be found with the molar ratio:

$$\begin{array}{rcc} & KCl & O_2 \\ \text{coefficients} \rightarrow & 1 & 3 \\ \hline \text{\# of moles} \rightarrow & 0.657 & x \end{array}$$

# of moles of O_2 (x) = 1.97 moles

C. Mass of O_2 = # of moles × molar mass = 1.97 ~~moles~~ × 32.0 g/~~mole~~ = 63.0 g

7. [572 g of H_2O]— $CH_4 + \cancel{2O_2} \rightarrow \cancel{CO_2} + 2H_2O$

255 g x g

A. # of moles of $CH_4 = \frac{\text{mass}}{\text{molar mass}} = \frac{255\ \cancel{g}}{16.0\ \cancel{g}/\text{mole}} = 15.9$ moles

B. # of moles of H_2O can be found with the molar ratio:

$$\begin{array}{rcc} & CH_4 & H_2O \\ \text{coefficients} \rightarrow & 1 & 2 \\ \hline \text{\# of moles} \rightarrow & 15.9 & x \end{array}$$

# of moles of H_2O (x) = 31.8 moles

C. Mass of H_2O = # of moles × molar mass = 31.8 ~~moles~~ × 18.0 g/~~mole~~ = 572 g

8. [416 g of H_2O]— $C_3H_8 + \cancel{5O_2} \rightarrow \cancel{3CO_2} + 4H_2O$

255 g x g

A. # of moles of $C_3H_8 = \frac{\text{mass}}{\text{molar mass}} = \frac{255\ \cancel{g}}{44.1\ \cancel{g}/\text{mole}} = 5.78$ moles

B. # of moles of H_2O can be found with the molar ratio:

$$\begin{array}{rcc} & C_3H_8 & H_2O \\ \text{coefficients} \rightarrow & 1 & 4 \\ \hline \text{\# of moles} \rightarrow & 5.78 & x \end{array}$$

# of moles of H_2O (x) = 23.1 moles

C. Mass of H_2O = # of moles × molar mass = 23.1 ~~moles~~ × 18.0 g/~~mole~~ = 416 g

Lesson 7–5 Review

1. [98.8 dm³ of hydrogen]— $3H_{2(g)} + \cancel{N_{2(g)}} \rightarrow 2NH_{3(g)}$

dm^3 50.0 g

A. # of moles of $NH_3 = \frac{\text{mass}}{\text{molar mass}} = \frac{50.0\ \cancel{g}}{17.0\ \cancel{g}/\text{mole}} = 2.94$ moles

B. # of moles of H_2 can be determined from the molar ratio:

$$\begin{array}{rcc} & H_2 & NH_3 \\ \text{coefficients} \rightarrow & 3 & 2 \\ \hline \text{\# of moles} \rightarrow & x & 2.94 \end{array}$$

# of moles of H_2 = 4.41 moles

C. Volume of H_2 = # of moles × molar volume = 4.41 ~~moles~~ × 22.4 dm³/~~mole~~ = 98.8 dm³

2. [35.7 g of ammonia]— $\cancel{3H_{2(g)}} + N_{2(g)} \rightarrow 2NH_{3(g)}$

23.5 dm³ x g

A. # of moles of $N_2 = \frac{\text{volume}}{\text{molar volume}} = \frac{23.5\ \cancel{dm^3}}{22.4\ \cancel{dm^3}/\text{mole}} = 1.05$ moles

B. # of moles of NH_3 can be determined from the molar ratio:

$$\begin{array}{rcc} & N_2 & NH_3 \\ \text{coefficients} \rightarrow & 1 & 2 \\ \hline \text{\# of moles} \rightarrow & 1.05 & x \end{array}$$

# of moles of NH_3 = 2.10 moles

C. Mass of NH_3 = # of moles × molar mass = 2.10 ~~moles~~ × 17.0 g/~~mole~~ = 35.7 g

3. [134 dm^3]— $3H_{2(g)} + N_{2(g)} \rightarrow 2NH_{3(g)}$

$x\ dm^3 \quad 56.0\ g$

A. # of moles of $N_2 = \dfrac{\text{mass}}{\text{molar mass}} = \dfrac{56.0\ \cancel{g}}{28.0\ \cancel{g}/\text{mole}} = 2.00$ moles

B. # of moles of H_2 can be determined from the molar ratio:

$$\begin{array}{lcc} & H_2 & N_2 \\ \text{coefficients} \rightarrow & 3 & 1 \\ \text{\# of moles} \rightarrow & x & 2.00 \end{array} \qquad \frac{3}{x} = \frac{1}{2.00}$$

# of moles of H_2 = 6.00 moles

C. Volume of H_2 = # of moles × molar volume = 6.00 $\cancel{\text{moles}}$ × 22.4 $dm^3/\cancel{\text{mole}}$ = 134 dm^3

4. [48.2 g of water]— $2H_2O \rightarrow 2H_2 + O_2$

$x\ g \quad 30.0\ dm^3$

A. # of moles of $O_2 = \dfrac{\text{volume}}{\text{molar volume}} = \dfrac{30.0\ \cancel{dm^3}}{22.4\ \cancel{dm^3}/\text{mole}} = 1.34$ moles

B. # of moles of H_2O can be determined from the molar ratio:

$$\begin{array}{lcc} & H_2O & O_2 \\ \text{coefficients} \rightarrow & 2 & 1 \\ \text{\# of moles} \rightarrow & x & 1.34 \end{array} \qquad \frac{2}{x} = \frac{1}{1.34}$$

# of moles of H_2O = 2.68 moles

C. Mass of H_2O = # of moles × molar mass = 2.68 $\cancel{\text{moles}}$ × 18.0 $g/\cancel{\text{mole}}$ = 48.2 g

5. [18.7 dm^3 of hydrogen]— $2H_2O \rightarrow 2H_2 + O_2$

$15.0\ g \quad x\ dm^3$

A. # of moles of $H_2O = \dfrac{\text{mass}}{\text{molar mass}} = \dfrac{15.0\ \cancel{g}}{18.0\ \cancel{g}/\text{mole}} = 0.833$ moles

B. # of moles of H_2 can be determined from the molar ratio:

$$\begin{array}{lcc} & H_2O & H_2 \\ \text{coefficients} \rightarrow & 2 & 2 \\ \text{\# of moles} \rightarrow & 0.833 & x \end{array} \qquad \frac{2}{0.833} = \frac{2}{x}$$

# of moles of H_2 = 0.833 moles

C. Volume of H_2 = # of moles × molar volume = 0.833 $\cancel{\text{moles}}$ × 22.4 $dm^3/\cancel{\text{mole}}$ = 18.7 dm^3

6. [8.00 g of oxygen]— $2H_2O \rightarrow 2H_2 + O_2$

$22.4\ dm^3 \quad x\ g$

A. # of moles of $H_2 = \dfrac{\text{volume}}{\text{molar volume}} = \dfrac{22.4\ \cancel{dm^3}}{22.4\ \cancel{dm^3}/\text{mole}} = 1.00$ moles

B. # of moles of O_2 can be determined from the molar ratio:

$$\begin{array}{lcc} & H_2 & O_2 \\ \text{coefficients} \rightarrow & 2 & 1 \\ \text{\# of moles} \rightarrow & 1.00 & x \end{array} \qquad \frac{2}{1.00} = \frac{1}{x}$$

# of moles of O_2 = 0.500 moles

C. Mass of O_2 = # of moles × molar mass = 0.500 $\cancel{\text{moles}}$ × 16.0 $g/\cancel{\text{mole}}$ = 8.00 g

7. [1.61 dm^3 of oxygen]— $2Mg + O_2 \rightarrow 2MgO$

$3.50\ g \quad x\ dm^3$

A. # of moles of $Mg = \dfrac{\text{mass}}{\text{molar mass}} = \dfrac{3.50\ \cancel{g}}{24.3\ \cancel{g}/\text{mole}} = 0.144$ moles

B. # of moles of O_2 can be determined from the molar ratio:

$$\begin{array}{lcc} & Mg & O \\ \text{coefficients} \rightarrow & 2 & 1 \\ \text{\# of moles} \rightarrow & 0.144 & x \end{array} \qquad \frac{2}{0.144} = \frac{1}{x}$$

# of moles of O_2 = 0.0720 moles

C. Volume of O_2 = # of moles × molar volume = 0.0720 ~~moles~~ × 22.4 dm^3/~~mole~~
= 1.61 dm^3

8. [2.52 g]— $2C_2H_6 + 7O_2 \rightarrow 4CO_2 + 6H_2O$

1.00 dm^3 x g

A. # of moles of C_2H_6 = $\frac{\text{volume}}{\text{molar volume}} = \frac{1.00\ dm^3}{22.4\ dm^3/\text{mole}}$ = 0.0446 moles

B. # of moles of H_2O can be determined from the molar ratio:

$$\begin{array}{lcc} & C_2H_6 & H_2O \\ \text{coefficients} \rightarrow & 2 & 6 \\ \text{\# of moles} \rightarrow & 0.0466 & x \end{array} \qquad \frac{2}{0.0466} = \frac{6}{x}$$

# of moles of H_2O = 0.140 moles

C. Mass of H_2O = # of moles × molar mass = 0.140 ~~moles~~ × 18.0 g/~~mole~~ = 2.52 g

Lesson 7–6 Review

1. [25 dm^3 of oxygen]— $C_3H_8 + 5O_2 \rightarrow$ ~~$4H_2O$~~ + ~~$3CO_2$~~

5.0 dm^3 x dm^3

propane oxygen

$$\frac{\text{coefficients} \rightarrow}{\text{volume} \rightarrow} \quad \frac{1}{5.0\ dm^3} = \frac{5}{x}$$

So, the volume of the oxygen (x) = 25 dm^3

2. [9.0 dm^3 of carbon dioxide]— ~~C_3H_8~~ + $5O_2 \rightarrow$ ~~$4H_2O$~~ + $3CO_2$

15 dm^3 x dm^3

O_2 CO_2

$$\frac{\text{coefficients} \rightarrow}{\text{volume} \rightarrow} \quad \frac{5}{15\ dm^3} = \frac{3}{x}$$

So, the volume of the CO_2 (x) = 9.0 dm^3

3. [120. dm^3 of water]— ~~C_3H_8~~ + ~~$5O_2$~~ $\rightarrow 4H_2O + 3CO_2$

x dm^3 90.0 dm^3

H_2O CO_2

$$\frac{\text{coefficients} \rightarrow}{\text{volume} \rightarrow} \quad \frac{4}{x} = \frac{3}{90.0\ dm^3}$$

So, the volume of the H_2O (x) = 120. dm^3

4. [6.0 dm^3 of nitrogen]— $3H_2 + N_2 \rightarrow$ ~~$2NH_3$~~

x dm^3 4.0 dm^3

H_2 N_2

$$\frac{\text{coefficients} \rightarrow}{\text{volume} \rightarrow} \quad \frac{3}{x} = \frac{2}{4.0\ dm^3}$$

So, the volume of the H_2O (x) = 6.0 dm^3

5. [5.0 dm^3 of oxygen]— $C_6H_{12}O_6 + 6O_2 \rightarrow 6H_2O + 6CO_2$

x dm^3 5.0 dm^3

O_2 CO_2

$$\frac{\text{coefficients} \rightarrow}{\text{volume} \rightarrow} \quad \frac{6}{x} = \frac{6}{5.0\ dm^3}$$

So, the volume of the O_2 (x) = 5.0 dm^3

Chapter 7 Examination

1. [C. 1.20×10^{24}]—Each mole of molecules would represent 6.02×10^{23} molecules. Two moles of water molecules would be $2(6.02 \times 10^{23}) = 1.20 \times 10^{24}$ molecules.
2. [D. 1.81×10^{24}]—Remember: Each water molecule contains 3 atoms. One mole of water molecules contains $3(6.02 \times 10^{23}) = 1.81 \times 10^{24}$ atoms. It is similar to asking how many tires there are in a dozen cars.
3. [B. 11]—If the question never mentions the word *mole*, then Avogadro's number doesn't enter the calculation. Each molecule of propane contains 3 carbon atoms and 8 hydrogen atoms, for a total of 11.
4. [A. 8]—As we mentioned in our previous answer, each molecule of propane contains 8 hydrogen atoms.
5. [D. 4.82×10^{24}]—Each molecule of propane has 8 atoms of hydrogen, so one mole of propane has $8(6.02 \times 10^{23}) = 4.82 \times 10^{24}$ atoms of hydrogen.
6. [C. 1.81×10^{24}]—Each molecule of propane contains 3 carbon atoms, so one mole of propane molecules contains $3(6.02 \times 10^{23}) = 1.81 \times 10^{24}$ atoms of carbon.
7. [A. 9.03×10^{24}]—5 moles of propane molecules represent $5(6.02 \times 10^{23}) = 3.01 \times 10^{24}$ molecules. If each of these contains 3 carbon atoms, the total number of carbon atoms would be $3(3.01 \times 10^{24}) = 9.03 \times 10^{24}$.
8. [C. 33.6 dm^3]—$1.50 \cancel{\text{moles}} \times \dfrac{22.4\ dm^3}{\cancel{\text{mole}}} = 33.6\ dm^3$
9. [D. 45.0 g]—$2.50 \cancel{\text{moles}} \times \dfrac{18.0\ g}{\cancel{\text{mole}}} = 45.0\ g$
10. [B. 3.01×10^{23} molecules]—$0.500 \cancel{\text{moles}} \times \dfrac{6.02 \times 10^{23}\ \text{molecule}}{\cancel{\text{mole}}}$
11. [2.08×10^{24} molecules]—$3.45 \cancel{\text{moles}} \times \dfrac{6.02 \times 10^{23}\ \text{molecule}}{\cancel{\text{mole}}} = 2.08 \times 10^{24}$ molecules
12. [0.748 dm^3]—$0.0334 \cancel{\text{moles}} \times \dfrac{22.4\ dm^3}{\cancel{\text{mole}}} = 0.748\ dm^3$
13. [239 g]—$5.43 \cancel{\text{moles}} \times \dfrac{44.0\ g}{\cancel{\text{mole}}} = 239\ g$
14. [5.81 moles]—$\dfrac{3.50 \times 10^{24} \cancel{\text{molecules}}}{6.02 \times 10^{23} \cancel{\text{molecules}}/\text{mole}} = 5.81$ moles
15. [2.57 moles]—$\dfrac{57.5 \cancel{dm^3}}{22.4 \cancel{dm^3}/\text{mole}} = 2.57$ moles
16. [2.51×10^{23} atoms]—# of moles of carbon $= \dfrac{\text{mass}}{\text{molar mass}} = \dfrac{5.00 \cancel{g}}{12.0 \cancel{g}/\text{mole}} =$

    $0.417 \cancel{\text{moles}} \times \dfrac{6.02 \times 10^{23}\ \text{atoms}}{\cancel{\text{mole}}} = 2.51 \times 10^{23}$ atoms
17. [13.6 g]—# of moles $= \dfrac{\text{\# of particles}}{\text{Avogadro's \#}} = \dfrac{4.56 \times 10^{23} \cancel{\text{molecules}}}{6.02 \times 10^{23} \cancel{\text{molecules}}/\text{mole}} = 0.757$ moles

    mass = # of moles × molar mass $= 0.757 \cancel{\text{moles}} \times \dfrac{18.0\ g}{\cancel{\text{mole}}} = 13.6\ g$
18. [32.0 dm^3]—# of moles $= \dfrac{\text{mass}}{\text{molar mass}} = \dfrac{39.9 \cancel{g}}{28.0 \cancel{g}/\text{mole}} = 1.43$ moles

    volume = # of moles × molar volume $= 1.43 \cancel{\text{moles}} \times \dfrac{22.4\ dm^3}{\cancel{\text{mole}}} = 32.0\ dm^3$

19. [48.0 g]— $\# \text{ of moles} = \dfrac{\text{volume}}{\text{molar volume}} = \dfrac{33.5 \text{ dm}^3}{22.4 \text{ dm}^3\text{/mole}} = 1.50 \text{ moles}$

$\text{mass} = \# \text{ of moles} \times \text{molar mass} = 1.50 \text{ moles} \times \dfrac{32.0 \text{ g}}{\text{mole}} = 48.0 \text{ g}$

20. [2.68×10^{23}]—$\# \text{ of moles} = \dfrac{\text{volume}}{\text{molar volume}} = \dfrac{10.0 \text{ dm}^3}{22.4 \text{ dm}^3\text{/mole}} = 0.446 \text{ moles}$

$\# \text{ of molecules} = 0.446 \text{ moles} \times \dfrac{6.02 \times 10^{23} \text{ molecule}}{\text{mole}} = 2.68 \times 10^{23} \text{ mol.}$

21. [98.8 dm^3]— $C_3H_8 + 5O_2 \rightarrow 3CO_2 + 4H_2O$

38.8 g x dm^3

A. $\# \text{ of moles } C_3H_8 = \dfrac{\text{mass}}{\text{molar mass}} = \dfrac{38.8 \text{ g}}{44.0 \text{ g/mole}} = 0.882 \text{ moles}$

B. # of moles of O_2 can be determined from the molar ratio:

	C_3H_8		O_2
coefficients →	1	=	5
# of moles →	0.882		x

# of moles of O_2 = 4.41 moles

C. $\text{Volume of } O_2 = \# \text{ of moles} \times \text{molar volume} = 4.41 \text{ moles} \times \dfrac{22.4 \text{ dm}^3}{\text{mole}} = 98.8 \text{ dm}^3$

22. [33.3 dm^3]— $C_3H_8 + 5O_2 \rightarrow 3CO_2 + 4H_2O$

x dm^3 20.0 dm^3

The volume of O_2 can be determined from the molar ratio:

	O_2		CO_2
coefficients →	5	=	3
volume →	x		20.0 dm^3

volume of O_2 = 33.3 dm^3

23. [66.1 g]— $C_3H_8 + 5O_2 \rightarrow 3CO_2 + 4H_2O$

40.0 g x g

A. $\# \text{ of moles } C_3H_8 = \dfrac{\text{mass}}{\text{molar mass}} = \dfrac{40.0 \text{ g}}{44.0 \text{ g/mole}} = 0.909 \text{ moles}$

B. # of moles of H_2O can be determined from the molar ratio:

	C_3H_8		H_2O
coefficients →	1	=	4
# of moles →	0.909		x

# of moles of H_2O = 3.67 moles

C. $\text{Mass of } H_2O = \# \text{ of moles} \times \text{molar mass} = 3.67 \text{ moles} \times \dfrac{18.0 \text{ g}}{\text{mole}} = 66.1 \text{ g}$

24. [43.1 g]— $2Mg + O_2 \rightarrow 2MgO$

30.0 g 12.0 dm^3 x g

A. $\# \text{ of moles of Mg} = \dfrac{\text{mass}}{\text{molar mass}} = \dfrac{30.0 \text{ g}}{24.3 \text{ g/mole}} = 1.23 \text{ moles}$

B. $\# \text{ of moles of } O_2 = \dfrac{\text{volume}}{\text{molar volume}} = \dfrac{12.0 \text{ dm}^3}{22.4 \text{ dm}^3\text{/mole}} = 0.536 \text{ moles (limiting)}$

C. The oxygen is the limiting reactant because there is more than twice as much magnesium than oxygen.

D. # of moles of MgO can be determined from the molar ratio:

	O_2		MgO
coefficients →	1	=	2
# of moles →	0.536		x

# of moles of MgO = 1.07 moles

E. $\text{Mass of MgO} = \# \text{ of moles} \times \text{molar mass} = 1.07 \text{ moles} \times \dfrac{40.3 \text{ g}}{\text{mole}} = 43.1 \text{ g}$

25. [41.5 g]—

$$2Mg + O_2 \rightarrow 2MgO$$
$$25.0\text{ g} \quad 28.0\text{ g} \quad x\text{ g}$$

A. $\#\text{ of moles Mg} = \dfrac{\text{mass}}{\text{molar mass}} = \dfrac{25.0\ \cancel{g}}{24.3\ \cancel{g}/\text{mole}} = 1.03\text{ moles (limiting)}$

B. $\#\text{ of moles } O_2 = \dfrac{\text{mass}}{\text{molar mass}} = \dfrac{28.0\ \cancel{g}}{32.0\ \cancel{g}/\text{mole}} = 0.875\text{ moles}$

C. The magnesium is the limiting reactant because you would need 2(0.875) = 1.75 moles of magnesium to react completely with the 0.875 moles of oxygen. We have less than that.

D. # of moles of MgO can be determined from molar ratio:

$$\begin{array}{lcc} & \text{Mg} & \text{MgO} \\ \dfrac{\text{coefficients} \rightarrow}{\#\text{ of moles} \rightarrow} & \dfrac{2}{1.03} & = \dfrac{2}{x} \end{array}$$

# of moles of MgO = 1.03 moles

E. Mass of MgO = # of moles × molar mass = $1.03\ \cancel{\text{moles}} \times \dfrac{40.3\text{ g}}{\cancel{\text{mole}}} = 41.5\text{ g}$

Gases

Lesson 8–1: Measuring Gas Pressure

The ***Kinetic Theory of Gases***, which is also called the ***Kinetic Molecular Theory*** (KMT), attempts to explain the behavior of gases in terms of the motion of their molecules. The KMT consists of the following points:

1. All gases are made up of individual particles (atoms and/or molecules) that are in constant motion. These particles move in straight lines until (as stated in Newton's second law of motion) acted on by an unbalanced force.
2. The collisions between the particles of the gas are considered to be perfectly "elastic," which means that, although kinetic energy may be transferred from one particle to another, the net kinetic energy is conserved.

As you can imagine, when studying a system as complex as a sample of gas, there are an unmanageable number of variables involved. A typical sample of gas might contain 10^{23} particles, all of which are in constant motion, interacting with each other by way of collisions and intermolecular forces. To be able to carry out calculations on these samples, it is necessary to "ignore" certain variables, which we can't really account for, in much the same way that physics students are often told to "ignore" wind resistance, or friction in general. When we study gases mathematically, we treat them as imaginary, or "ideal," gases.

Ideal gases are imaginary gases, which have these important properties:

1. Although a sample of a gas may occupy a significant volume, the actual molecules of an ideal gas take up no space. They are treated as "point masses." Many real gases can approach this

condition of ideal gases. For example, if you have a sample of hydrogen gas at high temperature and low pressure, the particles of the gas don't occupy a very significant amount of the space occupied by the sample. The lower the density of a real gas, the closer it will approach this characteristic of an ideal gas.

2. Ideal gases are treated as if there is no attraction between particles. Again, many real gases can approach this characteristic of ideal gases. Because intermolecular forces decrease over distance, less dense gases are more like ideal gases. Any characteristic that results in weaker intermolecular forces will make a real gas more like an ideal gas. For example, non-polar gases are more like ideal gases than polar gases, if all other variables are the same. Small molecules, with less subatomic particles and taking up less significant space, tend to come closest to approaching ideal gas characteristics.

Pressure is defined as a force exerted over a given area, as shown in the following formula:

$$\textbf{Pressure} = \frac{\textbf{force}}{\textbf{area}}$$

Notice that the formula for pressure looks similar to the formula for density. In this case the pressure is directly related to the force, so an increase in force leads to an increase in pressure. Pressure is inversely proportional to the surface area, so that an increase in the area will lead to a decrease in the pressure.

You have experienced the consequences of this formula in many real-life situations. Think of the shape of a nail. The head of a nail is relatively broad, meaning that the area that you strike with a hammer is relatively large. The formula for pressure shows us that, by increasing the area of the surface of the head of the nail, we decrease the pressure per unit of area. The tip of a nail, which is designed to plunge into a wood surface, is relatively small. The pressure formula shows us that when the area is small, we exert a relatively large pressure per unit of area.

Another example of where this formula comes into play involves a person standing on a frozen lake when the ice starts to crack. You may have heard that it is advisable to lie down on the surface of the ice, if you find yourself in this situation. The logic behind this can be seen in the pressure formula. By lying down on the surface of the ice, you will increase the area of your body that is in contact with the ice. Because the pressure that you exert on the surface of the ice is inversely proportional to the area of contact, you will be less likely to break through.

Gases exert pressure, because their particles bombard the surface of a material they come in contact with. At this moment, billions of gas particles, which make up the air around you, are striking your body every second. Under normal circumstances, your internal pressure keeps these collisions from having any noticeable effect on you. Sometimes the difference between the external air pressure and your internal pressure is evidenced by a discomfort in your eardrums. When your ears "pop," your body attempts to compensate for the difference in pressure.

The atmospheric pressure is measured with an instrument called a ***barometer***, which is also called a ***closed manometer***. When you watch your local weather report, you will often hear the meteorologist discuss the barometric pressure, which is a measure of the force exerted by the atmospheric gases per unit of area. The news often reports pressure in "inches of mercury," because a barometer, as are many thermometers, is filled with mercury. The SI unit for pressure is the Pascal, which is derived from Newtons and meters. Other units that are commonly used to measure pressure include the atmosphere (atm), torr, and millimeters of mercury (mm of Hg). You should make sure that you know how to convert between the various units of pressure. The necessary conversion factors are shown here.

Conversion Factors for Units of Pressure

1.0 atm = 101.3 kPa = 760 torr = 760 mm of Hg

1.0 kPa = 7.5 mm of Hg

If you want to convert inches of mercury to any of the other units, you will want to remember that 1 inch is approximately 25.4 mm. An example of such a conversion is shown in the following example:

Example 1

The local news reports the atmospheric pressure as 30.03 inches of mercury. How many kilopascals does this represent?

$$30.03 \text{ ~~inches~~ of Hg} = \frac{25.4 \text{ mm}}{\text{inch}} \times \frac{1 \text{ kPa}}{7.50 \text{ mm of Hg}} = 101.7 \text{ kPa}$$

This is also a good place to review the concept of "standard temperature and pressure." Because the volume of a gas is so susceptible to changes in temperature and pressure, you must always make note of the temperature and pressure conditions that the volume of the gas refers to. As you should recall from Chapter 7, if you collected a 2.50 L sample of a gas in your laboratory on a day when the temperature is low and the atmospheric pressure is high, you actually have more gas molecules in the sample than you would if you collected 2.50 L of the same gas when the temperature was high and the atmospheric pressure was low. Reporting the volume of the gas alone is never enough, so the practice is to calculate the volume that your gas sample would occupy at "standard temperature and pressure." ***Standard temperature and pressure (STP)*** has been set at 273 K and 101.3 kPa.

Lesson 8–1 Review

1. A ____________ is an instrument that is used to measure atmospheric pressure.
2. ____________ is a measure of the force exerted over a given area.
3. STP stands for standard __________ and __________.

Make the following conversions:

4. 2.0 atm = _____ kPa
5. 107.5 kPa = _____ mm of Hg
6. 790 torr = _____ atm
7. 1.04 atm = _____ mm of Hg
8. 33.2 inches of Hg = _____ atm
9. 754 Torr = _____ mm of Hg
10. Which of the following is equivalent to **standard** temperature?
    a) 0 °C
    b) 0 K
    c) 273 °C
    d) 273 K
    e) Both A and D are correct

Lesson 8–2: Boyle's Law

Did you ever squeeze a balloon and see bulges appear in a different area of it? Did you ever climb onto an inflatable float or air mattress and notice that as you sunk into it, it started to get firmer beneath you? If you did either of these two things, you were unknowingly experimenting with Boyle's Law. ***Boyle's Law*** states that the pressure and volume of a gas at constant temperature are inversely proportional. This means that if you change the volume of a gas, without changing the temperature or number of particles of the gas, you get an inversely proportional change in the gas sample's pressure. The relationship between the volume and pressure of a gas will be made clearer by looking at the formula and some examples of Boyle's Law.

Boyle's Law

$$P_1V_1 = P_2V_2$$

where P is the pressure and V is the volume

Boyle's Law can be used to solve for an unknown volume or pressure, when the temperature and the number of particles in the sample remain constant. Let's try a few examples.

Example 1

A sample of neon gas has a volume of 250 cm³ and a pressure of 86.7 kPa. At what volume would this gas exert 100. kPa of pressure? (Assume the temperature remains constant.)

We will start by listing what we have been given and identifying the unknown, which is Volume 2 (V_2). We will then take the original formula for Boyle's Law and isolate the unknown by dividing both sides by P_2. After dividing both sides of the equation by P_2 to isolate V_2, we get our working formula. We then substitute and solve, as shown here:

Given: $V_1 = 250\ cm^3$; $P_1 = 86.7\ kPa$; $P_2 = 100.\ kPa$

Find: V_2

Formula: $V_2 = \frac{P_1V_1}{P_2} = \frac{(86.7\ \cancel{kPa})(250\ cm^3)}{(100.\ \cancel{kPa})} = 216.75\ cm^3$

Answer rounded is **220 cm³**

Now, I will show an example where the second pressure is the unknown. I will do all of the work in one step, but if I lose you, you can refer back to Example 1 to see what I did.

Example 2

A student collects 435 cm³ of hydrogen gas at 2.3 atm of pressure. Assuming the temperature of the gas doesn't change, at what pressure would this gas occupy 500. cm³?

Given: $V_1 = 435\ cm^3$; $P_1 = 2.3\ atm$; $V_2 = 500.\ cm^3$

Find: P_2

Formula: $P_2 = \frac{P_1V_1}{V_2} = \frac{(2.3\ atm)(435\ \cancel{cm^3})}{(500.\ \cancel{cm^3})} = 2.001\ atm$

Answer rounded is **2.0 atm**

Eventually you will find that you can solve many of the easiest examples of Boyle's Law calculations in your head, just by understanding what it means to say that the pressure and volume of the gas are inversely proportional. Look at Example 3 as an illustration of what I mean.

Example 3

A student collects 4.0 L of chlorine gas at 2.0 atm of pressure. Assuming the temperature of the gas doesn't change, at what pressure would this gas occupy 8.0 L?

Given: $V_1 = 4.0$ L; $P_1 = 2.0$ atm; $V_2 = 8.0$ L

Find: P_2

Formula: $P_2 = \frac{P_1V_1}{V_2} = \frac{(2.0 \text{ atm})(4.0 \cancel{\text{L}})}{(8.0 \cancel{\text{L}})} = 1.0 \text{ atm}$

Can you see why you should be able to do this example in your head? When we say that two things are inversely proportional, it means that if you double one, you must divide the other by 2. If you divide one by a factor of 3, then you must triple the other. In Example 3 we wanted to double the volume, so we had to divide the pressure in half. Many of the Boyle's Law problems that you are likely to encounter, especially on standardized tests, can be solved with some quick "mental" math. If you want to be sure of your answer, the formula demonstrated in this lesson will always work.

Before we move on to the practice problems, I want to mention an error that some students make. Some students will read a word problem and assume that V_1 is the volume that is mentioned first and V_2 is the volume that is mentioned second. To avoid this problem, some teachers will refer to the volumes as $V_{initial}$ and V_{final}, or V_i and V_f. Regardless of the convention that you use in class, you should think of the gas in terms of "before" and "after" the change occurs. The pressure and volume before the change would be P_1 and V_1, or $P_{initial}$ and $V_{initial}$. The pressure and volume after the change would be P_2 and V_2, or P_{final} and V_{final}. Try the review problems, and keep this suggestion in mind.

Lesson 8–2 Review

1. _______________ Law states that the pressure and volume of a gas, at constant temperature, are inversely proportional to each other.
2. If you increase the pressure on a sample of gas at constant temperature, what will happen to its volume?
3. When you start with the original formula, $P_1V_1 = P_2V_2$, what do you get when you isolate P_1?

4. What factors are assumed to be constant, when doing a Boyle's Law calculation?
5. A sample of neon gas occupies 200 cm^3 at 4.0 atm. What would be the volume of this sample at 1.0 atm, assuming the temperature is held constant?
   a) 100 cm^3 b) 200 cm^3 c) 500 cm^3 d. 800 cm^3
6. A student collects a 4.5 dm^3 sample of hydrogen gas 1.0 atm. If the temperature remains constant, at what pressure would this sample have a volume of 9.0 dm^3?
   a) 0.50 atm b) 1.0 atm c) 2.0 atm d) 5.5 atm
7. A sample of oxygen gas has a volume of 4.0 L at standard temperature and pressure. If the temperature remains constant, what would be the volume of this gas at 202.6 kPa?
   a) 1.0 L b) 2.0 L c) 4.0 L d) 8.0 L
8. What volume would a 60 ml sample of neon gas at STP occupy if the pressure were tripled, while the temperature remains the same?
   a) 20 ml b) 180 ml c) 3.0 atm d) 760 torr
9. If a sample of hydrogen occupies 2.5 L at STP, how much space would it occupy at 120 kPa, assuming the temperature remains constant?
10. At what pressure would a sample of neon occupy 4.0 L, if it occupies 3.4 L at 720 mm of Hg? Assume temperature remains constant.

Lesson 8–3: Charles's Law

Like Boyle's Law, Charles's Law can explain many phenomenas that you have experienced in your real life. Maybe you have had the experience of inflating a pool float to the point where it seems quite firm, and then throwing it into a cold pool. Then you jump in the pool and climb on top of the pool float, which no longer seems nearly as firm as it did outside of the pool. Your first thought might be that the float has sprung a leak. It is probably more likely that you have simply experienced an example of ***Charles's Law***, which states that the volume of an ideal gas, at constant pressure, varies directly with its Kelvin temperature. In simpler terms, this tells us that when gases get hot, their volumes increase, and when gases get cold, their volumes decrease. Putting your float in the cold water just made the volume of the gas inside decrease.

The Formula for Charles's Law

$$\frac{V_1}{T_1} = \frac{V_2}{T_2}$$

It is very important that you work with Kelvin temperature, whenever you do a Charles's Law problem. If you work in Celsius you won't get the correct answer. You may recall that we discussed the problem with the Celsius scale in Chapter 2. You can have negative values for temperatures in the Celsius scale, which, when solving a Charles's Law problem for volume, would result in a negative value for volume. Can you imagine what a gas with a negative volume would be like? Neither can I, and that is why we must always convert our temperatures to Kelvin when solving Charles's Law problems.

Although you need to work in Kelvin, the temperature that you are given in a particular problem may be in Celsius degrees. This upsets some students, and they consider this an example of a "trick" question. There is, however, a logical reason for your teacher to give you some Charles's Law problems in Celsius degrees. Most students in chemistry laboratories work with thermometers marked with the Celsius scale. In these real-life situations, you will need to convert between temperature units before doing calculations. Some problems are designed to get you ready for these real-life experiences.

Now, let's look at an example of a problem that requires Charles's Law to solve.

Example 1

A student collects a 250. cm^3 sample of a gas at 21.0 °C. Assuming the pressure of the gas remains constant, at what temperature will the volume of this gas be 500. cm^3?

The first thing that we want to do is change the Celsius degrees to Kelvin, using the formula K = °C + 273 that we learned in Chapter 2. I like to do that right away, directly on the written problem, to avoid making mistakes. Because 21 + 273 = 294, I will change the temperature to 294 K, as shown below.

A student collects a 250. cm^3 sample of a gas at ~~21.0 °C~~ 294 K. Assuming the pressure of the gas remains constant, at what temperature will the volume of this gas be 500. cm^3?

Next, we list what we are given, identify the unknown, and write the original formula.

Given: V_1 = 250. cm^3; T_1 = 294 K; V_2 = 500. cm^3

Find: T_2

Formula: $\frac{V_1}{T_1} = \frac{V_2}{T_2}$

Note: To isolate the unknown (T_2) we must multiply both sides by T_2, divide both sides by V_1, and multiply both sides by T_1.

Original formula $\frac{V_1}{T_1} = \frac{V_2}{T_2}$

Multiplying both sides by T_2, we get $T_2 \times \frac{V_1}{T_1} = \frac{V_2}{\cancel{T_2}} \times \cancel{T_2}$

Dividing both sides by V_1, we get $\frac{\cancel{V_1} T_2}{T_1 \cancel{V_1}} = \frac{V_2}{V_1}$

Multiplying both sides by T_1, we get $\cancel{T_1} \times \frac{T_2}{\cancel{T_1}} = \frac{V_2 T_1}{V_1}$

Which leaves us with $T_2 = \frac{V_2 T_1}{V_1}$

Now, armed with this working formula, let's solve the problem.

Given: V_1 = 250. cm^3; T_1 = 294 K; V_2 = 500. cm^3

Find: T_2

Formula: $T_2 = \frac{V_2 T_1}{V_1} = \frac{(500.\ \cancel{cm^3})(294\ K)}{(250.\ \cancel{cm^3})} = \mathbf{588\ K}$

Notice that you could probably have solved this problem in your head as well, once you changed the temperature to Kelvin. Because the volume of the gas varies directly with its Kelvin temperature, if we want to double the volume (from 250. cm^3 to 500. cm^3), we needed to double the temperature (from 294 K to 588 K).

Now, let's try an example where the volume is the unknown.

Example 2

A student generates a sample of hydrogen gas, which occupies 1.2 L at 12.0 °C. Assuming that the pressure of the gas remained constant, what would be the volume of this gas sample at 67.0 °C?

You can clearly see that the final volume (V_2) is the unknown, because the question asks "what will be the volume of this gas?" Therefore, let's start this time by isolating V_2 in our original formula, by multiplying both sides by T_2.

Original formula $$\frac{V_1}{T_1} = \frac{V_2}{T_2}$$

Multiplying by T_2, we get $$T_2 \times \frac{V_1}{T_1} = \frac{V_2}{\cancel{T_2}} \times \cancel{T_2}$$

Which leaves us with $$V_2 = \frac{V_1T_2}{T_1}$$

Before we forget, let's make sure that we change the Celsius temperature in the problem to Kelvin. When we add 273 to 12.0 °C (T_1), we get 285 K. Adding 273 to 67.0 °C (T_2), we get 340 K. We can now go to our problem, identify the given and unknown, and solve.

> A student generates a sample of hydrogen gas, which occupies 1.2 L at ~~12.0 °C~~ 285 K. Assuming that the pressure of the gas remained constant, what would be the volume of this gas sample at ~~67.0 °C~~ 340 K?

Given: V_1 = 1.2 L; T_1 = 285 K; T_2 = 340 K

Find: V_2

Formula: $$V_2 = \frac{V_1T_2}{T_1} = \frac{(1.2\text{ L})(340\ \cancel{\text{K}})}{(285\ \cancel{\text{K}})} = \mathbf{1.4\ L}$$

Lesson 8–3 Review

1. ____________ Law states that the volume of a gas, at constant pressure, varies directly with its Kelvin temperature.
2. What factors are considered constant in Charles's Law calculations?
3. Which temperature scale must be used in Charles's Law calculations, and why?

Convert the following Celsius temperatures to Kelvin:

4. -132 °C 5. 0 °C 6. 32 °C 7. 96 °C

8. A student collects a sample of hydrogen gas, which has a volume of 350 cm^3 at 22.0 °C. If the pressure of the gas remains constant, what would be the volume of this gas at standard temperature?
9. A sample of argon gas has a volume of 5.00 L at STP. If the pressure remains constant, at what temperature would the volume of this gas reach 10.0 L?
10. A balloon has a volume of 0.855 L at 1.0 atm and 17.0 °C. If the pressure remains constant, what will be the volume of this gas at 0 °C?

Lesson 8–4: Dalton's Law

Dalton's Law of partial pressure states that the total pressure exerted by a mixture of gases is equal to the sum of each of the partial pressures of the gases in the mixture. Dalton's Law is expressed by the formula shown here:

$$P_{total} = P_1 + P_2 + ...P_x$$

How many "Ps" you show in the formula will depend upon the number of gases in the mixture. Look at how we use Dalton's Law to solve the following example.

Example 1

A mixture of gases consists of oxygen, nitrogen, and hydrogen with pressures of 1.2 atm, 0.8 atm, and 3.5 atm, respectively. Find the total pressure exerted by this mixture of gases.

$P_{total} = P_1 + P_2 + P_3 = 1.2 \text{ atm} + 0.8 \text{ atm} + 3.5 \text{ atm} = \mathbf{5.5\ atm}$

* If you are surprised by the way that I rounded in Example 1, review the rule for addition and subtraction found in Lesson 2–4.

All of the problems using only Dalton's Law will be very easy, but pay careful attention to the wording. Your teacher may make these problems slightly harder by making one of the partial pressures the unknown. The problem is still very easy, but it is possible that you could make a careless error if you read the question too quickly.

If one of the pressures, say P_2, is the unknown, simply isolate it on one side of the equation. Look at Example 2 to see how this is done.

Example 2

A mixture of two gases exerts a total pressure of 142.5 kPa. If the partial pressure exerted by one of the gases is 96.4 kPa, what is the partial pressure exerted by the other gas?

$$P_2 = P_{total} - P_1 = 142.5 \text{ kPa} - 96.4 \text{ kPa} = \mathbf{46.1 \text{ kPa}}$$

A variation on this type of problem involves making use of the molar ratio of the gases in the mixture. Take a look at Example 3.

Example 3

A mixture made up of 2.0 moles of nitrogen gas and 3.0 moles of hydrogen gas exerts a total pressure of 4.0 atm. What is the partial pressure exerted by the hydrogen gas?

As you can see, the gas is made up of 3.0 parts (moles) hydrogen and 2.0 parts nitrogen. The gas is then made up of 60% hydrogen, and therefore hydrogen is responsible for 60% of the total pressure exerted by the gas. Here is the work:.

pressure of hydrogen = total pressure × % of mixture that is hydrogen

= 4.0 atm × 60%

= 4.0 atm × .60

= **2.4 atm**

Another variation of this type of problem that you are likely to encounter involves mathematically removing the water vapor pressure from a sample of gas, in order to determine the pressure exerted by a dry gas. In the laboratory, you may collect a sample of a gas by bubbling it through a column of water. The resultant gas will be a mixture of the water vapor and whatever gaseous product that you intended to collect. In order to subtract out the water vapor pressure, you must determine the temperature of the gas sample, and then refer to a table such as the one shown on page 271.

Vapor Pressure of Water					
°C	mm of Hg	kPa	°C	mm of Hg	kPa
0	4.6	0.61	26	25.2	3.36
5	6.5	0.87	27	26.7	3.57
10	9.2	1.23	28	28.3	3.78
15	12.8	1.71	29	30.0	4.01
16	13.6	1.82	30	31.8	4.25
17	14.5	1.94	35	42.2	5.63
18	15.5	2.06	40	55.3	7.38
19	16.5	2.19	50	92.5	12.34
20	17.5	2.34	60	149.4	19.93
21	18.6	2.49	70	233.7	3.18
22	19.8	2.64	80	355.1	47.37
23	21.1	2.81	90	525.8	70.12
24	22.4	2.98	100	760.0	101.32
25	23.8	3.17	110	1074.6	143.28

Figure 8–4a

Let's see an example of how this chart would be used.

Example 4

A student collects hydrogen by displacing water. The mixture she collects exerts a total pressure of 743.3 mm of Hg at 25 °C. What is the pressure of just the dry hydrogen gas?

Working with the formula for Dalton's Law, we get the following:

$$P_2 = P_{total} - P_1 \text{ or}$$

$$P_{dry\ hydrogen} = P_{total} - P_{H_2O}$$

We will also need to know the pressure exerted by the water vapor at 25 °C, which, according to the Vapor Pressure of Water table, is 23.8 mm of Hg. Now, we are ready to solve the problem.

$$P_{\text{dry hydrogen}} = P_{\text{total}} - P_{H2O}$$

$$P_{\text{dry hydrogen}} = 743.3 \text{ mm of Hg} - 23.8 \text{ mm of Hg} = \mathbf{719.5 \text{ mm of Hg}}$$

The only challenge to this variation on Dalton's Law is that it is unlikely that anyone will remind you to use a table such as the one presented in this lesson (Vapor Pressure of Water). Just try to associate questions about water vapor pressure to this type of table. If a question talks about a gas "collected over water" or if it asks for the "pressure of the dry gas," then you probably will need a table of water vapor pressures.

Lesson 8–4 Review

1. A mixture of helium and hydrogen gases exerts a combined pressure of 2.5 atm. If the partial pressure of the hydrogen gas is 0.8 atm, what is the partial pressure exerted by the helium gas?
2. A mixture contains argon, neon, and hydrogen gases. If the partial pressures of these gases are 45.5 kPa, 23.6 kPa, and 85.3 kPa respectively, what is the total pressure exerted by the mixture?
3. A sample of hydrogen gas is generated and collected by displacing water at 30.0 °C. If the pressure of the resultant hydrogen and water vapor mixture is 745.3 mm of Hg, what is the pressure of the dry hydrogen gas?
4. A flask containing 2.0 moles of oxygen and 2.0 moles of carbon dioxide has a total pressure of 6.0 atm. What is the partial pressure of the oxygen gas?
5. What is the total pressure of a gaseous mixture containing helium, oxygen, and carbon dioxide, if the partial pressure exerted by each gas is 25.0 kPa?
6. A vessel contains a mixture of neon and argon with a total pressure of 1.8 atm. If the partial pressure of the neon gas is 0.5 atm, what is the pressure of the argon gas?

   a) 0.5 atm b) 1.3 atm c) 1.8 atm d) 2.3 atm

7. A sample of chlorine gas is collected in a bottle by displacing water at a temperature of 20.0 °C. The total pressure exerted by the mixture of chlorine and water vapor is 85.6 kPa. What is the pressure of the dry chlorine gas?
   a) 85.6 kPa b) 2.34 kPa c) 83.3 kPa d) 87.94 kPa

8. A mixture of gases is composed of 1.00 mole of carbon dioxide, 2.00 moles of helium, and 2.00 moles of oxygen. If the total pressure is 1.00 atm, what is the partial pressure of the carbon dioxide?
   a) 0.100 atm b) 0.200 atm c) 0.400 atm d) 1.00 atm

Lesson 8–5: Combined Gas Law

Although you learn about each of these gas laws in isolation, in reality, you often will need to use some combination of them in experimental situations. The term ***Combined Gas Law*** could really refer to any combination of gas laws, which you find yourself using. Often, however, the term is used to describe the combination of Boyle's Law and Charles's Law, which is often used to mathematically adjust the volume of a gas to STP conditions.

You see, it is unlikely that a student or scientist will actually be measuring the volume of a gas at standard temperature and pressure. Standard temperature is 0 °C, and that is colder than most of us would want to work at. Variations in temperature and pressure result in significant changes in the volume of a gas, and it is difficult to visualize the quantity of a gas at varying conditions. It is very common for this reason to mathematically determine the volume that a gas sample would occupy at STP. In order to do this, we combine Charles's and Boyle's Laws into the formula shown here.

The Combined Gas Law

$$V_2 = \frac{V_1 P_1 T_2}{P_2 T_1}$$

Keep in mind that the subscript numbers (as in V_1 and V_2) don't refer to the order in which the quantities appear in the word problem; rather, they refer to "before" and "after," or "initial" and "final." Let's go over an example of how this formula is used.

Example 1

A student generates and collects a 550 cm^3 sample of chlorine gas at 23 °C and 1.2 atm. What would be the volume of this sample of gas at STP?

As in the case of working with Charles's Law by itself, we must be sure to always do our calculations in Kelvin. Therefore, converting to Kelvin, by adding 273 to the Celsius temperature, is a priority. After that, we can list the givens, identify the unknown, and solve the equation.

A student generates and collects a 550 cm^3 sample of chlorine gas at ~~23 °C~~ 296 K and 1.2 atm. What would be the volume of this sample of gas at STP?

Given: $V_1 = 550\ cm^3$; $P_1 = 1.2$ atm; $T_1 = 296$ K, $P_2 = 1.0$ atm, $T_2 = 273$ K

Find: V_2

Formula: $$V_2 = \frac{V_1P_1T_2}{P_2T_1} = \frac{(550\ cm^3)(1.2\ \cancel{atm})(273\ \cancel{K})}{(1.0\ \cancel{atm})(296\ \cancel{K})} = \mathbf{610\ cm^3}$$

Of course, you might get called on to solve a problem where V_2 is not the unknown. The key to correctly answering these problems will be your ability to identify and isolate the unknown. It is an unfortunate fact that some students do poorly in chemistry, simply because their algebra skills are somewhat weak. If you identify such a weakness in yourself, don't let it affect your chemistry class. Get an algebra tutor, or look for *Master Math: Algebra* (also published by Career Press).

Let's look at another example.

Example 2

A student collects 4.85 L of hydrogen gas at 314 K and 101.8 kPa of pressure. The gas is then transferred to a 2.00 L vessel and allowed to cool to 294 K. What is the new pressure exerted by the gas?

In this example, P_2 is the unknown. In order to isolate P_2,we must multiply both sides of the original equation by P_2 and divide both sides by V_2, as shown here:

$$P_2 \times \frac{\cancel{V_2}}{\cancel{V_2}} = \frac{V_1P_1T_2}{\cancel{P_2}T_1V_2} \times \cancel{P_2}$$

This gives us:

$$P_2 = \frac{V_1P_1T_2}{T_1V_2}$$

Now we have the working formula necessary to solve the problem.

Given: V_1 = 4.85 L; P_1 = 101.8 kPa; T_1 = 314 K; V_2 = 2.00 L;
T_2 = 294 K

Find: P_2

Formula: $P_2 = \frac{V_1P_1T_2}{T_1V_2} = \frac{(4.85\ \cancel{L})(101.8\ kPa)(294\ \cancel{K})}{(314\ \cancel{K})(2.00\ \cancel{L})}$ = **231 kPa**

For our final example, let's try a problem where T_2 is the unknown.

Example 3

A sample of argon has a volume of 4.21 L at 34.0 °C and 1.13 atm. If the pressure is allowed to decrease to 1.00 atm, at what temperature will this gas occupy 5.00 L?

As always, we will start by changing the temperature to the Kelvin scale, by adding 273. Then we will list the given information and what we are asked to find.

A sample of argon has a volume of 4.21 L at ~~34.0 °C~~ 307 K and 1.13 atm. If the pressure is allowed to decrease to 1.00 atm, at what temperature will this gas occupy 5.00 L?

Given: V_1 = 4.21 L; T_1 = 307 K; P_1 = 1.13 atm; V_2 = 5.00 L;
P_2 = 1.00 atm

Find: T_2

Now, let's isolate the unknown in the equation.

Original formula: $V_2 = \frac{V_1P_1T_2}{P_2T_1}$

Multiply both sides by P_2T_1: $V_2P_2T_1 = \frac{V_1P_1T_2}{\cancel{P_2T_1}} \times \cancel{P_2T_1}$

Now, divide both sides by V_1P_1: $\frac{V_2P_2T_1}{V_1P_1} = \frac{\cancel{V_1P_1}T_2}{\cancel{V_1P_1}}$

Our formula becomes: $T_2 = \frac{V_2P_2T_1}{V_1P_1}$

We are ready to solve the problem.

A sample of argon has a volume of 4.21 L at ~~34.0 °C~~ 307 K and 1.13 atm. If the pressure is allowed to decrease to 1.00 atm, at what temperature will this gas occupy 5.00 L?

Given: V_1 = 4.21 L; T_1 = 307 K; P_1 = 1.13 atm; V_2 = 5.00 L;
P_2 = 1.00 atm

Find: T_2

Formula: $T_2 = \frac{V_2P_2T_1}{V_1P_1} = \frac{(5.00\ L)(1.00\ atm)(307\ K)}{(4.21\ L)(1.13\ atm)} = 322.66\ K = \mathbf{323\ K}$

When doing these practice problems, remember to change °C to Kelvin, and make sure that you label each quantity correctly. Pay attention to the initial (V_1, T_1, and P_1) and final (V_2, T_2, and P_2) conditions.

Lesson 8–5 Review

1. A sample of neon occupies 7.6 L at 23.0 °C and 1.12 atm. What would be the volume of this gas at STP?
2. A sample of hydrogen has a volume of 250. cm^3 at STP. What would be the volume of this gas at 359 K and 89.9 kPa?
3. A sample of oxygen has a volume of 5.56 dm^3 at 15.0 °C and 755 mm of Hg. If the gas were allowed to cool to -11.0 °C, at what pressure would it occupy 5.00 dm^3?
4. A student collects a 5.00×10^2 cm^3 sample of hydrogen gas at 22.5 °C and 103.3 kPa of pressure. How much space would this gas occupy at STP?
5. A sample of carbon dioxide occupies 2.75 L at 43.5 °C and 733 mm of Hg. How much space would this gas occupy at STP?

Lesson 8–6: Graham's Law

In order to smell something, molecules must come off that substance and enter your nose. When a woman who is wearing perfume enters a room, molecules of that perfume must leave her skin and travel to your nose in order for you to smell it. These molecules will not just leave her skin and travel in a straight path to your nose; rather, they ***diffuse***, or spread out in all directions. What factors determine the amount of time that elapses from when she enters the room and when you smell her perfume? A study of Graham's Law will allow you to answer this and other questions.

The first factor that will affect the rate of diffusion will be the temperature of the molecules. Remember that we defined temperature as the average kinetic energy of the molecules of a substance. Remember, also, that the formula for kinetic energy is:

$K.E. = 1/2\ MV^2$

where M is the mass and V is the velocity of the object

When we heat a substance, its temperature increases as the speed of its molecules increases. Why would a woman apply perfume to her neck and wrists? Because those are particularly hot areas of the body, which will heat the perfume faster and allow the molecules of the perfume to leave her body faster and travel faster.

How does the mass of the molecules affect the rate of diffusion? Let's compare the molecules of two different gases, say methane (CH_4) and helium (He), at equal conditions of temperature and pressure.

Comparison of 2 Different Gases at Equal Temperatures and Pressures

	CH_4 He
If they are at the same temperature, then	$1/2\ M_1V_1^2 = 1/2\ M_2V_2^2$
If we multiply both sides by 2, then	$M_1V_1^2 = M_2V_2^2$
Now, we rearrange the equation	$\frac{V_1^2}{V_2^2} = \frac{M_2}{M_1}$
Taking the square root of both sides, we get	$\frac{V_1}{V_2} = \sqrt{\frac{M_2}{M_1}}$

Now, the molecular mass of methane is 16.0 u and the mass of helium is 4.00 u. Filling these values in for M_1 and M_2, we get:

$$\frac{V_1}{V_2} = \sqrt{\frac{16u}{4u}} = \sqrt{4} = 2$$

Stating as a ratio of V_1 to V_2 we get $V_1:V_2 =$ **2:1**

This tells us that the less massive (helium) gas will diffuse twice as fast as the more massive (methane) gas. This makes sense, of course, because if two objects have the same kinetic energy, the more massive one must be moving slower. Graham's Law applies this concept to gases.

Graham's Law states that "under equal conditions of temperature and pressure, gases diffuse at a rate that is inversely proportional to the square roots of their molecular masses." If your teacher asks you which of a group of gases will diffuse most quickly under equal conditions, the answer will be the least massive gas. If the question involves the relative rate of diffusion, you would solve the problem as shown in the comparison just presented. Let's try a typical example.

Example 1

Calculate the ratio of the velocity of neon atoms to the velocity of oxygen molecules at the same temperature.

We will need the molecular mass of the oxygen, which is diatomic, and the atomic mass of the neon, which exists as monatomic particles. We can get this information from the periodic table.

$$\text{Neon (Ne)} = 20.2\ \text{u}$$
$$\text{Oxygen } (O_2) = 32.0\ \text{u}$$

Now, we just use the formula that we derived in the comparison of two different gases at equal temperatures and pressures. I always set it up so that the more massive particle goes on the top of the fraction under the radical sign.

$$\frac{V_{Ne}}{V_{O_2}} = \sqrt{\frac{M_{O_2}}{M_{Ne}}} = \sqrt{\frac{32.0\text{u}}{20.2\text{u}}} = \sqrt{1.584158} = 1.26$$

The less massive particles move faster, so we would say the neon diffuses 1.26 times as fast as the oxygen, or the ratio of $V_{O_2}:V_{Ne}$ = **1:1.26**

We will do one more example. Although the wording may seem quite different, it is still an example of a Graham's Law problem. Remember: We find the masses of the elements involved by looking them up on the periodic table.

Example 2

Calculate the relative rate of diffusion of propane (C_3H_8) molecules to methane (CH_4) molecules.

$$\frac{V_{CH_4}}{V_{C_3H_8}} = \sqrt{\frac{M_{C_3H_8}}{M_{CH_4}}} = \sqrt{\frac{44.1u}{16.0u}} = \sqrt{2.75625} = 1.66$$

The less massive particles move faster, so we would say the methane diffuses 1.66 times as fast as the propane, or the ratio of $V_{C_3H_8}$:V_{CH_4} = **1:1.66**

Try the following problems and check your answers at the end of the chapter before moving on to the next lesson.

Lesson 8–6 Review

1. Calculate the ratio of the velocity of carbon dioxide (CO_2) molecules to hydrogen (H_2) molecules at the same temperature.
2. Calculate the ratio of the velocity of argon (Ar) atoms to hydrogen (H_2) molecules at the same temperature.
3. Calculate the ratio of the velocity of methane (CH_4) molecules to helium (He) atoms at the same temperature.
4. Calculate the relative rate of diffusion of butane (C_4H_{10}) molecules to carbon monoxide (CO) molecules.
5. Calculate the relative rate of diffusion of oxygen (O_2) molecules to methane (CH_4) molecules.

Lesson 8–7: Ideal Gas Law

One of the most useful gas laws is the Ideal Gas Law, which shows the mathematical relationship between the pressure, volume, temperature, and number of particles of a gas. The formula for the Ideal Gas Law is shown here:

$$PV = nRT$$

where P = pressure; V = volume; n = number of moles;
T = temperature in Kelvin; and
R = the ideal gas law constant, 8.31 $dm^3 \times kPa/mol \times K$

The Ideal Gas Law can be used when any one of the four variables is missing, provided the other information is known. R, which is a constant, is the same for every calculation. The units that come with the constant, R, dictate the units that we must have for the other quantities for the calculation. For example, we must work in Kelvin for temperature and kilopascals for pressure. If we were asked to do a problem with atmospheres of pressure or Celsius degrees for temperature, we would need to make a conversion before we performed our Ideal Gas Law calculation.

For our first example using the Ideal Gas Law, let's calculate the molar volume of a gas at STP, which we assumed to be true in previous chapters.

Example 1

What volume in dm^3 would 1.00 mole of hydrogen gas occupy at standard temperature and pressure (STP)?

Given: P = 101.3 kPa; n = 1.00 mole;
R = 8.31 dm^3 × kPa/mol × K; T = 273 K

Find: V

Formula: $V = \frac{nRT}{P} = \frac{(1.00\ \cancel{mol})(8.31\ dm^3 \times \cancel{kPa/mol} \times \cancel{K})(273\ \cancel{K})}{(101.3\ \cancel{kPa})}$

$= \mathbf{22.4\ dm^3}$

So, the Ideal Gas Law can be used to verify our value for the volume occupied by one mole of any gas (molar volume) at STP. Remember that one dm^3 is equivalent to one liter, so these units are interchangeable.

For our next example, let's pretend that the value of the constant was unknown. How could we use the known and accepted value for the molar volume of a gas at STP to calculate the value of the constant, R?

Example 2

Assuming that the molar volume of a gas (22.4 dm^3/mole) at STP has been experimentally verified, use this information and the Ideal Gas equation to mathematically determine the value of R.

Given: P = 101.3 kPa; V = 22.4 dm^3; n = 1.00 mole; T = 273 K
Find: R

Formula: $R = \frac{PV}{nT} = \frac{(101.3\ kPa)(22.4\ dm^3)}{(1.00\ mol)(273\ K)} = \mathbf{8.31\ kPa \times dm^3/mol \times K}$

Hopefully, Example 2 shows you how the value for the constant, and the units that come with the constant, can be determined.

Suppose we collected a sample of chlorine gas in the laboratory, which was not at STP, and we wanted to determine the number of moles that the sample contained. With the calculations that we learned in earlier lessons, we would be able to adjust the volume of the gas to STP, using the Combined Gas Law. Then, we could use the known molar volume of a gas to determine the number of moles of chlorine gas we had. The Ideal Gas Law allows us to do this work in a single calculation, as shown in Example 3.

Example 3

A student collects a sample of chlorine gas, which occupies 0.750 dm^3 at a temperature of 297 K and a pressure of 103.2 kPa. How many moles of chlorine does this sample represent?

Given: V = 0.750 dm^3; T = 297 K; P = 103.2 kPa;
R = 8.31 $dm^3 \times$ kPa/mol × K

Find: n

Formula: $n = \frac{PV}{RT} = \frac{(103.2\ \cancel{kPa})(0.750\ \cancel{dm^3})}{(8.31\ \cancel{dm^3} \times \cancel{kPa}/mol \times \cancel{K})(297\ \cancel{K})} = \mathbf{0.0314\ moles}$

Just as a reminder, for our final problem let's try an example where you are required to make a couple of conversions before you can solve the problem. You will want to remember the formulas for temperature and pressure conversions, which we covered in earlier chapters.

Example 4

How much space would 3.45 moles of nitrogen gas occupy at a temperature of 18 °C and a pressure of 888 mm of Hg?

I believe in getting in the habit of making the necessary conversions as early as possible. I often make the conversions, and then cross out the original values and write the converted values right in the problem.

We would convert the temperature to Kelvin with this formula:

$$K = {}^{\circ}C + 273 = 18\ {}^{\circ}C + 273 = \mathbf{291\ K}$$

We would convert the pressure to kPa by making use of the conversion factor 1.00 kPa = 7.50 mm of Hg.

$$888 \text{ ~~mm~~ of Hg} \times \frac{1.00 \text{ kPa}}{7.50 \text{ ~~mm~~ of Hg}} = \mathbf{118\ kPa}$$

Now, substitute these values into the original problem and solve.

How much space would 3.45 moles of nitrogen gas occupy at a temperature of ~~18 °C~~ 291 K and a pressure of ~~888 mm of Hg~~ 118 kPa?

Given: n = 3.45 mol; T = 291 K; P = 118 kPa;
R = 8.31 dm^3 × kPa/mol × K

Find: V

Formula: $V = \frac{nRT}{P} = \frac{(3.45 \text{ mol})(8.31 \text{ dm}^3 \times \text{kPa/mol} \times \text{K})(291 \text{ K})}{118 \text{ kPa}}$

$= \mathbf{70.7\ dm^3}$

As you can see, using the Ideal Gas Law is really as simple as isolating the unknown. Don't be intimidated by the large number of units involved in the calculations. Simply isolate the unknown, substitute, and solve. Try the review problems, and check your answers at the end of the chapter.

Lesson 8–7 Review

In the Ideal Gas Law formula, shown here, identify what each letter stands for.

PV = nRT

1. P______________
2. V______________
3. n______________
4. T______________
5. R______________

Use the Ideal Gas Law to solve each of the following problems:

6. At what temperature would 2.5 moles of argon gas occupy 60.0 dm^3 at a pressure of 98.4 kPa?

7. How many moles of helium gas would occupy 33.5 dm^3 at 22.0 °C and 1.50 atm?

8. How much space would 6.0 moles of oxygen gas occupy at 112.0 kPa of pressure and 33.0 °C?

9. How many moles of neon gas would occupy 550 cm^3 at 112 K and 98.4 kPa of pressure?

10. At what temperature would 4.55 moles of carbon dioxide occupy 4.5 dm^3 at a pressure of 850 mm of Hg?

Chapter 8 Examination

Part I—Matching

Match the following gas laws to the descriptions and the formulas that follow. Answers can and will be used more than once.

a. Boyle's Law c. Dalton's Law e. Ideal Gas Law

b. Charles's Law d. Graham's Law

_____1. This law states that the volume of a gas at constant pressure varies directly with its Kelvin temperature.

_____2. This law states that under equal conditions of temperature and pressure, gases diffuse at a rate that is inversely proportional to the square roots of their molecular masses.

_____3. This law states that the total pressure exerted by a mixture of gases is equal to the sum of each of the partial pressures of the gases in the mixture.

_____4. This law states that the pressure and volume of a gas at constant temperature are inversely proportional to each other.

_____5. This law shows the relationship between the pressure, volume, temperature, and number of particles of a gas.

_____6. Mathematically, this law is represented by the formula $\frac{V_1}{V_2} = \sqrt{\frac{M_2}{M_1}}$.

_____7. Mathematically, this law is represented by the formula $P_{total} = P_1 + P_2 \ldots P_x$.

_____8. Mathematically, this law is represented by the formula PV = nRT.

_____9. Mathematically, this law is represented by the formula $P_1V_1 = P_2V_2$.

_____10. Mathematically, this law is represented by the formula $\frac{V_1}{T_1} = \frac{V_2}{T_2}$.

Part II—Multiple Choice

Select the best answer for each of the following questions.

11. Under which conditions will a real gas act most like an ideal gas?
    a) low temperature and high pressure
    b) high temperature and low pressure
    c) low temperature and low pressure
    d) high temperature and high pressure

12. Under equal conditions of temperature, pressure, and number of particles, which of the following will act most like an ideal gas?
    a) He b) CH_4 c) CO_2 d) H_2O

13. As the volume of a closed gas sample is decreased at constant temperature, ________.
    a) the pressure will increase and the mass will stay the same
    b) the pressure will increase and the mass will increase
    c) the pressure will decrease and the mass will decrease
    d) both the pressure and the mass will increase

14. As the temperature of a gas at constant pressure is decreased, ________.
    a) the volume will increase and the mass will stay the same
    b) the volume will decrease and the mass will stay the same
    c) the volume will stay the same and the mass will increase
    d) both the volume and the mass will increase

15. Several gases are physically mixed together in a closed vessel. The pressure of the resultant mixture of gases will be equal to ________.
    a) the pressure exerted by the most massive of the gases
    b) the pressure exerted by the least massive of the gases
    c) the product of all of the partial pressures of the individual gases
    d) the sum of all of the partial pressures of the individual gases

16. Under identical conditions of temperature and pressure, which of the following gases will diffuse most quickly?
    a) H b) Ne c) CO d) O_2
17. A sample of hydrogen gas is collected by displacing water at a temperature of 25.0 °C. The pressure of the resultant mixture is 85.5 kPa. What is the partial pressure exerted by the water vapor in the mixture?
    a) 3.17 kPa b) 85.5 kPa c) 88.6 kPa d) 82.3 kPa
18. A sample of hydrogen gas is collected by displacing water at a temperature of 25.0 °C. The pressure of the resultant mixture is 85.5 kPa. What is the partial pressure exerted by the dry hydrogen gas?
    a) 3.17 kPa b) 85.5 kPa c) 88.6 kPa d) 82.3 kPa
19. A sample of neon occupies 3.49 dm³ at 111 K and 0.98 atm. Which formula can be used to determine the volume that this gas would occupy at STP?
    a) $\frac{V_1}{V_2} = \sqrt{\frac{M_2}{M_1}}$ b) $\frac{V_1}{T_1} = \frac{V_2}{T_2}$ c) $V_2 = \frac{V_1P_1T_2}{P_2T_1}$ d) $P_1V_1 = P_2V_2$
20. What is the density of argon gas at STP?
    a) 22.4 dm³/mol b) 39.9 g/mol c) 1.12 g/dm³ d) 1.78 g/dm³

Part III—Calculations

Perform the following calculations.

21. How many moles of nitrogen (N_2) gas would occupy 3.75 dm³ at 23.0 °C and 1.45 atm?
22. A sample of argon gas occupies 350 cm³ at 750 mm of Hg. Assuming the temperature of the gas remains constant, what will be the volume of this gas at 1120 mm of Hg?
23. Calculate the relative rate of diffusion for argon and neon gas under identical conditions of temperature and pressure.
24. Calculate the volume that a sample of helium gas will occupy at 45.0 °C, if it occupies 0.993 L at 19.0 °C. Assume the pressure of the gas is constant.

25. A mixture of gases is made up of oxygen, hydrogen, and neon. If the partial pressures exerted by each of these gases is 0.23 atm, 0.33 atm, and 0.42 atm respectively, find the total pressure exerted by the mixture.

Answer Key

The actual answers will be shown in brackets, followed by the explanation. If you don't understand an explanation that is given in this section, you may want to go back and review the lesson that the question came from.

Lesson 8–1 Review

1. [barometer or closed manometer]—When you hear a meteorologist referring to "barometric pressure," he or she is talking about the pressure exerted by the gases in our atmosphere.
2. [pressure]—You may have used a pressure gauge to measure the pressure of the air in your bicycle or car tires.
3. [temperature, pressure]—Remember to report the temperature and pressure of a gas sample when you report its volume.
4. [2.0×10^2 kPa]— $2.0\ \cancel{\text{atm}} \times \frac{101.3\text{ kPa}}{\cancel{\text{atm}}} = 202.6\text{ kPa}$

   which rounds to 2.0×10^2 kPa
5. [806.3 mm of Hg]—$107.5\ \cancel{\text{kPa}} \times \frac{7.5\text{ mm of Hg}}{\cancel{\text{kPa}}} = 806.25\text{ mm of Hg}$

   which rounds to 806.3 mm of Hg
6. [1.0 atm]— $790\ \cancel{\text{Torr}} \times \frac{1.0\text{ atm}}{760\ \cancel{\text{Torr}}} = 1.03947\text{ atm}$

   which rounds to 1.0 atm
7. [790. mm of Hg]—$1.04\ \cancel{\text{atm}} \times \frac{760\text{ mm of Hg}}{1.0\ \cancel{\text{atm}}} = 790.4\text{ mm of Hg}$

   which rounds to 790. mm of Hg
8. [1.11 atm]—$33.2\ \cancel{\text{inches}}\text{ of }\cancel{\text{Hg}} \times \frac{25.4\ \cancel{\text{mm}}}{1.0\ \cancel{\text{inch}}} \times \frac{1.0\text{ atm}}{760\ \cancel{\text{mm}}\text{ of }\cancel{\text{Hg}}} = 1.109578\text{ atm}$

   which rounds to 1.11 atm
9. [754 mm of Hg]—Remember, 1 mm of Hg = 1 Torr.
10. [E. Both A and D are correct]—Standard temperature is equal to 0° on the Celsius scale, which is equal to 273 K (0 + 273 = 273).

Lesson 8–2 Review

1. [Boyle's]—As the volume of a gas sample goes up, the pressure that it exerts goes down.

2. [it will decrease]—When you exert a pressure on a balloon with your hands, you can decrease its volume.
3. [$P_1 = \frac{P_2V_2}{V_1}$]—Starting with the original formula, $P_1V_1 = P_2V_2$, we simply divide both sides by V_1.
4. [temperature and number of particles]—Temperature and number of particles are not represented in these problems, because they are assumed to remain unchanged.
5. [D. 800 cm^3]—If we divide the pressure (from 4.0 atm to 1.0 atm) by a factor of 4, then we must multiply the volume (from 200 cm^3 to 800 cm^3) by a factor of four.
6. [A. 0.50 atm]—To double the volume of the gas sample, we must halve the pressure.
7. [B. 2.0 L]—Standard pressure is 101.3 kPa, so changing the pressure to 202.6 kPa is doubling it. We must halve the volume, from 4.0 L to 2.0 L.
8. [A. 20 ml]—If we triple the pressure of the gas, the volume must be divided by a factor of three, giving us 20 ml (60 ml / 3 = 20 ml).
9. [2.1 L]—$V_2 = \frac{P_1V_1}{P_2} = \frac{(101.3\ \cancel{kPa})(2.5\ L)}{(120\ \cancel{kPa})} = 2.110416\ L$

   which rounds to 2.1 L
10. [610 mm of Hg]—$P_2 = \frac{P_1V_1}{V_2} = \frac{(720\ mm\ of\ Hg)(3.4\ \cancel{L})}{(4.0\ \cancel{L})} = 612\ mm\ of\ Hg$

    which rounds to 610 mm of Hg.

    Notice: 4.0 L is actually V_2, even though it appears first in the word problem. It may help to think of V_1 and V_2 as $V_{initial}$ and V_{final}.

Lesson 8–3 Review

1. [Charles's]—Heating a gas will cause it to expand, and cooling it will cause it to contract.
2. [pressure and the number of particles]—When we make statements such as "doubling the Kelvin temperature of a gas will cause its volume to double," we are assuming that the pressure and the number of particles of gas do not change.
3. [Kelvin, there are no negative values on the Kelvin scale]—If we do a Charles's Law calculation using the Celsius scale, we may end up with a negative value for the volume of the gas, which doesn't make sense.
4. [141 K]—(-132 + 273) = 141 K
5. [273 K]—(0 + 273) = 273 K
6. [305 K]—(32 + 273) 305 K
7. [369 K]—(96 + 273) = 369 K

8. [320 cm^3]—Change 22.0 °C to Kelvin and then solve, as shown below.

$$V_2 = \frac{V_1T_2}{T_1} = \frac{(350\ \text{cm}^3)(273\ \cancel{K})}{(295\ \cancel{K})} = 323.898\ \text{cm}^3$$

which rounds to 320 cm^3

9. [546 K]—Standard temperature is 273 K. Solve as shown here.

$$T_2 = \frac{V_2T_1}{V_1} = \frac{(10.0\ \text{L})(273\ \cancel{K})}{(5.00\ \cancel{K})} = 546\ \text{K}$$

10. [0.805 L]—Remember to change the temperatures to K, by adding 273.

$$V_2 = \frac{V_1T_2}{T_1} = \frac{(0.855\ \text{L})(273\ \cancel{K})}{(290.\ \cancel{K})} = 0.804879\ \text{L}$$

which rounds to 0.805 L.

Lesson 8–4 Review

1. [1.7 atm]—$P_{He} = P_{total} - P_H = 2.5\ \text{atm} - 0.8\ \text{atm} = 1.7\ \text{atm}$
2. [154.4 kPa]—$P_{total} = P_1 + P_2 + P_3 = 45.5\ \text{kPa} + 23.6\ \text{kPa} + 85.3\ \text{kPa} = 154.4\ \text{kPa}$
3. [713.5 mm of Hg]—$P_{dry\ hydrogen} = P_{total} - P_{H_2O}$ = 745.3 mm of Hg – 31.8 mm of Hg (from the Vapor Pressure of Water table) = 713.5 mm of Hg
4. [3.0 atm]—Gas is (2.0 moles/4.0 moles) = 50% oxygen.
   6.0 atm × 0.50 = 3.0 atm
5. [75.0 kPa]—$P_{total} = P_1 + P_2 + P_3 = 25.0\ \text{kPa} + 25.0\ \text{kPa} + 25.0\ \text{kPa} = 75.0\ \text{kPa}$
6. [B. 1.3 atm]—$P_{Ar} = P_{total} - P_{Ne} = 1.8\ \text{atm} - 0.5\ \text{atm} = 1.3\ \text{atm}$
7. [C. 83.3 kPa]—$P_{dry\ chlorine} = P_{total} - P_{H_2O} = 85.6\ \text{kPa} - 2.34\ \text{kPa} = 83.26\ \text{kPa}$, which rounds to 83.3 kPa
8. [B. 0.200 atm]—Carbon dioxide makes up (1.00 mole/5.00 mole) = 20.0% of gas. 1.0 atm × .200 = 0.200 atm

Lesson 8–5 Review

1. [7.9 L]—Work shown following:

   Given: $V_1 = 7.6$ L; $T_1 = 296$ K; $P_1 = 1.12$ atm; $T_2 = 273$ K; $P_2 = 1.00$ atm

   Find: V_2

   Formula: $V_2 = \frac{V_1P_1T_2}{P_2T_1} = \frac{(7.6\ \text{L})(1.12\ \cancel{\text{atm}})(273\ \cancel{K})}{(1.00\ \cancel{\text{atm}})(296\ \cancel{K})} = 7.85059 = 7.9\ \text{L}$

2. [370. cm^3]—Work shown following:

   Given: $V_1 = 250.$ cm^3; $T_1 = 273$ K; $P_1 = 101.3$ kPa; $T_2 = 359$ K; $P_2 = 89.9$ kPa

   Find: V_2

   Formula: $V_2 = \frac{V_1P_1T_2}{P_2T_1} = \frac{(250.\ \text{cm}^3)(101.3\ \cancel{\text{kPa}})(359\ \cancel{K})}{(89.9\ \cancel{\text{kPa}})(273\ \cancel{K})} = 370.4431\ \text{cm}^3 = 370.\ \text{cm}^3$

3. [764 mm of Hg]—Work shown following:

Given: $V_1 = 5.56\ dm^3$; $T_1 = 288\ K$; $P_1 = 755$ mm of Hg; $T_2 = 262\ K$; $V_2 = 5.00\ dm^3$

Find: P_2

Formula: $P_2 = \frac{V_1P_1T_2}{V_2T_1} = \frac{(5.56\ \cancel{dm^3})(755 \text{ mm of Hg})(262\ \cancel{K})}{(5.00\ \cancel{dm^3})(288\ \cancel{K})}$

$= 763.766 = 764$ mm of Hg

4. [471 cm³]—Work shown following:

Given: $V_1 = 5.00 \times 10^2\ cm^3$; $T_1 = 295.5\ K$; $P_1 = 103.3$ kPa; $T_2 = 273\ K$; $P_2 = 101.3$ kPa

Find: V_2

Formula: $V_2 = \frac{V_1P_1T_2}{P_2T_1} = \frac{(5.00 \times 10^2\ cm^3)(103.3\ \cancel{kPa})(273\ \cancel{K})}{(101.3\ \cancel{kPa})(295.5\ \cancel{K})}$

$= 471.0489 = 471\ cm^3$

5. [2.29 L]—Work shown following:

Given: $V_1 = 2.75$ L; $T_1 = 316.6\ K$; $P_1 = 733$ mm of Hg; $T_2 = 273\ K$; $P_2 = 760$ mm of Hg

Find: V_2

Formula: $V_2 = \frac{V_1P_1T_2}{P_2T_1} = \frac{(2.75\ L)(733\ \cancel{\text{mm of Hg}})(273\ \cancel{K})}{(760\ \cancel{\text{mm of Hg}})(316.6\ \cancel{K})}$

$= 2.287045 = 2.29$ L

Lesson 8–6 Review

1. [1:4.67]—Work shown following:

$$\frac{V_{H_2}}{V_{CO_2}} = \sqrt{\frac{M_{CO_2}}{M_{H_2}}} = \sqrt{\frac{44.0u}{2.02u}} = \sqrt{21.78218} = 4.67$$

2. [1:4.44]—Work shown following:

$$\frac{V_{H_2}}{V_{Ar}} = \sqrt{\frac{M_{Ar}}{M_{H_2}}} = \sqrt{\frac{39.9u}{2.02u}} = \sqrt{19.75248} = 4.44$$

3. [1:2]—Work shown following:

$$\frac{V_{He}}{V_{CH_4}} = \sqrt{\frac{M_{CH_4}}{M_{He}}} = \sqrt{\frac{16.0u}{4.00u}} = \sqrt{4.00} = 2.00$$

4. [1:1.44]—Work shown following:

$$\frac{V_{CO}}{V_{C_4H_{10}}} = \sqrt{\frac{M_{C_4H_{10}}}{M_{CO}}} = \sqrt{\frac{58.1u}{28.0u}} = \sqrt{2.075} = 1.44$$

5. [1:1.41]—Work shown following:

$$\frac{V_{CH_4}}{V_{O_2}} = \sqrt{\frac{M_{O_2}}{M_{CH_4}}} = \sqrt{\frac{32.0u}{16.0u}} = \sqrt{2.00} = 1.41$$

Lesson 8–7 Review

1. [pressure]—Make sure that your units of pressure match the units that come with the Ideal Gas Law constant.
2. [volume]—Make sure that your units of volume match the units that come with the Ideal Gas Law constant.
3. [moles]—Remember: If you are given the mass of a substance, you can convert to moles by dividing by the molar mass.
4 [temperature]—Make sure that your units of temperature match the units that come with the Ideal Gas Law constant.
5. [Ideal Gas Law constant]—Some texts may use a different unit for one of the quantities, changing the apparent value of this constant. You can, however, convert back to the constant that we use in this book.
6. [280 K]—Work shown following:

   Given: P = 98.4 kPa; V = 60.0 dm^3; n = 2.5 moles;
   R = 8.31 dm^3 × kPa/mole × K

   Find: T

   Formula: $T = \dfrac{PV}{nR} = \dfrac{(98.4\ \cancel{kPa})(60.0\ \cancel{dm^3})}{(2.5\ \cancel{mol})(8.31\ \cancel{dm^3} \times \cancel{kPa/mol} \times K)} = 284.188\ K = 280\ K$

7. [2.08 moles]—Work shown following:

   Convert: 22.0 °C + 273 = 295 K

   $1.50\ \cancel{atm} = \dfrac{101.3\ kPa}{1\ \cancel{atm}} = 152\ kPa$

   Given: P = 152 kPa; V = 33.5 dm^3; R = 8.31 dm^3 × kPa/mole × K;
   T = 295 K

   Find: n

   Formula: $n = \dfrac{PV}{RT} = \dfrac{(152\ \cancel{kPa})(33.5\ \cancel{dm^3})}{(8.31\ \cancel{dm^3} \times \cancel{kPa}/mol \times \cancel{K})(295\ \cancel{K})}$

   $= 2.0771\ moles = 2.08\ moles$

8. [140 dm^3]—Work shown following:

   Convert: 33.0 °C + 273 = 306 K

   Given: P = 112.0 kPa; n = 6.0 moles; R = 8.31 dm^3 × kPa/mole × K;
   T = 306 K

   Find: V

   Formula: $V = \dfrac{nRT}{P} = \dfrac{(6.0\ \cancel{mol})(8.31\ dm^3 \times \cancel{kPa/mol} \times \cancel{K})(306\ \cancel{K})}{(112.0\ \cancel{kPa})}$

   $= 136.22 = 140\ dm^3$

9. [0.058 moles]—Work shown following:

Convert: $550\ \cancel{cm^3} = \frac{1\ dm^3}{1000\ \cancel{cm^3}} = 0.55\ dm^3$

Given: P = 98.4 kPa; V = 0.55 dm^3; R = 8.31 dm^3 × kPa/mole × K;
T = 112 K

Find: n

Formula: $n = \frac{PV}{RT} = \frac{(98.4\ \cancel{kPa})(0.55\ \cancel{dm^3})}{(8.31\ \cancel{dm^3} \times \cancel{kPa}/mol \times \cancel{K})(112\ \cancel{K})}$

$= .0581485 = 0.058$ moles

10. [13 K]—Work shown following:

Convert: $850\ \cancel{mm\ of\ Hg} = \frac{1\ kPa}{7.50\ \cancel{mm\ of\ Hg}} = 113\ kPa$

Given: P = 113 kPa; V = 4.5 dm^3; n = 4.55 moles;
R = 8.31 dm^3 × kPa/mole × K

Find: T

Formula: $T = \frac{PV}{nR} = \frac{(113\ \cancel{kPa})(4.5\ \cancel{dm^3})}{(4.55\ \cancel{mol})(8.31\ \cancel{dm^3} \times \cancel{kPa}/\cancel{mol} \times K)}$

$= 13.44864\ K = 13\ K$

Chapter 8 Examination

1. [b. Charles's Law]
2. [d. Graham's Law]
3. [c. Dalton's Law]
4. [a. Boyle's Law]
5. [e. Ideal Gas Law]
6. [d. Graham's Law]
7. [c. Dalton's Law]
8. [e. Ideal Gas Law]
9. [a. Boyle's Law]
10. [b. Charles's Law]
11. [B. high temperature and low pressure]—When the gas sample is at its largest, the actual individual particle size will be of least significance. When the particles are far apart and moving fast, the intermolecular forces will be at their lowest.
12. [A. He]—Of the choices shown, helium will have the smallest particles and the least intermolecular attraction between particles, making it the most like an ideal gas.

13. [A. the pressure will increase and the mass will stay the same]—Boyle's Law tells us that the pressure will increase, but why would the mass change? Have you forgotten about conservation of mass?
14. [B. the volume will decrease and the mass will stay the same]—Charles's Law tells us that the volume will decrease, and the Law of Conservation of Mass tells us that the mass will remain the same.
15. [D. the sum of all of the partial pressures of the individual gases]—That is what Dalton's Law teaches us.
16. [A. H]—Graham's Law shows us that the least massive gas will have the greatest velocity, under identical conditions.
17. [A. 3.17 kPa]—We simply look up the water vapor pressure at 25.0 °C in the Vapor Pressure of Water table in Lesson 8–4.
18. [D. 82.3 kPa]—$P_{dry}\ H_2 = P_{total} - PH_2O$ = 85.5 kPa – 3.17 kPa = 82.33 kPa, which we must round according to the rule for addition and subtraction of significant digits.
19. [C. $V_2 = \frac{V_1P_1T_2}{P_2T_1}$]—This is the Combined Gas Law.
20. [D. 1.78 g/dm³]—$D = \frac{M}{V} = \frac{39.9\text{ g}}{22.4\text{ dm}_3} = 1.78125\text{ g/dm}^3 = 1.78\text{ g/dm}^3$
21. [0.224 moles]—We use the ideal gas equation. Work shown following:

Convert: 23.0 °C + 273 = 296 K

$$1.45\ \cancel{\text{atm}} = \frac{101.3\text{ kPa}}{1\ \cancel{\text{atm}}} = 147\text{ kPa}$$

Given: P = 147 kPa; V = 3.75 dm³; R = 8.31 dm³ × kPa/mole × K; T = 296 K

Find: n

Formula: $n = \frac{PV}{RT} = \frac{(147\ \cancel{\text{kPa}})(3.75\ \cancel{\text{dm}^3})}{(8.31\ \cancel{\text{dm}^3} \times \cancel{\text{kPa}}/\text{mol} \times \cancel{\text{K}})(296\ \cancel{\text{K}})}$

$= 0.224107\text{ moles} = .224\text{ moles}$

22. [234 cm³]—We use Boyle's Law. Work shown following:

$$V_2 = \frac{P_1V_1}{P_2} = \frac{(750\ \cancel{\text{mm of Hg}})(350\text{ cm}^3)}{(1120\ \cancel{\text{mm of Hg}})} = 234.375\text{ cm}^3 = 234\text{ cm}^3$$

23. [1:1.41]—We use Graham's Law. Work shown following:

$$\frac{V_{Ne}}{V_{Ar}} = \sqrt{\frac{M_{Ar}}{M_{Ne}}} = \sqrt{\frac{32.9\text{u}}{20.2\text{u}}} = \sqrt{1.975247} = 1.41$$

24. [1.08 L]—We use Charles's Law. Remember to convert both temperatures to Kelvin by adding 273. Also, read the problem carefully and note that the 45.0 °C (318 K) is really the final temperature.

$$V_2 = \frac{P_1T_2}{T_1} = \frac{(0.993\ \text{L})(318\ \cancel{\text{K}})}{(292\ \cancel{\text{K}})} = 1.08\ \text{L}$$

25. [0.98 atm]—We use Dalton's Law.

$$P_{total} = P_1 + P_2 + P_3$$
$$= 0.23\ \text{atm} + 0.33\ \text{atm} + 0.42\ \text{atm} = 0.98\ \text{atm}$$

24. [illegible] We use Charles' law. Remember to convert from temperature [illegible] to K [illegible] by adding 273. Also read the problem carefully and note that the [illegible] is 0°C (273 K) [illegible] the final temperature.

[illegible]

25. [illegible] Law

[illegible]

[illegible]

Solutions, Acids, and Bases

Lesson 9–1: Solutions

In Lesson 1–1, you learned that ***solutions*** are homogeneous mixtures of two or more substances physically mixed together in a uniform way. You also learned that there are various types of solutions, including solid, liquid, and gas solutions. In this section, we will concern ourselves with the types of solutions that are in the liquid phase—that is, solids, gases, or liquids dissolved within a liquid.

A solution is made up of two parts. The ***solute*** is the substance that gets dissolved into the solution. The ***solvent*** is the substance that does the "dissolving." A common example of a solution is salt water. If you mix up a glass of salt water, the salt would be the solute, and the water would be the solvent. Because of its ability to dissolve many substances well, water is sometimes called the "universal" solvent. Water is a polar substance, and polar substances dissolve other polar substances, as well as many ionic substances, well.

The ability of a substance to dissolve in another substance is called ***solubility***, which can be measured in a number of ways. When a substance does not dissolve in another substance, it is called ***insoluble***. Non-polar substances are often insoluble in polar substances. When both substances are liquids, and the liquids fail to mix, they might be called ***immiscible***, whereas liquids that do mix well are called ***miscible***.

In the chemistry laboratory activities that you carry out, you will probably be required to make several solutions, using water as a solvent. When we talk about the solubility of a solute in water, we normally speak in terms of the number of grams of a particular solute that will normally dissolve in a specific amount of water at a particular temperature. Table salt, for example, has a solubility of about 40g of NaCl/100g of water, at 90 °C.

This means that 100 g of water at 90 °C can usually hold about 40 g of dissolved table salt. If you add less than 40 g of NaCl to the 100 g of H_2O at 90 °C, you get an unsaturated solution. An ***unsaturated solution*** is one that is holding less solute than it normally can, at that temperature. If you were to add more salt to this unsaturated solution, it would dissolve, up to the maximum solubility, when you would get a saturated solution. A ***saturated solution*** is one that cannot dissolve any more of the given solute under the current conditions. Do you remember ever drinking hot chocolate and finding the "residue" on the bottom of the cup? This is evidence that your hot chocolate represented a saturated solution. When you add additional solute to a saturated solution, the additional solute will fall to the bottom of the vessel.

It is possible, under the right conditions, to produce a supersaturated solution. A ***supersaturated solution*** is one that is holding more dissolved solute than it should be able to under the current conditions. For example, if you create a solution that contains 60 g of NaCl/100 g of H_2O at 90 °C, it would be holding 20 more grams of NaCl than it should be able to. Creating a supersaturated solution can be a little tricky, and the resultant solution is unstable, but the practical applications of such solutions include producing rock candy.

Because the concentration of a solution is so variable, we need two ways to indicate how much solute is in a particular solution. There are several ways to measure the concentration of a solution, including molarity, molality, and mole fraction. The type of measurements you use will often depend upon the situation or on the calculations that you want to be able to carry out.

Molarity is probably the most common measurement for the concentration of a solution. ***Molarity*** is a measure of the number of moles of solute dissolved in every liter of solution. The formula for molarity is:

$$\text{Molarity (M)} = \frac{\text{moles of solute}}{\text{liters of solution}}$$

Molarity is measured in moles/L, but we use the symbol "M" for short, referring to the unit as "molar." So, for example, if we dissolved 3 moles of solute in 1 liter of solution we would write that we have a 3M solution, and we would say that the solution is "3 molar." Let's try a straightforward example.

Example 1

Calculate the molarity of a 2.0 L solution made with 3.0 moles of NaOH.

In this particular problem, you don't even need to pay attention to what solute is used, because the number of moles you have is indicated. The answer would turn out exactly the same if you were told that the solution was made with 3.0 moles of HCl or 3.0 moles of NaCl. Let's see the solution to the problem.

Given: Number of moles of solute = 3.0 moles; liters of solution = 2.0 L
Find: Molarity (M)

Formula	**Answer**
$\text{Molarity (M)} = \dfrac{\text{moles of solute}}{\text{liters of solution}}$	$= \dfrac{3.0 \text{ moles}}{2.0 \text{ L}} = 1.5 \text{ M}$

Try to visualize what this answer means, and you will have an easier time with these calculations. You may recall a time when you made a drink from a powder mix, and you altered the original recipe given on the container. Intuitively, you could figure out how to make your lemonade or iced tea twice as strong or half as strong. Molarity calculations work in much the same way. When we find that our answer is 1.5 M, it means that there are 1.5 moles of solute in every liter of solution. If you were to use twice as much solute, you would double the concentration of the solution, making it twice as strong, because 6.0 moles/2.0 L = 3.0 M. If you were to double the volume of the solution without adding more solute, you would dilute the solution to half its strength, because 3.0 moles/4.0 L = 0.75 M.

Slightly more difficult examples of molarity problems involve an extra calculation. Remember: You need to know the number of moles of solute in order to calculate the molarity of a solution. Sometimes you will start with the mass of the solute and you will need to determine the number of moles of solute that you are starting with by using a formula that you studied in Chapter 7:

$$\text{Number of moles of a substance (n)} = \frac{\text{mass of the sample}}{\text{molar mass of the substance}}$$

Let's try a molarity problem that involves two calculations.

Example 2

Determine the molarity of a 2.5 L solution made with 200.0 g of calcium chloride ($CaCl_2$).

Can you see how Example 2 differs from Example 1? In order to solve for molarity, we need to know how many moles of solute that we have. In Example 2, we're given the mass of the solute, instead of the number of moles. To solve this problem we need to:

1. Find the molar mass of the solute.
2. Divide the mass of our sample of solute by its molar mass, to get the number of moles of solute.
3. Divide the number of moles of solute by the number of liters of solution, in order to get the molarity.

For the sake of clarity, I will review each of these steps, before I summarize the solution.

1. **Find the molar mass of the solute, $CaCl_2$.** Remember: The process of finding the molar mass of a substance is the same as finding the molecular or formula mass of the substance. We look up the masses listed on the Periodic Table of Elements for each of the elements involved and multiply by the appropriate subscripts. The only difference is that you use the unit symbol "g" for grams, instead of "u" for atomic mass units.

   Calculating the Molar Mass of $CaCl_2$ — Ca = 40.1 g

   Cl_2 = + 71.0 g

   Molar mass of $CaCl_2$ = 111.1 g

2. **Divide the mass of our sample of solute by its molar mass, to get the number of moles of solute**. The mass of our original sample of $CaCl_2$ given in our problem is 200.0 g. We need to convert that into moles by dividing by the molar mass.

   $$\text{Number of moles (n)} = \frac{\text{mass of sample}}{\text{molar mass}} = \frac{200.0\ \text{g}}{111.1\ \text{g/mole}} = 1.800\ \text{moles}$$

3. **Divide the number of moles of solute by the number of liters of solution, in order to get the molarity.**

   $$\text{Molarity (M)} = \frac{\text{moles of solute}}{\text{liters of solution}} = \frac{1.800\ \text{moles}}{2.5\ \text{L/mole}} = 0.72\ \text{M}$$

Given: liters of solution = 2.5 L; mass of solute = 200.0 g

Find: A. moles of solute

B. molarity of solution

A. $$\text{Number of moles (n)} = \frac{\text{mass of sample}}{\text{molar mass}} = \frac{200.0\ \text{g}}{111.1\ \text{g/mole}} = 1.800\ \text{moles}$$

B. $$\text{Molarity (M)} = \frac{\text{moles of solute}}{\text{liters of solution}} = \frac{1.800\ \text{moles}}{2.5\ \text{L/mole}} = 0.72\ \text{M}$$

As with many of the calculations in chemistry, you should always be prepared to work with a different unknown. For example, let's suppose that you were given a problem where you knew the molarity and the volume of the solution that you needed to make, and you need to calculate the number of grams of solute that you need to use. Example 3 represents this type of problem.

Example 3

How many grams of NaCl would be required to produce a 3.0 M solution with a volume of 500. cm^3?

The first thing that I would do, in order to avoid confusion, is change the volume of the solution into liters. Dividing 500. cm^3 by 1000 cm^3/liter, you find that you have 0.500 L of solution. I would make the change right in the problem, so that I don't forget to do it later. I will also add the given information and the unknowns. You want to find the mass of NaCl, but to do that you will first need to find out how many moles of NaCl would be required to produce a 3.0 molar solution with a volume of 0.5 L.

Given: volume = 0.500 L; molarity = 3.0 M

Find: A. Number of moles (n) of NaCl

B. Mass of NaCl

Now, to solve this type of problem, you'll need to do the following:

1. Take the original formula for molarity and isolate the unknown, which in this case is the number of moles. Use this formula to calculate the number of moles of NaCl you will need.

   Original formula: $\text{Molarity (M)} = \frac{\text{moles of solute}}{\text{liters of solution}}$

   Isolating the number of moles of solute by multiplying both sides of the equation by liters of solution, you get:

   $$\text{Number of moles of solute} = \text{molarity} \times \text{liters of solution}$$

   Solving, you get:

   $$\text{Number of moles of solute} = 3.0\text{ M} \times 0.500\text{ L} = 1.5\text{ moles of NaCl}$$

2. Change the number of moles of NaCl to grams by multiplying by the molar mass of NaCl, which is 58.5 g/mole.

   Mass of sample = number of moles × molar mass

   Mass of sample = 1.5 ~~moles~~ of NaCl × 58.5 g/~~mole~~ = 87.75 g

   which you round to **88 g of NaCl**

Given: volume = 0.500 L; molarity = 3.0 M

Find: A. Number of moles (n) of NaCl

B. Mass of NaCl

A. number of moles of solute = 3.0 M × 0.500 L = **1.5 moles of NaCl**

B. mass of sample = number of moles × molar mass
= 1.5 ~~moles~~ × 58.5 g/~~mole~~ = 87.75 g

Final answer: **88 g of NaCl would be required**

Another useful way of indicating the concentration of a solution is called "molality." The ***molality*** of a solution is a measure of the number of moles of solute dissolved in each kilogram of solvent. The formula for molality is given here:

$$\text{Molality (m)} = \frac{\text{number of moles of solute}}{\text{kilograms of solvent}}$$

The fact that the words *molality* and *molarity* are so similar, and the respective formulas and symbols are also similar, can lead to confusion. The main difference is that molarity is based on the volume of the whole solution, whereas molality is based on the mass of just the solvent. Molality is measured in moles/kilogram, but it is given the derived unit "m," which stands for "molal."

Let's begin with a basic molality calculation.

Example 4

What would be the molality of a solution made from 6.0 moles of NaOH dissolved in 2.0 kg of water?

Given: number of moles of solute = 6.0 moles;
kilograms of solvent = 2.0 kg

Find: molality

Formula **Answer**

$$\text{Molality} = \frac{\text{number of moles of solute}}{\text{kilograms of solvent}} = \frac{6.0\text{ moles}}{2.0\text{ kg}} = \mathbf{3.0\ molal}$$

As with the molarity calculations, molality problems often incorporate the formula for determining the number of moles that a sample represents. Let's suppose you knew the mass of both the solute and the solvent that went into the solution. Would you be able to calculate the molality? Here's an example.

Example 5

Determine the molality of a solution made with 120. g of $CaCl_2$ and 1.5 kg of water.

As you can see, before you can calculate the molality of the solution, you will need to determine the number of moles of solute that you are starting with. You can do this by dividing the mass of the solute ($CaCl_2$) by its molar mass. You may recall from Example 2 that the molar mass of $CaCl_2$ is 111.1 g/mole. Once you know the number of moles of solute that you have, you will be able to divide that value by the mass of the solute (1.5 kg of water) in order to determine the molality. Let's see what this would look like.

Given: mass of solute = 120 g; kilograms of solvent = 1.5 kg

Find: A. number of moles of solute
B. molality of solution

$$\text{A. Number of moles of solute} = \frac{\text{mass of sample}}{\text{molar mass}} = \frac{120.0\text{ g}}{111.1\text{ g}}$$

$$= \mathbf{1.08\ moles\ of\ CaCl_2}$$

$$\text{B. Molality} = \frac{\text{moles of solute}}{\text{kilograms of solvent}} = \frac{1.08\text{ moles}}{1.5\text{ kg}} = \mathbf{0.72\ m\ or\ 0.72\ molal}$$

Working with a different unknown shouldn't really be any harder than the problem from Example 5. Let's try a problem where the mass of the solute is the unknown. Take a look at the problem and try to figure out the given information and what you need to find. Include the values that you will need to find in the order that you will need to find them.

Example 6

How many grams of strontium chloride ($SrCl_2$) must you add to 2.50 kg of water to produce a solution with a molality of 0.500 m?

Given: ______________________________

Find: ______________________________

You should have made note of the fact that you have been given the number of kilograms of the solvent (2.50 kg) and the molality (0.500 m) of the solution. You want to find the mass of the solute, but you need to find two other pieces of information before you can do that. First, you want to use the molality formula to determine the number of moles of solute that you need to produce a solution of the required concentration. Next, you want to determine the molar mass of the solute, so that you can use that information to change number of moles to mass.

Look at the original formula for molality, and figure out how you would isolate the unknown, which is the number of moles of solute.

$$\text{Molality (m)} = \frac{\text{number of moles of solute}}{\text{kilograms of solvent}}$$

Can you see that you will want to multiply both sides of the equation by "kilograms of solvent" as shown here?

$$\textit{kilograms of solvent} \times \textit{molality} = \frac{\textit{number of moles of solute}}{\textit{kilograms of solvent}} \times \textit{kilograms of solvent}$$

Crossing out the "kilograms of solvent" on the right-hand side of the equation, we get:

kilograms of solvent × *molality* = *number of moles of solute*

Rearranging the formula, and filling in the given information, we get:

number of moles of solute = *kilograms of solvent* × *molality*

number of moles of solute = 2.50 kg × 0.500 m = **1.25 moles of solute**

For the next part, you'll need to find the molar mass of strontium chloride ($SrCl_2$), by making use of the periodic table. The work is shown here:

Molar mass of strontium chloride ($SrCl_2$)

Sr = 87.6 g/mole

Cl_2 = + 71.0 g/mole

Molar mass of $SrCl_2$ = 158.6 g/mole which rounds to **159 g/mole**

Now you can determine the required mass of the solute by multiplying the number of moles of solute by the molar mass of the solute.

mass of solute = number of moles × molar mass = 1.25 ~~moles~~ × 159 g/~~mole~~
= 198.75 g, which rounds to **199 g of $SrCl_2$**

Here is a summary of this work:

Given: kilograms of solvent = 2.50 kg; molality of solution = 0.500 m

Find: A. number of moles of solute

B. molar mass of solute

C. mass of solute

A. number of moles of solute = kilograms of solvent × molaltiy (m)
= 2.50 kg × 0.500 m = **1.25 moles of $SrCl_2$**

B. molar mass of $SrCl_2$ = sum of the molar masses of the parts.
= 87.6 g/mole + 71.0 g/mole = 158.6 g/mole,
which rounds to **159 g/mole**

C. mass of solute = number of moles of solute × molar mass of solute
= 1.25 moles × 159 g /mole = 198.75 g,
which rounds to **199g**

Try the review questions and check your answers before moving on to Lesson 9–2.

Lesson 9–1 Review

1. Polar solutes tend to be soluble in ______________ solvents.
2. Polar solutes tend to be ______________ in non-polar solvents.
3. The ______________ of a solution is measured in moles of solute per liters of solution.
4. The ______________ of a solution is measured in moles of solute per kilograms of solvent.
5. The symbol "M" is used to indicate the __________ of a solution.
6. The symbol "m" is used to indicate the __________ of a solution.
7. Determine the molarity of a 3.0 L solution containing 2.0 moles of solute.
8. How many grams of NaOH would be required to produce a 1.25 M solution with a volume of 2.00 liters?
9. What would be the volume of a 2.0 M solution made with 8.0 moles of NaCl?
10. How many kilograms of water would be required to produce a 2.0 molal solution made with 111 g of $CaCl_2$?

Lesson 9–2: Properties of Acids and Bases

Many students have two very common misconceptions about acids and bases. First, they believe that all acids are extremely corrosive, capable of dissolving a person the way the acids that they see in movies are. Second, they believe that all bases are harmless, because they can be used to neutralize deadly acids. The truth is that not all acids are hazardous and not all bases are safe. All of the acids and bases that you encounter in the chemistry lab should be handled with care, especially if you are unsure of how strong they are.

Some acids are quite harmless, and most of us swallow acidic substances on a day-to-day basis. Soda and seltzer water contain carbonic acid. Orange juice, oranges, lemons, and limes contain citric acid. Vinegar contains acetic acid. So, not all acids are dangerous.

Soaps, detergents, baking powder, and antacids all contain relatively harmless bases, but that does not mean that you don't need to worry about any of the bases that you find in the home or in the lab. Drain cleaners, found in many homes, often contain a strong base called sodium hydroxide or lye. Not all bases are harmless!

If you were under these misconceptions, then you probably don't have a complete understanding of what acids and bases are. In this chapter, we will explore acids and bases in detail. We will begin with a summary of some of the properties and characteristics of acids and bases.

Properties and Characteristics of Acids and Bases

Acids

- Acids are electrolytes, meaning they dissolve in water to form a solution that conducts electricity.
- Acidic substances often taste tart or sour.
- Acids react with some metals to produce hydrogen gas.
- Acids react with carbonates to produce carbon dioxide.
- Acids cause indicators to change color. For example, they turn litmus paper red.
- Acidic solutions have a pH value of less than 7.
- Acidic solutions have a pOH value of greater than 7.
- In acids, $[H^+] > [OH^-]$.

Bases

- Bases are electrolytes, meaning they dissolve in water to form a solution that conducts electricity.
- Basic substances are slippery and taste bitter.
- Bases and acids neutralize each other in neutralization reactions.
- Bases react with fats to form compounds called soaps.
- Bases cause indicators to change color. For example, they turn litmus paper blue.
- Basic solutions have a pH value of greater than 7.
- Basic solutions have a pOH value of less than 7.
- In bases, $[OH^-] > [H^+]$.

Before we go any further, I would like to remind you that when you see a species in brackets, as in [H^+], it is referring to the concentration of the species, usually in moles/L. Therefore, if we say that [H^+] = 1.0×10^{-7}, we mean the concentration of hydrogen ions is .0000001 M. I would also like to point out that, in this section of the book, you should consider H_3O^+ to be synonymous with H^+. Many books describe acids in terms of hydrogen ions, [H^+], but others argue that hydrogen ions, which are essentially bare protons, won't exist for any significant length of time in water. They will attach to water molecules, forming hydronium ions, H_3O^+. You must not get confused when switching between chemistry books. Whether we discuss acids in terms of hydrogen ions (H^+) or hydroxide ions (H_3O^+), we are really talking about the same thing.

The pH and pOH scales are shorthand notations, in a sense. Instead of writing that the concentration of hydronium ions in a particular solution is 0.00001 M (or, 1×10^{-5} M, in scientific notation), we simply say that the pH of the solution is 5. How do we convert between the concentration of the ions and the pH or pOH scales? We simply use the following formulas.

$$pH = -\log[H_3O^+]$$
$$pOH = -\log[OH-]$$

To start with, try to use the pH formula to find the pH of a solution with a hydronium ion concentration of 1×10^{-5} M. I already told you that it comes out to a pH value of 5. Can you get that answer on your calculator? Of course, each calculator is different, so I can't instruct you on how to use your specific calculator, but I can give you some tips. First, check to see if your calculator has a "log" button. If it doesn't, you will need a different calculator to solve these problems. Second, to get the negative sign in front of the "log" in the formula, don't use the subtraction button. Look for another button that may have a smaller minus sign, perhaps in parentheses, such as (-).

We solve the problem as follows;

$$pH = -\log[H_3O^+] = -\log(1 \times 10^{-5}) = 5$$

A little trick that you should learn early on (especially if your calculator lacks a "log" button) is that when the coefficient in your concentration of hydronium ions is 1, as in 1×10^{-8}, the pH is equal to the absolute value of the exponent. The pOH scale is used in a similar way to indicate the concentration of hydroxide ions in a solution. A concentration of hydroxide ions of 1×10^{-3} would correspond to a 3 on the pOH scale. The following table summarizes the method of converting the concentrations of hydronium ions and hydroxide ions to the pH and pOH scales respectively.

Ion Concentrations and Corresponding Values on the pH and pOH Scales

$[H_3O^+]$	Value on pH Scale	$[OH^-]$	Value on pOH Scale
1×10^{-1} M	1	1×10^{-14} M	14
1×10^{-2} M	2	1×10^{-13} M	13
1×10^{-3} M	3	1×10^{-12} M	12
1×10^{-4} M	4	1×10^{-11} M	11
1×10^{-5} M	5	1×10^{-10} M	10
1×10^{-6} M	6	1×10^{-9} M	9
1×10^{-7} M	7	1×10^{-8} M	8
1×10^{-8} M	8	1×10^{-7} M	7
1×10^{-9} M	9	1×10^{-6} M	6
1×10^{-10} M	10	1×10^{-5} M	5
1×10^{-11} M	11	1×10^{-4} M	4
1×10^{-12} M	12	1×10^{-3} M	3
1×10^{-13} M	13	1×10^{-2} M	2
1×10^{-14} M	14	1×10^{-1} M	1

Of course, this only works when the coefficient in the concentration of ions is 1. If you have any other coefficient, as in $[H_3O^+] = 4.3 \times 10^{-5}$ M, we must use the pH formula, as shown here:

$$pH = -\log[H_3O^+] = -\log(4.3 \times 10^{-5}) = 4.4$$

One of the most interesting things about acids and bases is that they can react with each other in such a way as to render each other harmless. A ***neutralization reaction*** occurs when aqueous solutions of an acid and a base react with each other to produce a salt (ionic compound) and water. The general format for neutralization reaction is this:

acid + base → salt + water

So, a very corrosive base, such as NaOH, can react with a very strong base, such as HCl, and the products of the reaction will be salt and water! The following equation represents an example of a neutralization reaction:

$$HCl + NaOH \rightarrow NaCl + H_2O$$

hydrochloric acid + sodium hydroxide → sodium chloride + water

Of course, if you use a different acid or a different base you won't produce table salt, but you will end up with a different type of salt, as well as water. In order to better understand neutralization reactions, as well as the other properties and characteristics of acids and bases, we should examine acids and bases on the atomic level.

Keep in mind that even pure substances, such as water, are made up of particles that are in constant motion. These particles collide with each other as well, but a chemical reaction doesn't usually take place. However, for a certain number of particles, a reaction does take place! If two water molecules strike each other with the proper orientation and kinetic energy, a hydrogen ion can move from one molecule to another, as shown here:

$$\mathbf{H_2O + H_2O \rightarrow H_3O^+ + OH^-}$$

water + water → hydronium + hydroxide

In this way, two neutral water molecules can react and produce a positively charged hydronium (H_3O^+) ion and a negatively charged hydroxide (OH^-) ion. This type of reaction, called the self-ionization of water, is actually a relatively rare occurrence, but it does happen. You may have measured this occurrence in the laboratory, if you ever measured the pH of pure water. Pure water at 25 °C has a pH of 7, which means that the concentration of hydronium ions, $[H_3O^+]$ is 1.0×10^{-7} moles/liter (remember molarity?), or 1.0×10^{-7} M. This pure water, with a pH of 7, is said to be neutral, because it has the same number of hydronium ions and hydroxide ions, which makes sense, because of the chemical equation shown preceding this paragraph.

So, pure water at 25 °C has a H_3O^+ concentration of 1.0×10^{-7} M and an OH^- concentration of 1.0×10^{-7} M. The two types of ions balance each other out. Any aqueous solution in which the concentrations of these two ions are equal is said to be ***neutral***.

An ***acid*** solution has a greater concentration of hydronium ions than hydroxide ions. How does this happen? Let's suppose you added some concentrated HCl into your pure water sample. The HCl is a strong electrolyte, which means that it tends to ionize completely in aqueous solutions, according to the reaction shown here:

$$HCl_{(s)} \xrightarrow{water} H^+_{(aq)} + Cl^-_{(aq)}$$

The bare protons (H^+) that the acid releases are too reactive to exist as isolated particles for any real length of time. They will quickly attach themselves either to hydroxide ions to form neutral water molecules, or to neutral water molecules to form additional hydronium ions, according to the reaction shown here:

$$H^+ + H_2O \rightarrow H_3O^+$$

hydrogen ion + water → hydronium ion

The result of adding the acid to the water will be an increase in hydronium ions and a corresponding decrease in hydroxide ion. The product of the concentrations of hydronium ions and hydroxide ions is a constant, called the ***ion-product constant for water***, which is symbolized as k_w. The formula for this constant is shown here:

$$k_w = [H_3O^+] \times [OH^-] = 1.0 \times 10^{-14} \text{ M}$$

This equation shows us that the product of the concentrations of these ions must always be the same, 1.0×10^{-14}. This gives us the ability to calculate the concentration of one type of ion, when we know the concentration of the other type of ion. For example, if we know that the hydronium ion concentration of the HCl acid solution that we "created" was 1.0×10^{-2} M, we could solve for the concentration of OH^- ions, as shown here:

$$[OH^-] = \frac{k_w}{[H_3O^+]} = \frac{1.0 \times 10^{-14}}{1.0 \times 10^{-2}} = 1.0 \times 10^{-12}$$

To check your answer, make sure that the product of the concentration of hydronium ions and hydoxide ions is equal to 1.0×10^{-14}, as shown here:

$$k_w = [H_3O^+] \times [OH^-] = (1.0 \times 10^{-2}) \times (1.0 \times 10^{-12}) = 1.0 \times 10^{-14}$$

A basic solution is one that has a greater concentration of OH^- ions than H_3O^+ ions. How could this come to pass? Let's suppose that we started out with our pure water, with a pH of 7. Remember: This pure water has an equal number of hydronium ions and hydroxide ions. Then, we take some pure sodium hydroxide (NaOH) and dissolve it in our pure water. The sodium hydroxide will ionize according to this equation:

$$NaOH_{(s)} \xrightarrow{water} Na^+_{(aq)} + OH^-_{(aq)}$$

Some of these additional hydroxide ions may react with hydronium ions to form neutral water molecules, as shown in this equation:

$$H_3O^+ + OH^- \rightarrow H_2O + H_2O$$

The rest will go to increase the overall concentration of the hydroxide ions in the solution. If we measure the pH of our new solution, we can find the concentrations of hydronium and hydroxide ions, as well as another measure called pOH. Let's suppose that the pH of our solution is measured as 13. These would correspond to a concentration of hydronium ions of 1.0×10^{-13}. So, if $[H_3O^+] = 1.0 \times 10^{-13}$, then $[OH^-]$ must be equal to 1.0×10^{-1}, as shown by the following equation.

$$[OH^-] = \frac{k_w}{[H_3O^+]} = \frac{1.0 \times 10^{-14}}{1.0 \times 10^{-13}} = 1.0 \times 10^{-1}$$

A concentration of hydroxide ions of 1.0×10^{-1} corresponds to a pOH value of 1. In case you haven't guessed, just as $[H_3O^+] \times [OH^-] = 1.0 \times 10^{-14}$, pH + pOH = 14. If our solution has a pH of 13, then it must have a pOH of 1. If it had a pH of 6, then it would have a pOH of 8.

Let's try a few examples that will test the ideas you learned in this lesson.

Example 1

What are the $[H_3O^+]$ and $[OH^-]$ of a solution with a pH of 2?

We read this question as, "What are the concentration of hydronium ions and the concentration of hydroxide ions of a solution with a pH of 2?" A pH of 2 corresponds to a $[H_3O^+]$ of 1.0×10^{-2}. You can find the concentration of hydroxide ions in a number of ways. One way is to make use of the ion-product constant of water, k_w, with the following equation:

$$[OH^-] = \frac{k_w}{[H_3O^+]} = \frac{1.0 \times 10^{-14}}{1.0 \times 10^{-2}} = 1.0 \times 10^{-12}$$

Alternatively, you could have reasoned that if the pH is 2, than the pOH must be (14-2) 12, which corresponds to an $[OH^-]$ of 1.0×10^{-12}. Either way, you get the answers shown below:

Answer: $[H_3O^+] = 1.0 \times 10^{-2}$ and $[OH^-] = 1.0 \times 10^{-12}$

The next example can be solved by anyone who remembers that pH + pOH = 14.

Example 2

What is the pH of a solution with a pOH value of 3?

Answer: pH = 14 – pOH = 14 – 3 = **11**

Now, let's try an example that requires you to use the "log" button on your calculator. Try this with your calculator and see if you can get the same answer as shown.

Example 3

What is the pH of a solution with a $[H_3O^+]$ of 7.4×10^{-4}?

Answer: pH = $-\log[H_3O^+] = -\log(7.4 \times 10^{-4}) = 3.13076828$, which we round to **3.1**

What if you're given the concentration of the hydroxide (OH^-) ion instead of the hydronium (H_3O^+) ion? You make use of both of the formulas that we used in Example 2 and Example 3.

Example 4

What is the pH of a solution with a $[OH^-]$ concentration of 8.2×10^{-8}?

Answer: pH = $14 - \text{pOH} = 14 - (-\log[OH^-]) = 14 - (-\log(8.2 \times 10^{-8}))$ = **6.9**, after rounding

Lesson 9–2 Review

For questions 1–6, determine if each statement is describing an acid or a base.

1. It reacts with some metals to produced hydrogen gas.
2. It turns litmus paper red.
3. It has pH values greater than 7.
4. It often tastes sour or tart.
5. Sodium hydroxide (NaOH) is an example.
6. It turns litmus paper blue.
7. What would be the pOH value of a solution with a $[OH^-]$ of 1×10^{-5}?

8. What would be the pH value of a solution with a $[H_3O^+]$ of 5.9×10^{-3}?
9. What would be the pH value of a solution with a $[OH^-]$ of 3.5×10^{-5}?
10. What would be the H_3O^+ concentration of a solution with a pH of 5?

Lesson 9–3: Naming Acids

By now, you have probably noticed that the rules for naming acids seem to differ from the rules for naming other compounds. Perhaps you have heard your instructor refer to the compound with the formula HCl as "hydrochloric acid," but, according to the rules for naming compounds that we went over in Lesson 5–3, it should be called "hydrogen chloride." Following the rules from Lesson 5–3, the compound with the formula H_2SO_4 should be named "hydrogen sulfate," but you may have seen it referred to as "sulfuric acid." What is going on here? Are there special rules for naming acids? The answer is yes!

When a compound dissolves in water to form an acidic aqueous solution, it gets a special "acid" name. Fortunately, it is quite easy to derive the common names of these acids. I can summarize what you need to know in three easy rules.

1. If the acid is a binary compound, as in HCl, shorten the "hydrogen" to "hydro" and change the ending from "ide" to "ic." Then add the word *acid* at the end.

Naming the Acid With the Formula HCl

Hydrogen Chloride + ic + "acid" becomes Hydrochloric Acid

Following this rule, H_2S becomes "hydrosulfuric acid" and HI becomes "hydroiodic acid." Look at how the names change in the following table, and only move on to the next rule when you are sure that you see the pattern.

Names of the Binary Acids

Formula	Name of Solid Compound	Name of Acid in Solution
HBr	Hydrogen bromide	Hydrobromic acid
HCl	Hydrogen chloride	Hydrochloric acid
HF	Hydrogen fluoride	Hydrofluoric acid
HI	Hydrogen iodide	Hydroiodic acid
H_2S	Hydrogen sulfide	Hydrosulfuric acid

2. If the acid is a ternary acid, and it contains both chlorine and oxygen, move on to Rule #3. If the acid is any other ternary compound, delete the word *hydrogen* completely, and simply look at the name of the polyatomic ion that it contains. If the polyatomic ion ends in "ate," change the ending to "ic." If the polyatomic ion ends in "ite," change the ending to "ous." Add "acid" to the end of the name.

Naming the Acid With the Formula HNO_3

~~Hydrogen~~ Nitrate + ic + acid becomes nitric acid

The following table shows the names of several common ternary acids. Look at the names carefully, and try to see the pattern for the name changes. Remember: "Ate" becomes "ic" and "ite" becomes "ous."

Names of Ternary Acids

Formula	Name of Solid Compound	Name of Acid in Solution
HNO_3	Hydrogen Nitrate	Nitric Acid
H_2NO_2	Hydrogen Nitrite	Nitrous Acid
H_2SO_4	Hydrogen Sulfate	Sulfuric Acid
H_2SO_3	Hydrogen Sulfite	Sulfurous Acid

Okay, so you caught me oversimplifying Rule #2! When dealing with the ternary acids containing sulfur, we didn't really just cross out the "ate" and add "ic," as we did for the acids containing nitrogen. I guess that chemists thought that "sulfic acid" sounded too odd, so they added the "ic" to sulfur. At any rate, these names should be easy enough to remember.

3. Rule # 3 is just for ternary acids containing chlorine and oxygen. There are actually four different polyatomic ions that contain chlorine and oxygen, and there are four corresponding acids. We still use Rule #2 for two of the acids, so we look at the name of the polyatomic ions that they contain, and "ate" still becomes "ic" and "ite" becomes "ous." We must also add additional prefixes to two of the acids, so that we can tell them all apart. We simply take the prefixes of "hypo-" and "per-" that come with the polyatomic ions. Look at the following table and see if you can figure out the system for naming these ternary acids.

Names of Ternary Acids Containing Chlorine and Oxygen

Formula	Name of Solid Compound	Name of Acid in Solution
$HClO$	Hydrogen Hypochlorite	Hypochlorous Acid
$HClO_2$	Hydrogen Chlorite	Chlorous Acid
$HClO_3$	Hydrogen Chlorate	Chloric Acid
$HClO_4$	Hydrogen Perchlorate	Perchloric Acid

One method that some people use to remember the names of these four acids involves paying attention to the number of oxygen atoms in the formula. The acid that has four atoms of oxygen in its formula gets the prefix "per-," which is derived from the prefix "hyper-," meaning "greater than." The acid with only one oxygen atom in its formula gets the prefix "hypo-," meaning "less than."

Practice the rules for naming acids by answering the following review questions.

Lesson 9–3 Review

Name the following acids.

1. H_3PO_4 __________
2. HBr __________
3. H_2SO_4 __________
4. HClO __________
5. HNO_2 __________
6. $HClO_3$ __________

Write the correct formula for each of the following acids.

7. Hydrofluoric acid __________
8. Chlorous acid __________
9. Sulfurous acid __________
10. Nitric acid __________

Lesson 9–4: Theories of Acids and Bases

It may surprise you to find that, despite the reason that acids and bases are found in so many of the materials around us, there are several different theories defining what acids and bases are. The real difference between the theories has to do with how broad or limited our definitions of acids and bases should be. For example, should only compounds that ionize to release hydroxide ions (OH) be considered bases, or should other compounds that neutralize acids without releasing hydroxide ions be included in our definition? Examine the following theories.

Arrhenius Theory

The first acid-base theory that we will discuss is the Arrhenius Theory. Proposed by a Swedish chemist named Svante Arrhenius in 1887, this was the first acid-base theory, and it remains the most specific or limiting. According to the Arrhenius Theory, an ***acid*** is a substance that releases H^+ ions in aqueous solution. This would explain why the acids that you are familiar with, such as HCl and H_2SO_4, have hydrogen in their formulas. ***Arrhenius bases*** are only those substances that release OH^- ions in aqueous solutions. This would include common bases, such as NaOH and KOH, but would exclude other substances, such as NH_3, that also have the ability to neutralize acids.

Arrhenius Acids and Bases	
Arrhenius Acid	A substance that release H^+ ions in aqueous solutions. Examples incl› de HI, HBr, HCl, and H_2SO_4.
Arrhenius Base	A substance that relea› e OH^- ions in aqueous solutions. Examples include KOH, NaOH, $Ca(OH)_2$, and $Mg(OH)_2$.

Brønsted-Lowry Theory

The next theory of acids and bases is called the Brønsted-Lowry Theory, proposed by Danish chemist Johannes Brønsted and the English chemist Thomas Lowry, independently, in 1923. The definition of an acid, in this theory, sounds essentially the same as the Arrhenius acid. A ***Brønsted-Lowry acid*** is defined as a substance that donates a proton to another species. Now, you might say, "A proton is a hydrogen ion (H^+), so what is the difference between the two definitions for acids?" Most notably, this definition doesn't require that the acid be in an aqueous solution. So a substance that donates protons, even when in a solid or vapor phase, is still acting as a Brønsted-Lowry acid.

The Brønsted-Lowry definition of a base is even more inclusive, incorporating substances that would never be considered bases by the Arrhenius definition. A ***Brønsted-Lowry base*** is a substance that accepts protons from another species. This definition allows substances such as ammonia NH_3, which not only doesn't release OH^- ions, but doesn't even have oxygen in it, to be considered bases. Look at the following chemical equation, and you will see an example of ammonia acting as a base.

$$H_2O_{(l)} + NH_{3(aq)} \rightarrow OH^-_{(aq)} + NH_4^+{}_{(aq)}$$

water + ammonia → hydroxide ion + ammonium ion

As you can see, the ammonia is acting as a Brønsted-Lowry base, because it is accepting a proton from another species, the water. What is water acting like in this equation? If you said a Brønsted -Lowry acid, you are correct. Water is donating a proton to the ammonia, to become the hydroxide ion.

What do you think will happen to the hydroxide ion (OH^-) on the product side of this equation? This is a reversible reaction, so, as more and more hydroxide and ammonium ions are produced, some of them will have effective collisions with each other and start the reverse reaction. The reverse reaction can be represented by the following equation:

$$OH^-_{(aq)} + NH_4^+{}_{(aq)} \rightarrow H_2O_{(l)} + NH_{3(aq)}$$

Notice that we have an acid and a base in the reverse reaction as well. In this reaction, the hydroxide ion is accepting a proton from the ammonium ion to form water, so it is acting as a Brønsted -Lowry base. The ammonium ion, which is donating a proton to the hydroxide ion, is acting like a Brønsted -Lowry acid. When we write both sides of the reaction together, the equation looks like the following:

$$H_2O_{(l)} + NH_{3(aq)} \rightleftarrows OH^-_{(aq)} + NH_4^+{}_{(aq)}$$

So, a reversible reaction of this type shows two acids and two bases. We will label the acids and bases, but first we must introduce two more definitions. A ***conjugate acid*** is the substance that is left after a base accepts a proton. A ***conjugate base*** is the substance that is left after an acid donates a proton. Using these definitions, let's label the equation.

$$\underset{\text{acid}}{H_2O_{(l)}} + \underset{\text{base}}{NH_{3(aq)}} \rightleftarrows \underset{\text{conjugate base}}{OH^-_{(aq)}} + \underset{\text{conjugate acid}}{NH_4^+{}_{(aq)}}$$

Typically, the acid and base will be labeled on the left side of the equation. In our example, H_2O is the acid, because it is donating a proton on the left side of the equation. OH^- is the conjugate base, because it is the species that is left over when the acid donates a proton. The NH_3 is acting as a base on the left side of the equation, because it is accepting a proton. The NH_4^+ is the conjugate acid of the base, because it is the species that is formed when the base accepts a proton.

An acid and its conjugate base, or a base and its conjugate acid, are often called "conjugate acid-base pairs." Conjugate acid-base pairs are two substances that are related by the gain or loss of a single proton (H^+). In our example, water (H_2O) and the hydroxide ion (OH^-) make up one conjugate acid-base pair, and ammonia (NH_3) and the ammonium ion (NH_4^+) make up the other conjugate acid-base pair.

The following summarizes some of the important terms presented in this section:

Brønsted -Lowry Acids and Bases	
Brønsted-Lowry Acid	A substance that donates a proton to another substance. Examples include HCl, H_2SO_4, NH_3, and H_2O.
Brønsted-Lowry Base	A substance that accepts a proton from another substance. Examples include KOH, LiOH, NH_3, and H_2O.
Conjugate Acid	The substance that is left after a base accepts a proton. For example, NH_4^+ is left after NH_3 accepts a proton.
Conjugate Base	The substance that is left after an acid donates a proton. For example, OH^- is left, after H_2O donates a proton.
Conjugate Acid-Base Pairs	Two substances that are related by the gain or loss of a single proton. An example of a pair is NH_3 and NH_4^+.

If you took a good look at the information just presented, you might think that I made a couple of mistakes. Did you notice that I included water (H_2O) and ammonia (NH_3) as examples of both Brønsted-Lowry acids and Brønsted-Lowry bases? This is no error. Water and ammonia are examples of amphoteric substances. An ***amphoteric substance*** is a substance that acts as an acid in some cases and as a base in other cases. Look at the following sets of reactions to see examples of these substances acting as both acids and bases.

1. Water acting as an acid (proton donor)

$$\underset{\text{acid}}{H_2O_{(l)}} + \underset{\text{base}}{NH_{3(aq)}} \rightarrow \underset{\text{conjugate base}}{OH^-_{(aq)}} + \underset{\text{conjugate acid}}{NH_4^+{}_{(aq)}}$$

2. Water acting as a base (proton acceptor)

$$\underset{\text{acid}}{H_2SO_4} + \underset{\text{base}}{H_2O} \rightarrow \underset{\text{conjugate base}}{HSO_4} + \underset{\text{conjugate acid}}{H_3O^+}$$

3. Ammonia acting as an acid (proton donor)

$$\underset{\text{base}}{OH^-_{(aq)}} + \underset{\text{acid}}{NH_{3(aq)}} \rightarrow + \underset{\text{conjugate base}}{NH^-_{2\,(aq)}} + \underset{\text{conjugate acid}}{H_2O_{(l)}}$$

4. Ammonia acting as a base (proton acceptor)

$$\underset{\text{acid}}{HCl} + \underset{\text{base}}{NH_3} \rightarrow \underset{\text{conjugate acid}}{NH_4^+} + \underset{\text{conjugate base}}{Cl^-}$$

I purposely switched the order of the acid and the base in #3 to point out a potential mistake that some students make. If your instructor seems to be in the habit of always writing the acid and base in the same order, don't try to memorize that order to avoid understanding the definitions. Even if that trick helped you on one of your quizzes, how is it going to help you on a standardized test, which your instructor doesn't create? Don't try to look for patterns that will help you avoid learning the material correctly.

Before we go on to the third, and last, theory of acids and bases, let's try a couple examples that are typical for this material.

Example 1

What would be the conjugate base for each of the following acids?

A. HF B. HNO_3 C. H_2SO_4 D. NH_4^+

Remember: We define the conjugate base as the substance that is left over when an acid loses a single proton. The key to solving this problem is to remember that we are removing a bare proton (H^+), which means that each acid will change in two ways. First, we will take one "H" off the formula. Second, we must remember to subtract +1 from each formula, because the hydrogen is leaving its electron behind. Substances with a charge of +1 become (+1 – 1 = 0) neutral. Neutral substances become (0 – 1 = -1) negatively charged. So, HF becomes F-; HNO_3 becomes NO_3^-; H_2SO_4 becomes HSO_4^-; and NH_4^+ becomes NH_3. If it helps, you can think of each change as a balanced reaction, as shown here:

$$H_2SO_4 \rightarrow HSO_4^- + H^+$$

Notice that both mass and charge are conserved in the reaction.

Answers:

A. F- B. NO_3^- C. HSO_4^- D. NH_3

The procedure for the next example is exactly the opposite. The conjugate acid of a base is the substance that is left behind after the base has gained a proton. Remember to add both the H and the + to each substance, and then check your answers.

Example 2

What would be the conjugate acid for each of the following bases?

A. OH^- B. H_2O C. NH_2^- D. NH_3

Adding an H and a +1 charge to each substance, OH^- becomes H_2O; H_2O becomes H_3O^+; NH_2^- becomes NH_3; and NH_3 becomes NH_4^+. Again, write out an equation that is balanced for atoms and for charge, if you don't see one of these answers. For example:

$$NH_2^- + H^+ \rightarrow NH_3$$

Answers:

A. H_2O B. H_3O^+ C. NH_3 D. NH_4^+

Another common type of question is shown in Example 3.

Example 3

Identify the two conjugate acid-base pairs shown in this equation:

$$HCl + NH_3 \rightleftarrows NH_4^+ Cl^-$$

This question should be easy enough to solve, provided that you recall the definition of a conjugate acid-base pair. A conjugate acid-base pair is a pair of substances that differ from each other by a single (H^+) proton. Clearly, when HCl donates a single proton, it becomes Cl^-. That makes HCl and Cl^- our first conjugate acid-base pair. When NH_3 accepts a proton, it becomes NH_4^+, making them our second conjugate acid-base pair.

Answer: The two conjugate acid-base pairs are (HCl and Cl^-) and (NH_3 and NH_4^+).

Lewis Acid-Base Theory

The final acid-base theory that we shall consider was proposed by chemist Gilbert Lewis in the early 1920s. The Lewis Theory is the most general, including more substances under its definitions than the other theories of acids and bases. A ***Lewis acid*** is a substance that accepts a pair of electrons to form a covalent bond. A ***Lewis base*** is a substance that provides a pair of electrons to form a covalent bond. In order for a substance to act as a Lewis base, it must have a pair of unshared electrons in its valence shell. An example of this is seen when a hydrogen ion attaches to the unpaired electrons of oxygen in a water molecule, as shown here:

H^+ Acting as a Lewis Acid and H_2O Acting as a Lewis Base

$$H^+ + H_2O \rightarrow H_3O^+$$

The bond formed between the hydrogen ion and the water, in this example, is a coordinate covalent bond. A ***coordinate covalent bond*** is formed between two atoms, where one atom provides both of the electrons that become shared. In the Lewis Acid-Base Theory, all interactions between acids and bases will involve coordinate covalent bonding.

Lewis Acids and Bases

Lewis Acid	A substance that accepts a pair of electrons to form a covalent bond. Examples include H^+ and BF_3.
Lewis Base	A substance that donates a pair of electrons to form a covalent bond. Examples include Cl^-, NH_3, and H_2O.

Lesson 9–4 Review

1. A(n) ______________ acid is a substance that donates a proton to another substance.
2. A(n) ______________ substance acts as an acid in some situations and a base in other situations.
3. A ______________ base is the substance that is left over, after an acid donates a proton.
4. Write the conjugate base of each of the following Brønsted-Lowry acids.

   a) HI b) H_3PO_4 c) H_3 d) H_2O

5. Write the conjugate acid for each of the following Brønsted-Lowry bases.
   a) OH^- b) Cl^- c) NH_3 d) H_2O
6. Identify the two Brønsted-Lowry acids in the following reaction:
   $$HCl + H_2O \rightleftarrows H_3O^+ + Cl^-$$
   a) HCl and H_2O b) $H_3O^+ + Cl^-$ c) H_2O and Cl^- d) HCl and H_3O^+
7. Identify the two Brønsted-Lowry bases in the following reaction:
   $$NH_4^+ + OH^- \rightleftarrows NH_3 + H_2O$$
   a) NH_4^+ and OH^- c) OH^- and NH_3
   b) $NH_3 + H_2O$ d) OH^- and H_2O
8. Identify a conjugate acid-base pair from the following reaction:
   $$F^- + HSO_4^- \rightleftarrows HF + SO_4^{-2}$$
   a) HF and F^- b) F^- and HSO_4^- c) HF and SO_4^{-2} d) F^- and SO_4^{-2}
9. Which of the following would be considered a Lewis base but not an Arrhenius base?
   a) KOH b) NH_3 c) NaOH d) $Ca(OH)_2$
10. In the following reaction, how would you describe HF?
    $$HF + SO_4^{-2} \rightarrow F^- + HSO_4$$
    a) amphoteric c) a proton acceptor
    b) a proton donor d) a Lewis base

Lesson 9–5: Strengths of Acids and Bases

The strengths of acids and bases vary quite a bit, as suggested by the wide variety of their uses. We consume weaker acids, and stronger acids in our stomach are used to break up the food we eat. We wash our hands with mild bases and use strong bases to break up the clogs in our drains. What is it that makes an acid or base "strong" or "weak"? It certainly isn't the number of H^+ ions in an acid formula that determines its strength, because hydrochloric acid (HCl) is considered much stronger than phosphoric acid (H_3PO_4). It does, however, have everything to do with the number of protons that the acid is able to donate.

Remember: Not all electrolytes dissociate completely in water. We said that a "strong" electrolyte ionizes to a greater degree than a "weak" electrolyte. The same idea is true for acids and bases, which, as you know, are electrolytes. The strongest acids are the ones that are considered to ionize completely, donating the greatest number of protons to the solution. The strongest bases are those that accept protons most readily, forming positive ions in the process.

An interesting aspect of the Brønsted-Lowry Theory of acids and bases is that the strongest acids have the weakest conjugate bases and the weakest acids have the strongest conjugate bases. For example, hydrochloric acid (HCl) is considered a very strong acid, because it ionizes almost completely, donating a great number of protons. The conjugate base of hydrochloric acid (Cl^-) is a very weak base, accepting relatively few protons. This makes sense, because if Cl accepted protons readily to reform HCl, then how could the HCl ionize to such a high degree?

The degree to which an acid ionizes is shown by its ionization constant, represented by the ionization constant for the acid, K_a. A high value for K_a represents a strong acid, whereas a low value represents a weak acid. The strength of a base is also indicated by an ionization constant, represented by K_b. The stronger the base, the higher the value its ionization constant will show. So, the acids with the lowest values for K_a are the weakest, but they will have the strongest conjugate bases, with the highest values for K_b.

You can find reference tables in your textbook, listing the values for K_a for many different acids. For the purpose of this book, I will give you K_a values for the acids that we use in specific questions, as they are needed.

Example 1

The K_a values of three acids at 1 atm and 29 K are listed here.

K_a of HF = 3.5×10^{-4}

K_a of $H_3PO_4 = 7.5 \times 10^{-3}$

K_a of $H_2O = 1.0 \times 10^{-14}$

A. List the three acids in order from strongest to weakest.

B. List the conjugate bases of the three acids in order from strongest to weakest.

The key to solving this problem correctly is to remember that we are looking at numbers in scientific notation and that these numbers all have negative exponents. Some students only look at the coefficients, and try to use them to judge the size of the number. What you really need to do is convert these numbers to standard notation, and then see which is the largest.

K_a of HF = 3.5×10^{-4} = 0.00035

K_a of H_3PO_4 = 7.5×10^{-3} = 0.0075

K_a of H_2O = 1.0×10^{-14} = 0.00000000000001

The largest of these numbers is 0.0075, making the H_3PO_4 our strongest acid. The number 0.00035 is the next largest, so the HF would be the next strongest acid. The water, with a K_a value of 0.00000000000001, is our weakest acid, which shouldn't come as much of a surprise. We can use this information to answer the first part of Example 1.

A. List the three acids in order from strongest to weakest.

Answer: A. **H_3PO_4, HF, H_2O**

In order to answer part B of the question, we need to find the conjugate bases of all three acids. You will remember from Lesson 9–4 that the conjugate base of an acid is the substance that is left after the acid donates a proton. When an acid loses a proton, we take one "H" away from its formula, and we decrease its charge by +1. So,

H_3PO_4 becomes H_2PO_4

HF becomes F^-

H_2O becomes OH^-

In order to arrange them from strongest base to weakest base, we need only remember that the stronger an acid is, the weaker its conjugate base will be. Therefore, the order of our bases will be the opposite of the order of our acids.

A. List the three acids in order from strongest to weakest.

B. List the conjugate bases of the three acids in order of strongest to weakest.

Answer: A. **H_3PO_4, HF, H_2O**
B. **OH^-, F^-, $H_2PO_4^-$**

When you are asked to solve problems involving acids and bases, you need to pay attention to their strengths, even if the word *strength* isn't mentioned in the problem. Strong acids and bases are treated as if they ionize completely, so the concentration of the ions in an aqueous solution of a strong acid or base is directly related to the molarity of the acid or base. If a question asks for the concentration of H_3O^+ ions in a 0.001 molar solution of HCl, the answer would be 0.001 M, because each molecule of HCl, a strong acid, is thought to provide one hydronium ion.

Example 2

Determine the pH and the pOH of a 0.01 M solution of NaOH, a strong base.

The key to this problem is that we are dealing with a strong base. The fact that it is strong means that the NaOH will ionize completely, releasing enough ions so that the concentration of hydroxide ions will be 0.01 M.

Given: [OH-] = 0.01 M **Find:** A. pOH B. pH

Answer: A. pOH = -log(0.01) = **2**

B. pH = 14 – pOH = 14 – 2 = **12**

Lesson 9–5 Review

1. The ionization constants for four bases are shown here. Select the constant that represents the strongest base.

   a) $K_b = 1.0 \times 10^{-2}$ c) $K_b = 1.0 \times 10^{-5}$

   b) $K_b = 1.0 \times 10^{-3}$ d) $K_b = 1.0 \times 10^{-7}$

2. The ionization constants for four acids are shown here. Select the constant that represents the strongest acid.

   a) $K_a = 1.0 \times 10^{-9}$ c) $K_a = 1.0 \times 10^{-5}$

   b) $K_a = 1.0 \times 10^{-6}$ d) $K_a = 1.0 \times 10^{-2}$

3. The ionization constants for four bases are shown here. Which of the bases represented would have the strongest conjugate acid?

   a) $K_b = 1.0 \times 10^{-6}$ c) $K_b = 1.0 \times 10^{-3}$

   b) $K_b = 1.0 \times 10^{-4}$ d) $K_b = 1.0 \times 10^{-1}$

Lesson 9–6: Acid-Base Titrations

One common laboratory activity, which makes use of the information that we have covered in this chapter, is an acid-base titration reaction. A ***titration*** reaction is a neutralization reaction carried out using a solution with a known concentration (called a ***standard solution***), in order to find the concentration of an unknown solution. By determining the volume of the ***standard solution*** required to neutralize a specific volume of the unknown solution, we can mathematically determine the concentration of the unknown. An indicator is used to signal the time when neutralization occurs, which is called the ***end point*** or the ***equivalence point*** of the titration.

The formula we use for these calculations is shown here:

$$V_aM_a = V_bM_b$$

V_a = the volume of the acid
M_a = the molar concentration of the H^+ ions in the acid
V_b = the volume of the base
M_b = the molar concentration of the OH^- ions in the base

As you can imagine, we can use this formula to solve for any of the four possible variables, provided that we know the other three values. Let's try a simple example to start.

Example 1

A student carries out an acid-base titration in the laboratory and finds that it takes 4.00×10^2 ml of a standard 1.0 M solution of NaOH to neutralize 1.00×10^2 ml of an HCl solution with an unknown concentration. What would be the concentration of the HCl solution?

Let's begin by listing the information that we know and the original formula. Please don't be confused by the use of scientific notation in the problem. It is simply used to indicate a certain number of significant digits.

Given: $V_b = 4.00 \times 10^2$ ml; $M_b = 1.0$ M; $V_a = 1.00 \times 10^2$ ml
Find: M_a
Original Formula: $V_aM_a = V_bM_b$

We isolate for the unknown, M_a, by dividing both sides of our original equation by V_a. Then, all we need to do is substitute, solve, and round to the correct number of significant digits.

Given: $V_b = 4.00 \times 10^2$ ml; $M_b = 1.0$ M; $V_a = 1.00 \times 10^2$ ml
Find: M_a

$$M_a = \frac{V_bM_b}{V_a} = \frac{(4.00 \times 10^2 \text{ ml})(1.0 \text{ M})}{(1.00 \times 10^2 \text{ ml})} = \mathbf{4.0\ M}$$

As you can see, titration problems are just a matter of simple algebra. As much as I hate to make things more complicated, just when they seemed so easy, I have to add one small detail. Hydrochloric acid is a ***monoprotic acid***, meaning that it theoretically yields one mole of hydrogen ions for every mole of acid. When dealing with HCl, one mole of acid generates one mole of H^+. Similarly, one mole of NaOH yields one mole of OH^-.

What happens when we are dealing with a ***polyprotic acid***, such as H_2SO_4, or a base that releases more hydroxide ions per mole, such as $Ca(OH)_2$?

One mole of H_2SO_4 theoretically yields two moles of H^+, and we must work that into our titration problems involving this acid. Some teachers use a concept called normality to handle these problems, stating that the ***normality*** of an acid is equal to the molarity of the acid, multiplied by the number of H^+ ions it releases per mole. So, a 2.0 M solution of H_2SO_4 would actually be 4.0 normal, as shown here:

normality = molarity × # of moles of ions/mole = 2.0 M × 2 = 4.0 normal

Some teachers, including myself, believe that the concept of normality adds an unnecessary extra level of complication, not to mention a new definition, to titration problems. We ask our students to remember that when we define M_a as the molar concentration of the ions in the acid solution, we really mean that it is equal to the molarity of the solution, multiplied by the number of moles of H^+ ions that it yields per mole. In this way, a 2.0 M solution of H_2SO_4 would have an M_a value of 4.0 M. A 2.0 M solution of H_3PO_4 would have an M_a value of 6.0 M. The same concept applies to bases, so a 4.0 M solution of $Ca(OH)_2$ would have an M_b value of 8.0 M.

If your instructor uses the concept of normality, then I would advise you to solve the following problems employing that concept. That way, you will be practicing for the exam that he or she will give you. Notice that, regardless of which method you use, you will still get the same answer that I do for the practice problems. Let's do an example involving sulfuric acid, H_2SO_4.

Example 2

A student conducts an acid-base titration laboratory activity and finds that it takes 0.50 L of 4.0 M H_2SO_4 standard solution to completely neutralize 2.0 L of NaOH solution. What is the concentration of the NaOH solution?

As with our previous example, we will carry out several steps in this problem. We will list what information we have been given, and what we have been asked to find, and we will show the original formula that we use to solve this type of problem. For the molar concentration of the H^+ ions in the acid (M_a) we will multiply the 4.0 M H_2SO_4 solution by 2 moles of H^+ per mole of H_2SO_4, for a total of 8.0 M.

Given: $V_a = 0.50$ L; $M_a = 8.0$ M; $V_b = 2.0$ L

Find: M_b

Original Formula: $V_aM_a = V_bM_b$

We can isolate the unknown, M_b, by dividing both sides of the equation by V_b. Then we substitute, solve, and round, as shown here:

Given: $V_a = 0.50$ L; $M_a = 8.0$ M; $V_b = 2.0$ L

Find: M_b

$$M_b = \frac{V_aM_a}{V_b} = \frac{(0.50 \cancel{L})(8.0\ M)}{(2.0 \cancel{L})} = \mathbf{2.0\ M}$$

Of course, the titration formula can be used to determine the volume of a solution with a known concentration needed to neutralize another volume of a known concentration. Because this will be our third example with this formula, I will risk solving Example 3 in one step. The only potentially confusing aspect of this next problem will be that it involves the base $Ca(OH)_2$. Because $Ca(OH)_2$ releases two moles of OH^- for every mole of the base (as indicated by the subscript of "2" on the outside of the parentheses), we will multiply the molarity of the base by 2 in order to get the molar concentration of the ions.

Example 3

How many liters of a 1.0 M $Ca(OH)_2$ solution would be required to completely neutralize 3.0 L of a 2.0 M HCl solution?

Given: $M_b = 2.0$ M; $V_a = 3.0$ L; $M_a = 2.0$ M

Find: V_b

$$V_b = \frac{V_aM_a}{M_b} = \frac{(3.0\ L)(2.0 \cancel{M})}{2.0 \cancel{M}} = \mathbf{3.0\ L}$$

Try the following practice problems and check your answers before moving on to the chapter examination.

Lesson 9–6 Review

For the following titration reactions, assume the ionization of the acids and bases are complete.

1. How many liters of 3.0 M H_2SO_4 would be required to completely neutralize 2.0 L of 3.0 M NaOH?
2. A student finds that it takes 150 ml of standard 3.0 M NaOH solution to neutralize 450 ml of HCl. What is the molarity of the acid?

3. In a laboratory experiment, a student neutralizes 25 ml of NaOH with an unknown concentration by reacting it with 25 ml of a standard 2.0 M H_2SO_4 solution. What is the molarity of the base?

4. A student carries out an acid-base titration in the laboratory, and finds that it takes 1.0 L of a standard 4.0 M HCl solution to neutralize 3.0 L of a $Ca(OH)_2$ solution with an unknown concentration. Determine the molarity of the base.

Chapter 9 Examination

Part I—Matching

Match the following terms to the definitions that follow. Not all of the terms will be used.

a. neutralization	f. Lewis base	k. hydronium ion
b. amphoteric	g. Lewis acid	l. titration
c. indicator	h. Arrhenius base	m. conjugate acid-base pair
d. Brønsted-Lowry acid	i. Arrhenius acid	n. conjugate base
e. Brønsted-Lowry base	j. hydroxide ion	o. conjugate acid

1. A(n) ______________ is a substance that donates a proton to another substance.
2. A(n) ______________ donates a pair of electrons to form a covalent bond.
3. A(n) ______________ releases hydroxide ions in an aqueous solution.
4. A(n) ______________ is an organic substance that changes color at certain pH values.
5. A(n) ______________ is the substance that is left after an acid loses a proton.

Part II—Multiple Choice

For each of the following questions, select the best answer.

6. When the H_3O^+ ion concentration of a solution is greater than the OH^- ion concentration, the solution is said to be __________.

   a) acidic b) basic c) amphoteric d) Arrhenius

7. Which of the following substances would be considered an Arrhenius acid?
   a) NaF b) HNO_3 c) CaI_2 d) $Ba(OH)_2$
8. Which of the following would be considered an Arrhenius base?
   a) NaF b) HNO_3 c) CaI_2 d) $Ba(OH)_2$
9. Which of the following could be considered a Brønsted-Lowry base, but *not* an Arrhenius base?
   a) NH_3 b) NaOH c) KCl d) H_2SO_4
10. A solution of sulfuric acid contains ________.
    a) more H_3O^+ than OH^- c) equal amounts of each ion
    b) more OH^- than H_3O^+ d) no free ions
11. As the concentration of OH^- ions in a solution at 298 K increases, k_w ________.
    a) increases c) remains the same
    b) decreases d) approaches a value of 1000
12. What is the H_3O^+ concentration in an aqueous solution with a pH value of 5?
    a) 5.0×10^1 M b) 1.0×10^5 M c) 5.0×10^{-1} M d) 1.0×10^{-5} M
13. What is the OH^- concentration in an aqueous solution with a pH value of 2?
    a) 2.0×10^1 M b) 1.0×10^{12} M c) 1.0×10^{-12} M d) 1.0×10^{-2} M
14. A solution has a hydronium ion concentration of 5.64×10^{-4} M. The pH value of this solution is closest to ________.
    a) 3.2 b) 4.0 c) 5.6 d) 10
15. A solution has a hydroxide ion concentration of 2.8×10^{-11} M. The pH value of this solution is closest to ________.
    a) 2.8 b) 3.4 c) 10.6 d) 14
16. Calculate the pOH value of a solution with a hydronium ion concentration of 1.0×10^{-4} M.
    a) 1 b) 4 c) 10 d) 14
7. A solution that turns phenolphthalein pink would best be described as ________.
   a) acidic b) basic c) amphoteric d) neutral

18. The ionization constants for four bases are shown here. Select the constant that represents the strongest base.
a) $K_b = 1.0 \times 10^{-3}$
b) $K_b = 1.0 \times 10^{-5}$
c) $K_b = 1.0 \times 10^{-7}$
d) $K_b = 1.0 \times 10^{-9}$

19. The ionization constants for four bases are shown here. Which of the bases represented would have the strongest conjugate acid?
a) $K_b = 1.0 \times 10^{-3}$
b) $K_b = 1.0 \times 10^{-5}$
c) $K_b = 1.0 \times 10^{-7}$
d) $K_b = 1.0 \times 10^{-9}$

20. What is the conjugate base of the acid NH_4^+ ?
a) NH_3^+ b) NH_3 c) NH_3^- d) NH_2^{+2}

Part III—Calculations

Perform the following calculations. Round your answers to the correct number of significant digits and include units.

21. What volume of a 3.0 M solution of NaOH would be required to completely neutralize 2.0 L of a 3.0 M solution of H_2SO_4?

22. What volume of 2.0 M NaOH would be required to completely neutralize 250 ml of 1.0 M HCl?

23. What is the pH and pOH of a solution with a concentration of OH^- ions of 0.000001 M?

24. What is the pH and pOH of a solution with a concentration of H_3O^+ of 0.001 M?

25. What is the pH and pOH of a solution with a concentration of H_3O^+ of 0.000073 M?

Answer Key

The actual answers will be shown in brackets, followed by the explanation. If you don't understand an explanation that is given in this section, you may want to go back and review the lesson that the question came from.

Lesson 9-1 Review

1. [polar]—Remember: "Likes dissolve likes."
2. [insoluble]—Remember: "Likes dissolve likes."
3. [molarity]—Molarity (M) $= \frac{\text{moles of solute}}{\text{liters of solution}}$

4. [molality]— Molality $= \frac{\text{moles of solute}}{\text{kilograms of solvent}}$
5. [molarity]—We read M as "molar."
6. [molality]—We read m as "molal."
7. [0.67 M]— Molarity (M) $= \frac{\text{moles of solute}}{\text{liters of solution}} = \frac{2.0 \text{ moles}}{3.0 \text{ L}} = 0.67 \text{ M}$
8. [100. g]—First, find out how many moles of NaOH you need. Then multiply by the molar mass of NaOH, which is (23.0 g + 16.0 g + 1.01 g) = 40.0 g/mole.
   # moles of solute = molarity × liters of solution = 1.25 M × 2.00 L = 2.50 moles
   Mass = # of moles × molar mass = 2.50 ~~moles~~ × 40.0 g/~~mole~~ = 100. g
9. [4.0 L]—liters of solution $= \frac{\text{moles of solute}}{\text{molarity}} = \frac{8.0 \text{ moles}}{2.0 \text{ M}} = 4.0 \text{ L}$
10. [0.5 kg]—The molar mass of $CaCl_2$ is 111.1 g (40.1 g + 71.0 g = 111.1 g), so you start with 1.00 mole of it. The kilograms of solvent that you need can be determined with the following formula:

$$\text{kilograms of solvent} = \frac{\text{moles of solute}}{\text{molality}} = \frac{1.0 \text{ moles}}{2.0 \text{ m}} = 0.5 \text{ kg}$$

Lesson 9–2 Review

1. [acid]—A common method for producing hydrogen gas in a chemistry lab involves zinc reacting with an acid, such as HCl, in a single replacement reaction.
2. [acid]—Litmus paper is a common indicator found in the chemistry lab.
3. [base]—Neutral substances have pH values of 7; above that is considered basic.
4. [acid]—Of course, you would never taste a strong or an unknown acid. We are talking about weak acids such as lemon juice.
5. [base]—Sodium hydroxide is found in some types of drain cleaners.
6. [base]—Remember: Litmus paper is only one of several types of indicators that you may encounter as you study chemistry.
7. [5]—We solve for pOH as shown here:
   $pOH = -\log[OH^-] = -\log(1 \times 10^{-5}) = 5$
8. [2.2]—This time, you really need to use the formula. Practice using your calculator correctly, by making sure that you can find this answer.
   $pH = -\log[H_3O^+] = -\log(5.9 \times 10^{-3}) = 2.229147988 = 2.2$
9. [9.5]—Again, you are given the concentration of the hydroxide ions, rather than the concentration of the hydronium ions. The easiest way to solve this is shown.
   $pH = 14 - pOH = 14 - (-\log[OH^-]) = 14 - (-\log(3.5 \times 10^{-5})) = 9.544068044 = 9.5$
10. [1×10^{-5} M]—See the Ion Concentrations chart on page 306.

Lesson 9–3 Review

1. [phosphoric acid]—This is a ternary acid, and the PO_4^{3-} is the phosphate ion, so "ate" becomes "ic."
2. [hydrobromic acid]—This is a binary acid, so we add the prefix of "hydro-" and the ending of "ide" becomes "ic."
3. [sulfuric acid]—This is a ternary acid, and the SO_4^{2-} is the sulfate ion, so "ate" becomes "ic."
4. [hypochlorous acid]—This is a ternary acid that contains chlorine, and the least number of oxygen atoms per formula unit.
5. [nitrous acid]—This is a ternary acid, and the NO_2^- is the nitrite ion, so the "ite" becomes "ous."
6. [chloric acid]—This is a ternary acid that contains chlorine. The ClO_3^- is the chlorate ion, so "ate" becomes "ic."
7. [HF]—The prefix of "hydro-" tells us that it is a binary compound beginning with hydrogen. The "fluoric" must be derived from "fluoride."
8. [$HClO_2$]—The acid doesn't start with "hydro-" so it must be a ternary acid. The "chloro" tells us that it contains chlorine. The ending of "ous" tells us that the polyatomic ion must end in "ite," like the chlorite ion, ClO_2^-.
9. [H_2SO_3]—The name doesn't start with "hydro-," so it must be a ternary acid, and it must contain a polyatomic ion. The ending of "ous" means that the polyatomic ion must end in "ite," like sulfite, SO_3^{2-}.
10. [HNO_3]—The name doesn't start with "hydro-," so it must be a ternary acid, and it must contain a polyatomic ion. The ending of "ic" means that the polyatomic ion must end in "ate," like nitrate, NO_3^-.

Lesson 9–4 Review

1. [Brønsted-Lowry]—This definition is slightly broader than the Arrhenius definition, as it isn't limited to aqueous solutions.
2. [amphoteric]—Water and ammonia are common examples.
3. [conjugate]—OH^- is the conjugate base of water.
4. We simply remove H^+ from each of the formulas.

   A. [I^-] B. [$H_2PO_4^-$] C. [NH_2^-] D. [OH^-]
5. We add H^+ to each of the formulas.

   A. [H_2O] B. [HCl] C. [NH_4^+] D. [H_3O^+]
6. [D. HCl and H_3O^+]—Brønsted-Lowry acids are proton donors.
7. [C. OH^- and NH_3]—Brønsted-Lowry bases are proton acceptors.
8. [A. HF and F^-]—Conjugate acid-base pairs differ from each other by a single proton.

9. [B. NH_3]—An Arrhenius base must contain the hydroxide (OH^-) ion.
10. [B. a proton donor]—HF donates a proton to become F^-.

Lesson 9–5 Review

1. [A. $K_b = 1.0 \times 10^{-2}$]—The strongest base will have the highest value for K_b. Remember that the exponents are all negative, so the largest negative exponent shows the smallest value.
2. [D. $K_a = 1.0 \times 10^{-2}$]—The strongest acid will have the highest value for K_a.
3. [A. $K_b = 1.0 \times 10^{-6}$]—Remember that the weakest base will have the strongest conjugate acid, so we look for the lowest value for K_b.

Lesson 9–6 Review

1. [1.0 L]—The sulfuric acid (H_2SO_4) releases 2 moles of H^+ per mole of acid, so multiply the molarity of the acid by 2.

   Given: $M_a = 6.0$ M; $V_b = 2.0$ L; $M_b = 3.0$M **Find:** V_a

   $$V_a = \frac{V_b M_b}{M_a} = \frac{(2.0 \text{ L})(3.0 \text{ M})}{(6.0 \text{ M})} = 1.0 \text{ L}$$

2. [1.0 M]—Work shown following.

   Given: $V_a = 450$ ml; $V_b = 150$ ml; $M_b = 3.0$ M **Find:** M_a

   $$M_a = \frac{V_b M_b}{V_a} = \frac{(150 \text{ ml})(3.0 \text{ M})}{(450 \text{ ml})} = 1.0 \text{ M}$$

3. [4.0 M]—The sulfuric acid (H_2SO_4) releases 2 moles of H^+ per mole of acid, so multiply the molarity of the acid by 2.

   Given: $V_a = 25$ ml; $M_a = 4.0$ M; $V_b = 25$ ml **Find:** M_b

   $$M_b = \frac{V_a M_a}{V_b} = \frac{(25 \text{ ml})(4.0 \text{ M})}{(25 \text{ ml})} = 4.0 \text{ M}$$

4. [0.67 M]—Now, the $Ca(OH)_2$ will release 2 moles of OH^- per mole of base. When we find the molarity, we will divide by 2, because this base is twice as effective as a base that only releases 1 mole of OH^- per mole of base.

   Given: $V_a = 1.0$ L; $M_a = 4.0$ M; $V_b = 3.0$ L **Find:** M_b

   $$M_b = \frac{V_a M_a}{V_b} = \frac{(1.0 \text{ L})(4.0 \text{ M})}{(3.0 \text{ L})} = 1.33 \text{ M}$$

   $$\frac{1.33 \text{ M}}{2} = 0.67 \text{ M}$$

Chapter 9 Examination

1. [d. Brønsted-Lowry acid]—Proton donors are Brønsted-Lowry acids.
2. [f. Lewis base]—Remember that they can neutralize H^+ ions.
3. [h. Arrhenius base]—Each OH^- can neutralize an H^+ ion.
4. [c. indicator]—Litmus paper and phenolphthalein are examples of indicators.

5. [n. conjugate base]—NH_2^- is the conjugate base of NH_3.
6. [A. acidic]—Remember: Some texts will concentrate on H^+ ions.
7. [B. HNO_3]—Arrhenius acids must have H^+ ions to release.
8. [D. $Ba(OH)_2$]—Arrhenius bases must have OH^- ions to release.
9. [A. NH_3]—Ammonia can't release OH^- ions the way an Arrhenius base can, but it can accept protons as a Brønsted-Lowry base.
10. [A. more H_3O^+ than OH^-]—Having more H_3O^+ than OH^- is a characteristic of any acidic solution.
11. [C. remains the same]—k_w is a constant for a given temperature.
12. [D. 1.0×10^{-5} M]—Remember: $-\log(1.0 \times 10^{-5}) = 5$.
13. [C. 1.0×10^{-12} M]—If the pH is 2, then the concentration of hydronium (H_3O^+) ions is 1.2×10^{-2} M. We can determine the concentration of OH^- ions by remembering that $[H_3O^+] \times [OH^-] = 1.0 \times 10^{-14}$.

$$[OH^-] = \frac{k_w}{[H_3O^+]} = \frac{1.0 \times 10^{-14}}{1.0 \times 10^{-2}} = 1.0 \times 10^{-12}$$

14. [A. 3.2]— $pH = -\log[H_3O^+] = -\log(5.64 \times 10^{-4}) = 3.248720896$, which rounds to 3.2
15. [B. 3.4]—This question gives us the concentration of hydroxide (OH^-) ions, so we must solve as shown here:

$$pH = 14 - (-\log[OH^-]) = 14 - (-\log(2.8 \times 10^{-11})) = 3.4, \text{ after rounding}$$

16. [C. 10]—Remember that pH + pOH = 14. We can solve for pH as follows: $pH = -\log[H_3O^+] = -\log(1.0 \times 10^{-4}) = 4$. Because the question asks for the pOH value, we solve pOH = 14 – pH = 14 – 4 = 10.
17. [B. basic]—Phenolphthalein is an indicator that turns pink in basic solutions. It is often used to signal the endpoint of a titration.
18. [A. $K_b = 1.0 \times 10^{-3}$]—Simply look for the greatest value for K_b.
19. [D. $K_b = 1.0 \times 10^{-9}$]—The weakest base has the strongest conjugate acid.
20. [B. NH_3]—$NH_4^+ \rightarrow NH_3 + H^+$. The conjugate base is the species that is left after the acid gives up one H^+.
21. [4.0 L]—The H_2SO_4 will release 2 moles of H^+ per mole of acid, so we will multiply the molarity of the acid by 2 and solve as shown here:

Given: $M_a = 6.0$ M; $V_a = 2.0$ L; $M_b = 3.0$ M **Find:** V_b

$$V_b = \frac{V_a M_a}{M_b} = \frac{(2.0\text{ L})(6.0\text{ M})}{(3.0\text{ M})} = 4.0\text{ L}$$

22. [130 ml]—Work shown following:

Given: $M_a = 1.0$ M; $V_a = 250$ ml; $M_b = 2.0$ M **Find:** V_b

$$V_b = \frac{V_a M_a}{M_b} = \frac{(250\text{ ml})(1.0\text{ M})}{(2.0\text{ M})} = 125\text{ ml} = 130\text{ ml, after rounding}$$

23. [pH = 8; pOH = 6]—Because we are given the concentration of OH^- ions, it is easier to find the pOH first.

$$pOH = -\log[OH^-] = -\log(0.000001) = 6$$

$$pH = 14 - pOH = 14 - 6 = 8$$

24. [pH = 3; pOH = 11]—We are given the concentration of H_3O^+ ions this time, so it is easier to find the pH first.

$$pH = -\log[H_3O^+] = -\log(0.001) = 3$$

$$pOH = 14 - pH = 14 - 3 = 11$$

25. [pH = 4.1; pOH = 9.9]—Work shown following:

$$pH = -\log[H_3O^+] = -\log(0.000073) = 4.1. \; pOH = 14 - 4.1 = 9.9$$

Glossary

Accuracy: How close a measurement or the answer to a calculation is to an accepted value.

Activation energy: The energy required to get a reaction started.

Alkali metal: An element (other than hydrogen) found in the first column of the periodic table.

Alkaline earth metal: An element found in the second column of the periodic table.

Anion: A negative ion.

Aqueous: A solution in which water is the solvent.

Arrhenius acid: A substance that releases H^+ ions in aqueous solution.

Arrhenius base: A substance that releases OH^- ions in aqueous solution.

Atom: The smallest complete unit of an element.

Atomic mass: A weighted average of the masses of the naturally occurring isotopes of an element.

Atomic number: A number equal to the number of protons in the nucleus of an atom.

Avogadro's number: 6.02×10^{23}.

Barometer: An instrument that can be used to measure the pressure of gases.

Binary compound: A compound made up of two, and only two, elements.

Boyle's Law: States that the pressure and volume of a gas at constant temperature are inversely proportional to each other.

Brønsted-Lowry acid: A substance that donates a proton to another substance.

Brønsted-Lowry base: A substance that accepts a proton from another substance.

Cation: A positive ion.

Change-of-phase operation: A physical change that takes place as a substance changes phase or state. (Examples include melting and evaporation.)

Change-of-state operation: See Change-of-Phase Operation.

Charles's Law: States that the volume of an ideal gas at constant pressure varies directly with its Kelvin temperature.

Chemical change: A change that results in the production of a new substance. (An example would be the rusting of iron to produce iron oxide.)

Chemical energy: Energy "stored" within a substance, due to its bonding states.

Chemical equation: A group of chemical formulas and numbers that can be used to represent a chemical reaction.

Chemical property: A property of a substance that pertains to how it reacts chemically.

Combustion reaction: A chemical reaction in which a hydrocarbon reacts with oxygen to produce carbon dioxide and water.

Compound: A substance made from two or more elements chemically combined.

Concentration: The "strength" of a solution, in terms of the relative number of "parts" of solvent to solute.

Condensation: The process by which a vapor becomes a liquid.

Coordinate covalent bond: A bond formed when both electrons that become a shared pair of electrons between two atoms are provided by only one of the atoms.

Covalent bond: A bond that results from electrons being shared between two atoms.

Dalton's Law: States that the total pressure exerted by a mixture of gases is equal to the sum of all of the partial pressures exerted by the gases in the mixture.

Decomposition reaction: A chemical reaction in which a complex substance breaks up into two or more simpler substances.

Density: A measure of the amount of matter in a given unit of volume. Density is often measured in g/cm^3, but any unit of mass over any unit of volume will do.

Deposition: The process by which a vapor becomes a solid.

Diatomic: A two-atom form.

Diatomic element: An element that is found in two-atom molecules in its elemental form. The seven diatomic elements are fluorine, chlorine, bromine, iodine, hydrogen, oxygen, and nitrogen.

Dipole: Another term for a polar molecule.

Electrolyte: A substance that dissociates in water to produce a current-carrying solution.

Electron: A negatively charged subatomic particle found in the "cloud" region of an atom.

Electron affinity: A measure of the attraction of an element for additional electrons.

Electron cloud: The region of the quantum mechanical atom where the electrons may be found.

Electron configuration: A notation that shows information about the likely locations of the electrons in an atom.

Electronegativity: A relative measure of an atom's attraction for bonding electrons.

Element: A substance made up entirely of one type of atom, with the same atomic number.

Elemental notation: A notation that summarizes information about an isotope.

Elemental symbol: One or more letters that represent a shorthand notation for an element.

Empirical formula: A formula that shows the simplest whole number ratio for the elements that make up a compound.

Endothermic reaction: A reaction that takes in energy from its surroundings.

Energy: The ability to do work.

Excess reactant: A reactant that doesn't get totally consumed in a chemical reaction, as there is plenty of it.

Exothermic reaction: A reaction that releases energy, usually in the form of heat and light.

Extensive property: A property of a substance that depends upon the size of the sample, such as mass and volume.

Factor-label method: A method used to convert between units of different quantities.

Family: One of the several groups of elements found in the same column on the periodic table.

Formula mass: The mass of one formula unit of an ionic compound.

Formula unit: The simplest whole number ratio of cations to anions found in a compound.

Fusion: The process by which a solid becomes a liquid. Also, nuclear fusion is the process by which two or more lighter nuclei combine to form a heavier nucleus.

Graham's Law: States that under equal conditions of temperature and pressure, gases diffuse at a rate that is inversely proportional to the square roots of their molecular masses.

Gravitational potential energy: The "stored" energy that an object has due to its mass and its height above a reference point.

Heterogeneous mixture: Two or more substances physically mixed together, without uniform distribution of particles.

Homogeneous mixture: Two or more substances physically mixed together, with uniform distribution of particles. Also called a solution.

Hydrocarbon: A compound that contains hydrogen and oxygen.

Intensive property: A property of a substance that is not dependent upon the size of the sample, such as density and color.

Intermolecular force: A force between the individual molecules of a substance.

Ion: An atom, or a group of covalently bonded atoms, that has obtained a charge by either gaining or losing one or more electrons.

Ionic bond: A bond formed from the electrostatic attraction between ions with different charges.

Ionic compound: Two or more elements held together by ionic bonds.

Ionization energy: The energy required to remove the most loosely held electron from an atom.

Isomers: Two or more compounds that have the same molecular formula but different structural formulas.

Isotopes: Atoms of the same element with different numbers of neutrons and different mass numbers.

Kelvin: A temperature scale based on the concept of "absolute zero." There are no negative values in this scale.

Kernel: A term referring to the nucleus of an atom and all of the electrons other than those in the valence shell.

Kinetic energy: The energy that an object has due to its motion.

Lewis acid: A substance that accepts a pair of electrons to form a covalent bond.

Lewis base: A substance that provides a pair of electrons to form a covalent bond.

Lewis dot notation: A notation that shows detail about an atom's valence shell.

Limiting reactant: A reactant that is totally consumed in a chemical reaction, causing the reaction to stop.

Malleability: A physical property of many metals, describing the ability to be hammered into thin sheets.

Mass: The amount of matter that an object contains.

Mass number: The total number of protons and neutrons in the nucleus of an atom.

Matter: Anything that has both mass and volume.

Metal: An element with relatively few valence electrons that tends to form a positive ion when forming an ion.

Metalloid: An element with some of the characteristics associated with metals and some associated with nonmetals. Also called semimetals.

Mixture: Two or more substances physically mixed together.

Molality: A measure of the number of moles of solute dissolved in every kilogram of solvent.

Molarity: A measure of the number of moles of solute dissolved in every liter of solution.

Molar mass: The mass of one mole of a substance.

Molar volume: The amount of space occupied by one mole of any gas at STP. Equal to 22.4 dm^3/mole.

Mole: A group of 6.02×10^{23} items.

Molecular compound: Two or more elements held together by covalent bonds.

Molecular formula: A formula that represents the particles that make up a compound, as they actually exist.

Molecular mass: The mass of one molecule of a substance.

Molecule: A group of atoms that are held together with covalent bonds.

Monatomic: A single atom.

Monatomic element: An element that exists in a single atom form.

Monatomic ion: A single atom that has gained or lost additional electrons.

Negative ion: An atom that has gained one or more electrons.

Neutron: A neutrally charged particle found in the nucleus of an atom. It has essentially the same mass as a proton.

Nonelectrolyte: A substance that dissolves in water to produce a solution that does not conduct electricity.

Nonmetal: An element that is a poor conductor of heat and electricity, which will tend to form a negative ion when forming an ion.

Non-polar covalent bond: A bond formed when atoms share one or more pairs of electrons relatively equally.

Non-polar molecule: A molecule that either has symmetrical molecular geometry, or it only has non-polar covalent bonds, or both.

Nuclear charge: The charge on the nucleus of an atom, which is equal to the number of protons the nucleus contains.

Orbital: A space that can be occupied by up to two electrons.

Orbital notation: A notation that displays information about the electrons of an atom and the orbits they occupy.

Organic compound: A compound containing carbon.

Oxidation: When a substance loses electrons, or appears to lose electrons, as its oxidation number appears to increase algebraically.

Oxidation number: A number indicating the charge, or the apparent charge, of an atom.

Oxidizing agent: The substance that appears to cause an oxidation to occur, by accepting or appearing to accept electrons.

Period: A horizontal row on the periodic table.

Physical change: A change that does not result in the production of a new substance. (Breaking glass is an example of a physical change.)

Physical property: A property of a substance that is readily observable or easy to measure, such as color, luster, and mass.

Polar covalent bond: A bond formed when atoms share one or more pairs of electrons, but one atom attracts the electrons more strongly, resulting in unequal sharing.

Polar molecule: A molecule with a positive and negative side, as a result of its bonds and geometry.

Polyatomic ion: Two or more atoms covalently bonded together that obtain a net charge. (Some examples include hydroxide [OH^-] and nitrate [NO_3^-] ions.)

Positive ion: An atom that has lost one or more electrons.

Potential energy: The "stored" energy that an object has due to its position relative to some reference point.

Precipitate: A solid deposit that forms when a substance comes out of a solution.

Proton: A positively charged particle found in the nucleus of an atom. The number of protons in the nucleus of an atom will determine its identity.

Redox reaction: An oxidation-reduction reaction, where one substance is oxidized and another is reduced.

Reducing agent: The substance that appears to cause a reduction to occur, by donating electrons, or appearing to donate electrons.

Reduction: When a substance gains electrons or appears to gain electrons, as its oxidation number is algebraically reduced.

Saturated solution: A solution that is currently holding as much solute as it can under current conditions.

SI: An abbreviation for the International System of Measurements.

Single replacement reaction: A chemical reaction in which a free element replaces a similar element from a compound.

Solidification: The process by which a liquid becomes a solid.

Solubility: The ability of a substance to dissolve into another substance.

Solute: The part of a solution that gets dissolved into the solvent.

Solution: Another name for a homogeneous mixture.

Solvent: The part of a solution into which the solute is dissolved.

Sublimation: The process by which a solid changes to the gas phase directly, without passing through the liquid phase.

Substance: A form of matter with constant composition.

Supersaturated solution: A solution that is holding more solute then it normally can under current conditions.

Synthesis reaction: A chemical reaction in which two or more simpler substances form a more complex substance.

Temperature: A measure of the average kinetic energy of the particles of a substance.

Ternary compound: A compound containing three or more elements.

Unsaturated solution: A solution that can hold more solute at current conditions.

Valence electron: An electron that is found in the outer energy level of an atom.

Vaporization: The process by which a liquid becomes a vapor.

Volume: The amount of space that an object occupies.

Weight: A measure of the attraction between two objects, usually the Earth and another object, due to gravity.

Work: A force exerted over a distance.

Index

M

N

About the Author

Greg Curran has been teaching science for fifteen years. He currently teaches chemistry and AP physics at Fordham Preparatory School, where he also serves as the chair of the science department. He also teaches online science classes through the University of Wisconsin, Cambridge College, the Academy of Teaching Excellence, and the JASON Academy. He is the author of the popular *Science Help Online* Website, which is designed to help students who are learning chemistry (*www.fordhamprep.com/gcurran/*). Greg has a M.S. in Education from Fordham University and a B.A. in biology from SUNY Purchase. He lives in Garrison, New York, with his wife, Rosemarie, and children, James, Amanda, and Jessica.